ENERGY

SHORTAGE, GLUT, OR ENOUGH?

ENERGY

SHORTAGE, GLUT, OR ENOUGH?

Doug Dupler

INFORMATION PLUS® REFERENCE SERIES
Formerly published by Information Plus, Wylie, Texas

Detroit
New York
San Francisco
London
Boston
Woodbridge, CT

ENERGY: SHORTAGE, GLUT, OR ENOUGH?

Doug Dupler, *Author*

The Gale Group Staff:

Editorial: John F. McCoy, *Project Manager and Series Editor*; Andrew Claps, *Series Associate Editor*; Jason M. Everett, *Series Associate Editor*; Michael T. Reade, *Series Associate Editor*; Rita Runchock, *Managing Editor*; Luann Brennan, *Editor*

Image and Multimedia Content: Barbara J. Yarrow, *Manager, Imaging and Multimedia Content*; Robyn Young, *Project Manager, Imaging and Multimedia Content*

Indexing: Lynne Maday, *Indexing Specialist*; Amy Suchowski, *Indexing Specialist*

Permissions: Julie Juengling, *Permissions Specialist*; Maria Franklin, *Permissions Manager*

Product Design: Michelle DiMercurio, *Senior Art Director*; Kenn Zorn, *Product Design Manager*

Production: Evi Seoud, *Assistant Manager, Composition Purchasing and Electronic Prepress*; NeKita McKee, *Buyer*; Dorothy Maki, *Manufacturing Manager*

ISBN 0-7876-5103-6 (set)
ISBN 0-7876-5392-6 (this volume)
ISSN 1534-1585 (this volume)
Printed in the United States of America
10 9 8 7 6 5 4 3 2 1

TABLE OF CONTENTS

PREFACE

Energy: Shortage, Glut, or Enough? is the latest volume in the ever-growing *Information Plus Reference Series*. Previously published by the Information Plus company of Wylie, Texas, the *Information Plus Reference Series* (and its companion set, the *Information Plus Compact Series*) became a Gale Group product when Gale and Information Plus merged in early 2000. Those of you familiar with the series as published by Information Plus will notice a few changes from the 1999 edition. Gale has adopted a new layout and style that we hope you will find easy to use. Other improvements include greatly expanded indexes in each book, and more descriptive tables of contents.

While some changes have been made to the design, the purpose of the *Information Plus Reference Series* remains the same. Each volume of the series presents the latest facts on a topic of pressing concern in modern American life. These topics include today's most controversial and most studied social issues: abortion, capital punishment, care for the elderly, crime, health care, the environment, immigration, minorities, social welfare, women, youth, and many more. Although written especially for the high school and undergraduate student, this series is an excellent resource for anyone in need of factual information on current affairs.

By presenting the facts, it is Gale's intention to provide its readers with everything they need to reach an informed opinion on current issues. To that end, there is a particular emphasis in this series on the presentation of scientific studies, surveys, and statistics. This data is generally presented in the form of tables, charts, and other graphics placed within the text of each book. Every graphic is directly referred to and carefully explained in the text. The source of each graphic is presented within the graphic itself. The data used in these graphics is drawn from the most reputable and reliable sources, in particular from the various branches of the U.S. government and from major independent polling organizations. Every effort was made to secure the most recent information available. The reader should bear in mind that many major studies take years to conduct, and that additional years often pass before the data from these studies is made available to the public. Therefore, in many cases the most recent information available in 2001 dated from 1998 or 1999. Older statistics are sometimes presented as well, if they are of particular interest and no more-recent information exists.

Although statistics are a major focus of the *Information Plus Reference Series* they are by no means its only content. Each book also presents the widely held positions and important ideas that shape how the book's subject is discussed in the United States. These positions are explained in detail and, where possible, in the words of those who support them. Some of the other material to be found in these books includes: historical background; descriptions of major events related to the subject; relevant laws and court cases; and examples of how these issues play out in American life. Some books also feature primary documents, or have pro and con debate sections giving the words and opinions of prominent Americans on both sides of a controversial topic. All material is presented in an even-handed and unbiased manner; the reader will never be encouraged to accept one view of an issue over another.

HOW TO USE THIS BOOK

The United States is the world's largest consumer of energy in all its forms. Gasoline and other fossil fuels power its cars, trucks, trains, and aircraft. Electricity generated by burning oil, coal and natural gas—or from nuclear or hydroelectric plants—runs America's lights, telephones, televisions, computers, and appliances. Without a steady, affordable, and massive amount of energy, modern America could not exist. This book presents the latest information on America's energy consumption and production, compared with years past. Controversial issues such as America's dependence on foreign oil; the possibility of exhausting fossil fuel supplies; and the harm done to the environment by mining, drilling, and pollution; are explored.

Energy: Shortage, Glut, or Enough? consists of ten chapters and three appendixes. Each of the major elements of the U.S. energy system—such as coal, nuclear power, renewable energy sources, and electricity generation—has a chapter devoted to it. For a summary of the information covered in each chapter, please see the synopses provided in the Table of Contents at the front of the book. Chapters generally begin with an overview of the basic facts and background information on the chapter's topic, then proceed to examine sub-topics of particular interest. For example, Chapter 5: Energy Reserves—Oil, Gas, Coal, and Uranium begins with a description of the different ways in which natural resource reserves are measured and their reliability. The chapter then moves into an examination of U.S. reserves of oil, followed by gas, coal, and uranium. Statistics on known and estimated reserves are presented, as are projections of how long these reserves will last if they continue to be consumed at current rates. Then the chapter presents similar statistics on worldwide reserves of oil, coal, gas, and uranium. Readers can find their way through a chapter by looking for the section and sub-section headings, which are clearly set off from the text. Or, they can refer to the book's extensive index, if they already know what they are looking for.

Statistical Information

The tables and figures featured throughout *Energy: Shortage, Glut, or Enough?* will be of particular use to the reader in learning about this topic. These tables and figures represent an extensive collection of the most recent and valuable statistics on energy production and consumption; for example: the amount of coal mined in the United States in a year and how it is used, the rate at which energy consumption is increasing in the United States, and the percentage of U.S. energy that comes from renewable sources. Gale believes that making this information available to the reader is the most important way in which we fulfill the goal of this book: To help readers understand the topic of energy and reach their own conclusions about controversial issues related to energy use and conservation in the United States.

Each table or figure has a unique identifier appearing above it, for ease of identification and reference. Titles for the tables and figures explain their purpose. At the end of each table or figure, the original source of the data is provided.

In order to help readers understand these often complicated statistics, all tables and figures are explained in the text. References in the text direct the reader to the relevant statistics. Furthermore, the contents of all tables and figures are fully indexed. Please see the opening section of the index at the back of this volume for a description of how to find tables and figures within it.

In addition to the main body text and images, *Energy: Shortage, Glut, or Enough?* has three appendices. The first is the Important Names and Addresses directory. Here the reader will find contact information for a number of organizations that study energy. The second appendix is the Resources section, which is provided to assist the reader in conducting his or her own research. In this section, the author and editors of *Energy: Shortage, Glut, or Enough?* describe some of the sources that were most useful during the compilation of this book. The final appendix is this book's index. It has been greatly expanded from previous editions, and should make it even easier to find specific topics in this book.

COMMENTS AND SUGGESTIONS

The editor of the *Information Plus Reference Series* welcomes your feedback on *Energy: Shortage, Glut, or Enough?* Please direct all correspondence to:

Editor
Information Plus Reference Series
27500 Drake Rd.
Farmington Hills, MI, 48331-3535

ACKNOWLEDGEMENTS

The editors wish to thank the copyright holders of the excerpted material included in this volume and the permissions managers of many book and magazine publishing companies for assisting us in securing reproduction rights. We are also grateful to the staffs of the Detroit Public Library, the Library of Congress, the University of Detroit Mercy Library, Wayne State University Purdy/Kresge Library Complex, and the University of Michigan Libraries for making their resources available to us. Following is a list of the copyright holders who have granted us permission to reproduce material in this volume of Energy: Shortage, Glut, or Enough?. *Every effort has been made to trace copyright, but if omissions have been made, please let us know.*

COPYRIGHTED MATERIAL IN *ENERGY: SHORTAGE, GLUT, OR ENOUGH?* ***WAS REPRODUCED FROM THE FOLLOWING SOURCES:***

Coal: Ancient Gift Serving Modern Man, table. Copyright by American Coal Foundation. Reproduced by permission of the American Coal Foundation.

Council on Environmental Quality, table. Reproduced by permission of publisher.

Flavin, Christopher, chart. From an illustration in *Reassessing Nuclear Power: The Fallout from Chernoby from Natural Resources Defense Council: World Watch Paper 75*. World Watch Institute, 1987. Copyright © 1987 by World Watch Institute. Reproduced by permission of publisher. For on-line information see http://www.worldwatch.org.

Mahoney, Katherine and Linda K. Murakami, table, 1993. From *Farewell to Arms: Cleaning Up Nuclear Weapons Facilities.* National Conference of State Legislatures, 1993. Copyright © 1993 by National Conference of State Legislatures. Reproduced by permission.

OCRWM Bulletin, charts, Summer/Fall, 1994. Office of Civilian Radioactive Waste Management, 1994. Copyright © by 1994 Office of Civilian Radioactive Waste Management. Reproduced by permission.

U.S. General Accounting Office, April, 2000, charts. From *Nuclear Safety: Concerns with the Continuing Operation of Soviet-Designed Nuclear Power Reactors;* September, 1999, charts. From *Low-Level Radioactive Wastes: States are Not Developing Disposal Facilities;* 1998, chart. From *Department of Energy: Problems and Progress in Managing Plutonium;* 1998, chart. From *Nuclear Waste: Understanding of Waste Migration at Hanford is Inadequate for Key Decisions;* 1993, 1994 charts. From *Geothermal Energy: Outlook Limited for Some Uses but Promising for Geothermal Heat Pumps;* February, 2000, chart. From *Energy Policy Act of 1992: Limited Progress in Acquiring Alternative Fuel Vehicles and Reaching Fuel Goals.* All courtesy of the U.S. General Accounting Office.

CHAPTER 1

AN ENERGY OVERVIEW

Energy is essential to life. Living creatures draw on energy flowing through the environment and convert it to forms they can use. The most fundamental energy flow for living creatures is the energy of sunlight, and the most important conversion is the act of primary production, in which plants and phytoplankton convert sunlight into biomass by photosynthesis. Earth's web of life, including human beings, rests on this foundation.

— U.S. Energy Information Administration, 1998

A HISTORICAL PERSPECTIVE

Over the centuries, people have found ways to expand the amount of energy available, such as harnessing animals or inventing machines to tap the power of wind or water. Industrialization of the modern world was accompanied by the widespread and growing use of fossil fuels, such as coal, oil, and natural gas.

The result was one of the most profound social changes in history in the space of only a few generations. In the mid-1800s, most Americans lived in rural areas and worked in agriculture. The country ran mainly on wood fuel. A hundred years later, after the nation had become the world's largest producer and consumer of fossil fuels, most Americans were city dwellers, and the United States had roughly tripled its per capita use of energy and become a global superpower.

The United States has always been a resource-abundant nation. However, it was not until after the mid-1800s that the total work output of all types of engines surpassed that of work animals. The westward expansion helped change things. As railroads grew to the west and to the mountains, coal found along railroad rights of way began to replace wood as the major fuel source. As industrialization proceeded, petroleum and natural gas replaced coal. The United States has relied heavily on these three fossil fuels—coal, petroleum, and natural gas.

For much of its history, the United States was mostly self-sufficient in energy, although small amounts of coal were imported from Britain in colonial times. Through the 1950s production and consumption were nearly equal. Following that, however, consumption outpaced domestic production, and by the 1970s, the gap had widened and continues to do so. That gap has been made up by energy imports from other countries.

In 1973 the United States moved to support Israel in its war with its Arab neighbors leading several of the oil-exporting nations of the Middle East to cut off exports of oil. This caused the price of oil to climb to between $14 and $19 per barrel, up from $2 and $3 a year earlier. The United States then faced a severe energy shortage, leading to blackouts of cities and industries, and the closing of factories, schools, and lines at gasoline service stations. The energy problem quickly became an energy crisis.

Energy is not the same issue that it was during the 1970s and the early 1980s. Then, energy prices soared and people worried about shortages. The mid-1980s brought a surplus of oil as a result of increased pumping by the Organization of Petroleum Exporting Countries (OPEC), and a slowing of the economies of the industrially developed nations. Consequently, oil prices dropped, production declined, consumption grew, and imports increased.

Since 1980 energy scarcities have no longer been seen as a "crisis." The emphasis of the 1970s on making the United States energy-independent dwindled and virtually disappeared. Tax credits for solar, wind, and other energy conservation approaches were eliminated during the 1980s. During the Reagan Administration the policy was to let the economic marketplace decide the nation's energy policy—a practice that continued into the 1990s—and the marketplace was dominated by fossil fuel producers. Nonetheless, the willingness of the United States to go to war against Iraq when it invaded Kuwait indicated American commitment to protect its oil resources. The collapse of the Asian economies in the mid-1990s led to a further drop in the demand for energy, and petroleum reached its lowest cost ever in the late 1990s. OPEC reacted to this by curtailing production, which sharply boosted prices in 2000.

An emerging factor in energy policy has been the environment. Concerns about pollution and ozone depletion have led to automobile and industrial emissions regulation. Governments around the world have become concerned about global warming and other environmental degradation. Concern about environmental issues in general, however, has dropped during times when oil seems once again plentiful and cheap.

A PROBLEM FOR THE GOVERNMENT OR THE MARKETPLACE?

In 1977 President Jimmy Carter (a Democrat) described the energy problem as one that could only be "effectively addressed by a Government that accepts responsibility for dealing with it comprehensively and by a public that understands its seriousness and is ready to make necessary sacrifices." On the other hand, the Republican administration of Ronald Reagan downplayed the importance the Carter Administration had placed on government responsibility for dealing with the energy problem.

The Reagan Administration sharply cut federal programs for energy and opposed government intervention in energy markets. For example, the administration refused to tax energy imports which may have stimulated domestic production and conservation. President Reagan believed that the expansion of the federal government's role in energy policy was counterproductive and misguided. His administration pursued a course of transferring the center of the decision-making process to the states, the private sector, and individuals. Reagan's decontrol of petroleum and natural gas reflected his commitment to reducing federal regulation.

The subsequent Republican administration of President George Bush continued the Reagan policy of limiting government regulation of the energy industry. In 1991 Bush unveiled a long-awaited energy policy that promised to reduce U.S. dependence on foreign oil by increasing domestic oil production and nuclear power. President Bush's aim was to rely on "the power of the marketplace, the common sense of the American people, and the responsible leadership of government and industry." He planned to achieve this by, among other proposals, producing additional oil from environmentally sensitive areas, encouraging pipeline construction, simplifying the construction permit process of nuclear power plants, and increasing competition in the production of electricity. His proposals did not include government-directed conservation efforts or tax incentives.

Conservationists disagreed with President Bush, objecting to increased offshore drilling, especially in the coastal plain of the Arctic National Wildlife Refuge (ANWR) in Alaska. (See Chapter 5 for additional discussion.) They also wanted to see automobile fuel mileage increased and conservation methods stressed, rather than increasing the use of nuclear power.

The Democratic administration of Bill Clinton brought government involvement back into energy and environmental issues, although major efforts were stalemated by a Republican Congress. The Clinton Administration increased funds for alternative energy research, mandated new energy efficiency measures, and proceeded with the enforcement of emission standards to clean up the environment. At the same time, it opened up several areas for oil exploration that had previously been out of bounds, such as some Alaskan and offshore areas, and backed off on raising CAFE (Corporate Average Fuel Economy) standards in automobiles. (See Chapter 10 for additional discussion.)

DOMESTIC ENERGY USAGE

Domestic Production

American energy production has leveled off in recent years, remaining almost steady since 1996. Low oil prices throughout the 1990s have led to a situation where drilling costs can often be higher than potential returns. In fact, the drop in oil prices contributed to a decline in the prices of other fossil fuels.

Total energy production in the U.S. in 1999 reached 72.5 quadrillion Btu. (See Table 1.1 and Figure 1.1. Note: one quadrillion Btu equals approximately the energy produced by 170 million barrels of crude oil. Large production and consumption figures are given in these units to make it easier to compare the various types of energy, which come in different forms. "Btu" stands for British thermal units.) Figure 1.2 shows that over the past 25 years, the energy production from coal has generally been increasing, the energy production from oil has generally been declining, and the energy from natural gas, after falling in the 1970s and early 1980s, has been increasing. Energy from nuclear power has increased, while hydroelectric and biofuel power have remained relatively steady.

Domestic Consumption

Energy consumption more than doubled from 1949 to 1973, increasing from 30 to 74 quadrillion Btu. Meanwhile, the economy grew at about the same rate, so the increased consumption of energy reflected the growth in the economy—as the nation grew, it used more fuel, mainly more petroleum and natural gas. However, after the huge 1973 oil price increases, energy consumption fluctuated, eventually returning to 1973 levels by 1986. Following the drop in crude oil prices in 1986, U.S.

TABLE 1.1

Energy production by source, 1950–99

(quadrillion btu)

Year	Fossil fuels: Coal	Fossil fuels: Natural gas (dry)	Fossil fuels: Crude oil [1]	Fossil fuels: Natural gas plant liquids	Fossil fuels: Total fossil fuels	Nuclear electric power	Hydroelectric pumped storage [2]	Renewable energy: Conventional hydroelectric power	Renewable energy: Geothermal	Renewable energy: Wood and waste [3]	Renewable energy: Solar	Renewable energy: Wind	Renewable energy: Total renewable energy	Total
1950	14.060	6.233	11.447	0.823	32.563	0	(4)	1.415	0	1.562	0	0	2.978	35.540
1960	10.817	12.656	14.935	1.461	39.869	0.006	(4)	1.608	0.001	1.320	0	NA	2.929	42.804
1970	14.607	21.666	20.401	2.512	59.186	0.239	(4)	2.634	0.011	[R]1.429	0	NA	[R]4.074	[R]63.499
1975	14.989	19.640	17.729	2.374	54.733	1.900	(4)	3.155	0.070	[R]1.427	0	NA	[R]4.722	[R]61.355
1980	18.598	19.908	18.249	2.254	59.008	2.739	(4)	2.900	0.110	[R]2.483	0	NA	[R]5.493	[R]67.240
1981	18.377	19.699	18.146	2.307	58.529	3.008	(4)	2.758	0.123	2.590	0	NA	5.471	67.007
1982	18.639	18.319	18.309	2.191	57.458	3.131	(4)	3.266	0.105	[R]2.615	0	NA	[R]5.985	[R]66.574
1983	17.247	16.593	18.392	2.184	54.416	3.203	(4)	3.527	0.129	2.831	0	(s)	6.488	64.106
1984	19.719	18.008	18.848	2.274	58.849	3.553	(4)	3.386	0.165	2.880	0	(s)	6.431	68.832
1985	19.325	16.980	18.992	2.241	57.539	4.149	(4)	2.970	0.198	[R,5]2.862	0	(s)	[R,5]6.030	[R,5]67.718
1986	19.509	16.541	18.376	2.149	56.575	4.471	(4)	3.071	0.219	[R,5]2.840	0	(s)	[R,5]6.131	[R,5]67.177
1987	20.141	17.136	17.675	2.215	57.167	4.906	(4)	2.635	0.229	[R]2.822	0	(s)	[R]5.686	[R]67.759
1988	20.738	17.599	17.279	2.260	57.875	5.661	(4)	2.334	0.217	[R,5]2.940	0	(s)	[R,5]5.491	[R,5]69.028
1989	21.346	17.847	16.117	2.158	57.468	5.677	(4)	[R,6]2.856	[R,6]0.327	[R,6]3.050	[R,6]0.059	[R,6]0.024	[R,6]6.316	[R,6]69.461
1990	22.456	18.362	15.571	2.175	58.564	[R]6.162	-0.036	[R,7]3.049	[R]0.348	[R]2.665	0.063	[R]0.032	[R]6.157	[R]70.847
1991	21.594	18.229	15.701	2.306	57.829	[R]6.580	-0.047	[R]3.022	[R]0.353	[R]2.679	0.066	[R]0.032	[R]6.152	[R]70.513
1992	21.629	18.375	15.223	2.363	57.590	[R]6.608	-0.043	2.618	0.361	[R]2.826	0.068	0.030	[R]5.903	[R]70.058
1993	20.249	18.584	14.494	2.408	55.736	[R]6.520	-0.042	2.893	0.375	[R]2.782	0.071	0.031	[R]6.152	[R]68.366
1994	22.111	19.348	14.103	2.391	57.952	6.838	-0.035	2.685	0.370	2.914	0.072	0.036	6.077	70.833
1995	22.029	19.101	13.887	2.442	57.458	7.177	-0.028	3.209	0.321	[R]3.044	0.073	0.033	[R]6.679	[R]71.287
1996	22.684	[R]19.363	13.723	2.530	[R]58.299	7.168	-0.032	[R]3.594	0.339	[R]3.104	0.075	0.035	[R]7.147	[R]72.582
1997	23.211	19.394	13.658	2.495	58.758	6.678	-0.042	[R]3.720	[R]0.327	[R]2.982	0.074	[R]0.034	[R]7.138	[R]72.532
1998	[R]23.719	[R]19.288	[R]13.235	[R]2.420	[R]58.662	7.157	-0.046	[R]3.347	[R]0.334	[R]2.991	0.074	[R]0.031	[R]6.778	[R]72.550
1999[P]	23.328	19.295	12.544	2.506	57.673	7.733	-0.063	3.226	0.327	3.514	0.076	0.038	7.181	72.523

[1] Includes lease condensate.
[2] Represents total pumped storage facility production minus energy used for pumping.
[3] Values are estimated. For all years, includes wood consumption in all sectors. Beginning in 1970, includes electric utility waste consumption. Beginning in 1981, includes industrial sector waste consumption, and transportation sector use of ethanol blended into motor gasoline. Beginning in 1989, includes expanded coverage of nonutility wood and waste consumption.
[4] Through 1989, pumped storage is included in conventional hydroelectric power.
[5] Not all data were available; therefore, values were interpolated.
[6] There is a discontinuity in this time series between 1988 and 1989 due to the expanded coverage of renewable energy beginning in 1989.
[7] There is a discontinuity in this time series between 1989 and 1990; beginning in 1990, pumped storage is removed.

R=Revised. P=Preliminary. (s)=Less than 0.0005 quadrillion Btu. NA=Not available.

SOURCE: *Annual Energy Review 1999,* Energy Information Administration, Washington, D.C., 2000

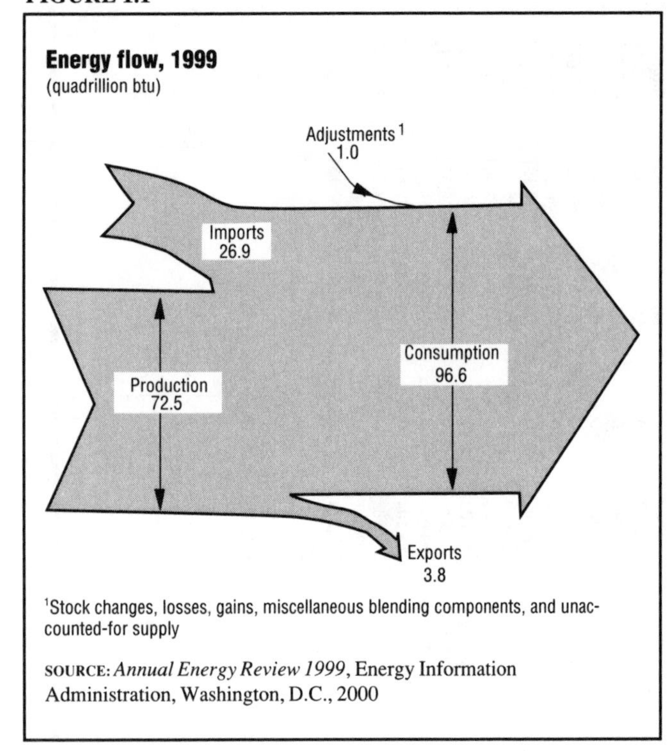

FIGURE 1.1

Energy flow, 1999
(quadrillion btu)

[1]Stock changes, losses, gains, miscellaneous blending components, and unaccounted-for supply

SOURCE: *Annual Energy Review 1999*, Energy Information Administration, Washington, D.C., 2000

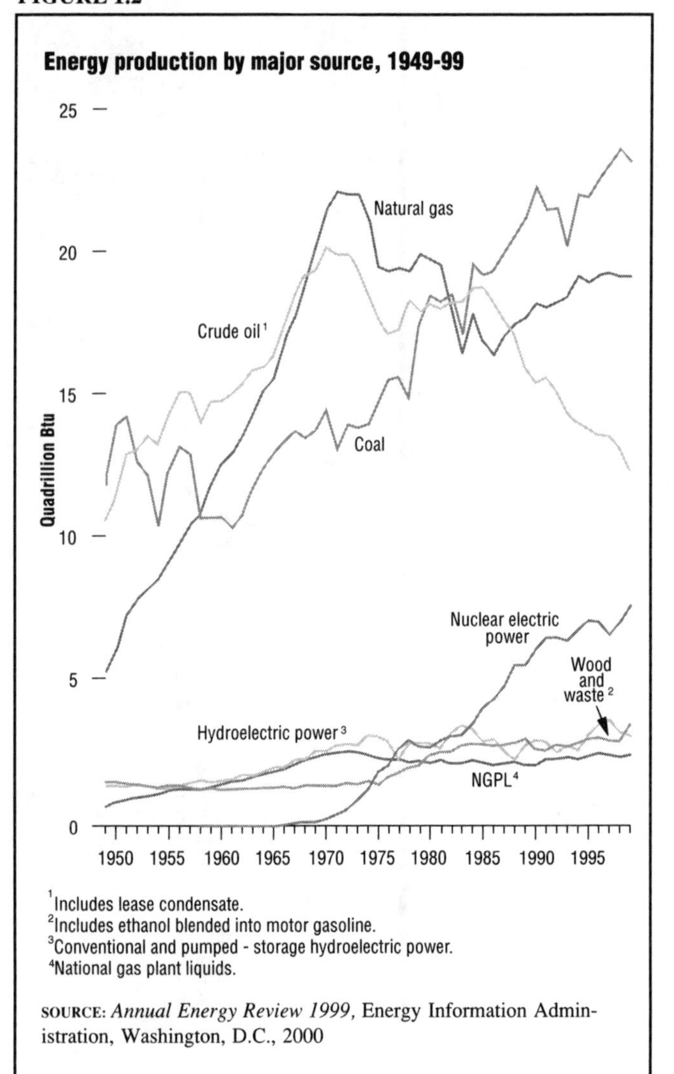

FIGURE 1.2

Energy production by major source, 1949-99

[1]Includes lease condensate.
[2]Includes ethanol blended into motor gasoline.
[3]Conventional and pumped - storage hydroelectric power.
[4]National gas plant liquids.

SOURCE: *Annual Energy Review 1999,* Energy Information Administration, Washington, D.C., 2000

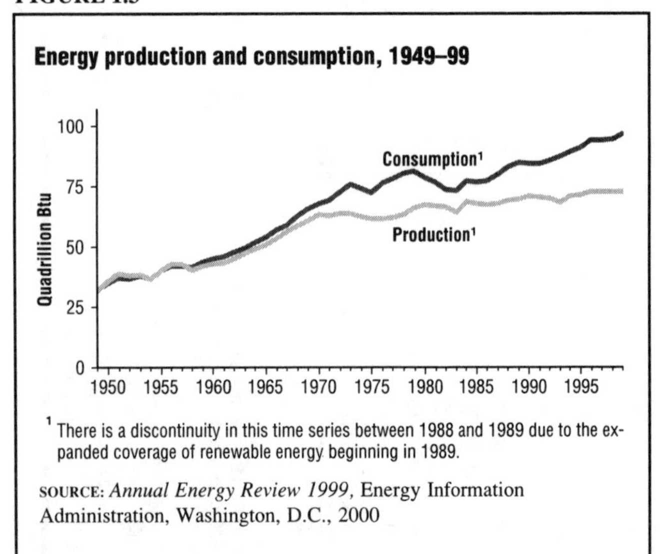

FIGURE 1.3

Energy production and consumption, 1949–99

[1] There is a discontinuity in this time series between 1988 and 1989 due to the expanded coverage of renewable energy beginning in 1989.

SOURCE: *Annual Energy Review 1999,* Energy Information Administration, Washington, D.C., 2000

imports of oil began to rise, and energy consumption increased, reaching an all time high of 97 quadrillion Btu in 1999. (See Figure 1.3.) According to U.S. government census data, the U.S. population expanded by 79 percent from 1949 to 1997, while energy consumption grew by 209 percent during the same period.

Before the 1973 oil crisis, U.S. energy consumption grew faster than economic output. Since 1973 energy consumption has continued to increase, but less vigorously than the increase of economic output, as Americans have become more efficient and use less energy to accomplish more. Energy consumption has shifted slightly away from petroleum and natural gas toward electricity generated by other fuels. In 1973 petroleum and natural gas accounted for 77 percent of total energy consumption; by 1999 their share had dropped to 62 percent (39 percent petroleum and 23 percent natural gas).

Coal, which in 1973 accounted for 17 percent of the energy consumed, accounted for 22 percent in 1999, or 21.7 quadrillion Btu out of a total of 96.6 quadrillion Btu. (See Figure 1.4 for energy production and consumption flows in 1999.) Nuclear power, which contributed barely 1 percent of the nation's consumption in 1973, accounted for 8 percent in 1999. Renewable energy sources (hydroelectric, solar, biofuels, and wind energy) accounted for 8 percent of energy consumed. (See the separate following chapters for further discussion of each energy source.)

ENERGY IMPORTS AND EXPORTS

Since 1958 the United States has consumed more energy than it has produced and has made up the difference by importing energy. Imports (mainly oil) grew

FIGURE 1.4

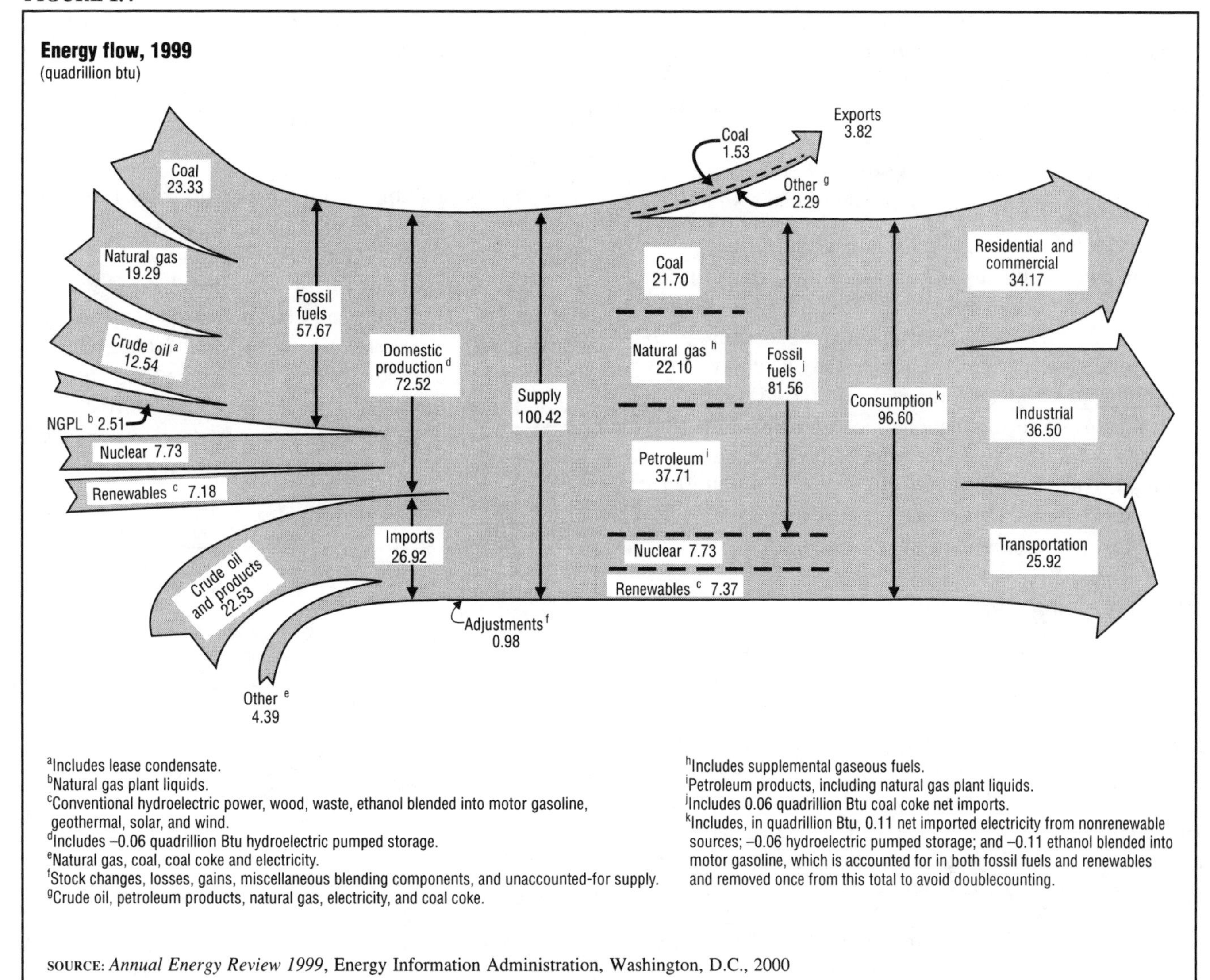

SOURCE: *Annual Energy Review 1999*, Energy Information Administration, Washington, D.C., 2000

rapidly from 1953 through 1973 as the United States built up its economy using inexpensive oil. In 1973 net imports of petroleum reached almost 13 quadrillion Btu.

Although the Arab oil embargo of 1973–74, coupled with increased oil prices, momentarily slowed the growth in petroleum imports, the general increase continued, exceeding 18 quadrillion Btu in 1977. That year, U.S. dependence on petroleum imports rose to 46.5 percent of the nation's oil consumption, the highest as of that date. Despite the lesson of 1973, it took a second round of price increases in 1979–80, accompanied by lengthy and frustrating lines at the gas station, to convince Americans that they had to either become less dependent on imported oil or more conserving of resources, or both. Oil imports declined in 1985, and U.S. dependence on foreign oil decreased sharply to 27.3 percent of oil consumption. (See Figure 2.9, Chapter 2.)

Nonetheless, when the price of crude oil dropped again, the demand returned, and U.S. dependence on foreign sources of oil increased. When Iraq invaded Kuwait in 1990 and appeared to threaten the international flow of oil, the United States quickly showed that it intended to guarantee the flow of oil to America and other industrialized nations by challenging Saddam Hussein and eventually declaring war. By 1999 imported oil accounted for a record 49.7 percent of U.S. oil consumption. (See Figure 2.9, Chapter 2.) The idea of "energy independence," considered important two decades ago, is a distant goal now, and demand continues to grow.

Since 1950 the United States has produced more coal than it has consumed and has been an exporter of coal to other nations. In 1999 coal exports totaled 1.53 quadrillion Btu, approximately 40 percent of U.S. energy exports. (See Figure 1.5 and Table 1.2.)

FOSSIL FUEL PRICES

Fossil fuel (crude oil, natural gas, and coal) production prices increased dramatically through the 1970s into the early 1980s. Prices began to level off in the mid-1980s, and in some instances, to drop very sharply—36 percent in 1986 alone. Crude oil, the most expensive of the fossil fuels, tumbled (in real dollars that account for inflation) from $8.78 per million Btu in 1981 to $2.56 in 1999, after an all-time low of $1.82 in 1998. The price of natural gas sank from $3.37 per million Btu in 1983 to $1.78 in 1999. Coal dropped from $2.11 per million Btu in 1975 to 80 cents per million Btu in 1999. (See Figure 1.6 and Table 1.3.)

To indicate how marked this decline in fuel prices was, the composite value of all fossil fuel prices (in real dollars) dropped by two-thirds, from $4.40 in 1981 to $1.55 in 1999. These huge drops meant economic problems in fuel-producing American states such as Texas, Louisiana, Oklahoma, Montana, West Virginia, and Ohio, and in energy-exporting nations such as many of the Middle East nations, Nigeria, Indonesia, Venezuela, and Trinidad. On the other hand, they were a windfall for industries that used a lot of energy, such as airlines, trucking companies, steel mills, and electric utilities.

ENERGY USE BY SECTOR

Energy use can be classified by three main sectors: the residential and commercial sector, the industrial sector, and the transportation sector. Historically, industry has been the largest consuming sector of the economy, followed by the residential and commercial sector and transportation. In 1999 industry used about 37 quadrillion Btu, compared to 34 quadrillion Btu in the residential and commercial sector and 26 quadrillion Btu in transportation. (See Figure 1.7.)

Within the sectors, energy sources have changed over time. For example, in the residential and commercial sector, coal was the leading source until about 1951 but disappeared after that in favor of petroleum and then natural gas. In transportation, however, reliance on petroleum has been steady since 1949. (See Table 1.4.)

INTERNATIONAL ENERGY USAGE

World Production

World production of primary energy rose from 248 quadrillion Btu in 1973 to 382 quadrillion Btu in 1998. (See Table 1.5.) The Energy Information Administration (EIA) of the U.S. Department of Energy reported that the world's total output of primary energy has increased at an average annual rate of 1.8 percent since 1987. Crude oil has been the most heavily used type of energy, accounting for 37 percent of world energy production. Coal production contributed 23 percent; natural gas, 22 percent; and hydroelectric power, nuclear power, and renewables provided the remainder.

FIGURE 1.5

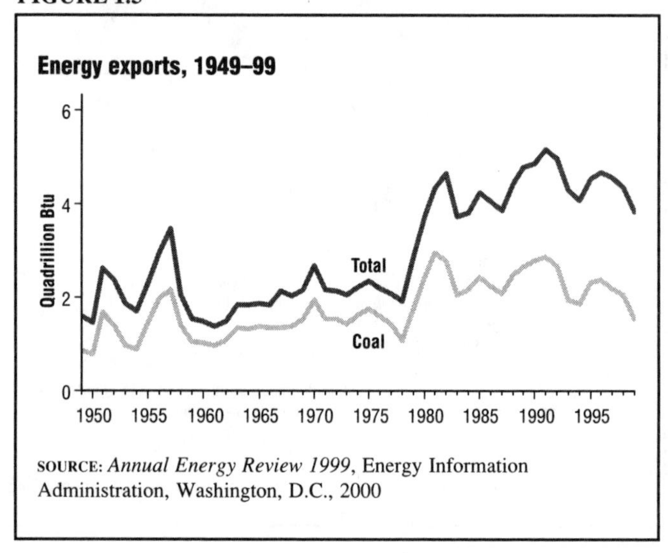

SOURCE: *Annual Energy Review 1999*, Energy Information Administration, Washington, D.C., 2000

FIGURE 1.6

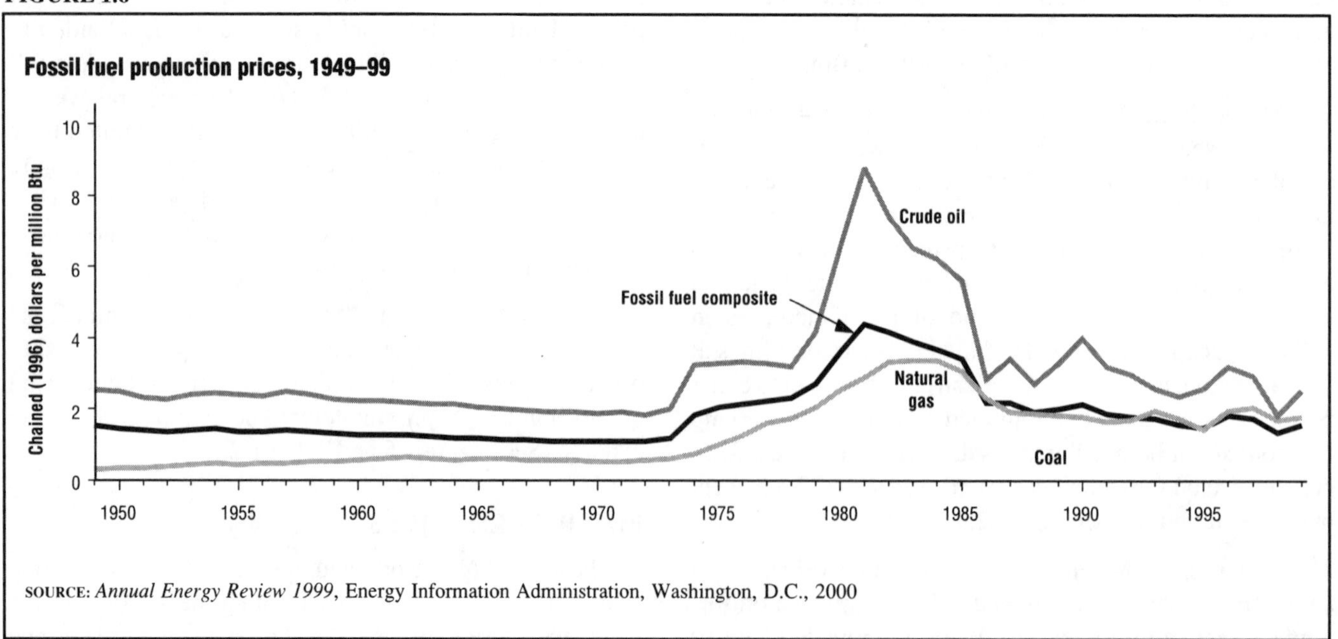

SOURCE: *Annual Energy Review 1999*, Energy Information Administration, Washington, D.C., 2000

TABLE 1.2

Energy imports, exports, and net imports, 1950–99
(quadrillion Btu)

Year	Imports: Coal	Imports: Natural gas (dry)	Imports: Petroleum[1]	Imports: Other[2]	Imports: Total	Exports: Coal	Exports: Natural gas (dry)	Exports: Petroleum	Exports: Other[2]	Exports: Total	Net Imports: Coal	Net Imports: Natural gas (dry)	Net Imports: Petroleum[1]	Net Imports: Other[2]	Net Imports: Total
1950	0.01	0.00	1.89	0.04	1.93	0.79	0.03	0.64	0.01	1.47	-0.78	-0.03	1.24	0.03	0.47
1960	0.01	0.16	4.00	0.06	4.23	1.02	0.01	0.43	0.02	1.48	-1.02	0.15	3.57	0.04	2.74
1970	(s)	0.85	7.47	0.07	8.39	1.94	0.07	0.55	0.11	2.66	-1.93	0.77	6.92	-0.04	5.72
1975	0.02	0.98	12.95	0.16	14.11	1.76	0.07	0.44	0.08	2.36	-1.74	0.90	12.51	0.08	11.75
1980	0.03	1.01	14.66	0.28	15.97	2.42	0.05	1.16	0.09	3.72	-2.39	0.96	13.50	0.18	12.25
1981	0.03	0.92	12.64	0.39	13.97	2.94	0.06	1.26	0.06	4.33	-2.92	0.86	11.38	0.33	9.65
1982	0.02	0.95	10.78	0.35	12.09	2.79	0.05	1.73	0.06	4.63	-2.77	0.90	9.05	0.28	7.46
1983	0.03	0.94	10.65	0.41	12.03	2.04	0.06	1.57	0.05	3.72	-2.01	0.89	9.08	0.36	8.31
1984	0.03	0.85	11.43	0.46	12.77	2.15	0.06	1.54	0.05	3.80	-2.12	0.79	9.89	0.40	8.96
1985	0.05	0.95	10.61	0.49	12.10	2.44	0.06	1.66	0.08	4.23	-2.39	0.90	8.95	0.41	7.87
1986	0.06	0.75	13.20	0.43	14.44	2.25	0.06	1.67	0.08	4.06	-2.19	0.69	11.53	0.36	10.38
1987	0.04	0.99	14.16	0.57	15.76	2.09	0.05	1.63	0.08	3.85	-2.05	0.94	12.53	0.49	11.91
1988	0.05	1.30	15.75	0.47	17.56	2.50	0.07	1.74	0.10	4.42	-2.45	1.22	14.01	0.37	13.15
1989	0.07	1.39	17.16	0.34	18.96	2.64	0.11	1.84	0.18	4.77	-2.57	1.28	15.33	0.15	14.19
1990	0.07	1.55	17.12	R0.22	R18.95	2.77	0.09	1.82	R0.18	R4.87	-2.70	1.46	15.29	0.03	R14.09
1991	0.08	1.80	16.35	R0.27	R18.50	2.85	0.13	2.13	R0.04	R5.16	-2.77	1.67	14.22	R0.22	R13.34
1992	0.10	2.16	16.97	R0.35	R19.58	2.68	0.22	2.01	R0.05	R4.96	-2.59	1.94	14.96	R0.31	R14.62
1993	R0.20	2.40	18.51	R0.39	R21.50	1.96	0.14	2.12	R0.06	R4.28	R-1.76	2.25	16.40	0.32	R17.22
1994	R0.22	2.68	R19.24	R0.58	R22.73	1.88	0.16	1.99	R0.05	R4.08	R-1.66	2.52	17.26	R0.53	R18.65
1995	R0.24	2.90	18.86	R0.55	R22.54	2.32	0.16	1.99	R0.07	R4.54	R-2.08	2.74	16.87	R0.47	R18.00
1996	R0.20	3.00	20.27	0.52	R23.99	2.37	0.16	2.06	R0.07	R4.66	R-2.17	2.85	18.21	R0.45	R19.33
1997	0.19	3.06	R21.74	R0.52	R25.52	2.19	0.16	2.10	R0.12	R4.57	-2.01	2.90	R19.64	R0.40	R20.94
1998	0.22	R3.22	R22.91	R0.50	R26.86	R2.05	0.16	R1.97	R0.16	R4.34	R-1.83	R3.06	R20.94	R0.34	R22.51
1999P	0.23	3.64	22.53	0.52	26.92	1.53	0.16	1.96	0.17	3.82	-1.31	3.48	20.57	0.36	23.10

[1] Includes imports into the Strategic Petroleum Reserve, which began in 1977.
[2] Coal coke and small amounts of electricity transmitted across U.S. borders with Canada and Mexico.

R=Revised. P=Preliminary. (s)=Less than 0.005 quadrillion Btu and greater than -0.005 quadrillion Btu.

Notes: Includes trade between the United States (50 States and the District of Columbia) and its territories and possessions.

SOURCE: *Annual Energy Review 1999*, Energy Information Administration, Washington, D.C., 2000

TABLE 1.3

Fossil fuel production prices, 1949–99
(dollars per million btu)

	Coal [1]		Natural gas [2]		Crude oil [3]		Fossilf fuel composite [4]		
Year	Nominal	Real [5]	Nominal	Real [5]	Nominal	Real [5]	Nominal	Real [5]	Percent change [6]
1949	0.21	[R]1.22	0.05	[R]0.31	0.44	[R]2.54	0.26	[R]1.52	—
1950	0.21	[R]1.19	0.06	[R]0.36	0.43	[R]2.48	0.26	[R]1.46	-3.8
1951	0.21	[R]1.13	0.06	[R]0.34	0.44	[R]2.33	0.26	[R]1.38	[R]-5.3
1952	0.21	[R]1.10	0.07	[R]0.38	0.44	[R]2.30	0.26	[R]1.37	-0.7
1953	0.21	[R]1.08	0.08	[R]0.42	0.46	[R]2.40	0.27	[R]1.42	[R]3.2
1954	0.19	[R]0.99	0.09	[R]0.46	0.48	[R]2.46	0.28	[R]1.42	0.5
1955	0.19	[R]0.94	0.09	[R]0.45	0.48	[R]2.42	0.27	[R]1.37	[R]-3.9
1956	0.20	[R]0.97	0.10	[R]0.48	0.48	[R]2.35	0.28	[R]1.36	[R]-0.8
1957	0.21	[R]0.99	0.10	[R]0.47	0.53	[R]2.52	0.30	[R]1.42	[R]4.1
1958	0.20	[R]0.94	0.11	[R]0.50	0.52	[R]2.40	0.29	[R]1.35	[R]-4.7
1959	0.20	[R]0.91	0.12	[R]0.54	0.50	[R]2.28	0.29	[R]1.31	[R]-3.1
1960	0.19	[R]0.87	0.13	[R]0.57	0.50	[R]2.24	0.28	[R]1.28	[R]-2.4
1961	0.19	[R]0.85	0.14	[R]0.60	0.50	[R]2.22	0.29	[R]1.28	0.0
1962	0.19	[R]0.82	0.14	[R]0.64	0.50	[R]2.20	0.29	[R]1.27	[R]-0.7
1963	0.18	[R]0.80	0.14	[R]0.63	0.50	[R]2.16	0.28	[R]1.23	[R]-2.8
1964	0.18	[R]0.79	0.14	[R]0.58	0.50	[R]2.13	0.28	[R]1.19	[R]-3.5
1965	0.18	[R]0.77	0.14	[R]0.61	0.49	[R]2.07	0.28	[R]1.16	[R]-1.9
1966	0.19	[R]0.77	0.14	[R]0.59	0.50	[R]2.03	0.28	[R]1.14	-1.7
1967	0.19	[R]0.76	0.14	[R]0.58	0.50	[R]2.00	0.28	[R]1.13	-1.6
1968	0.19	[R]0.74	0.14	[R]0.54	0.51	[R]1.93	0.28	[R]1.08	[R]-3.8
1969	0.21	[R]0.76	0.15	[R]0.56	0.53	[R]1.93	0.30	[R]1.08	0.0
1970	0.27	[R]0.92	0.15	[R]0.53	0.55	[R]1.89	0.32	[R]1.09	[R]0.9
1971	0.30	[R]1.00	0.16	[R]0.53	0.58	[R]1.91	0.34	[R]1.11	[R]1.8
1972	0.33	[R]1.04	0.17	[R]0.54	0.58	[R]1.84	0.35	[R]1.10	[R]-1.3
1973	0.37	[R]1.09	0.20	[R]0.60	0.67	[R]2.00	0.40	[R]1.18	[R]7.7
1974	0.69	[R]1.87	0.27	[R]0.74	1.18	[R]3.23	0.68	[R]1.85	55.8
1975	0.84	[R]2.11	0.40	[R]1.00	1.32	[R]3.30	0.82	[R]2.05	[R]11.1
1976	0.86	[R]2.02	0.53	[R]1.26	1.41	[R]3.34	0.90	[R]2.13	[R]3.9
1977	0.88	[R]1.96	0.72	[R]1.61	1.48	[R]3.28	1.01	[R]2.24	[R]5.0
1978	0.98	[R]2.04	0.84	[R]1.73	1.55	[R]3.22	1.12	[R]2.31	[R]3.3
1979	1.06	[R]2.02	1.08	[R]2.07	2.18	[R]4.17	1.42	[R]2.71	[R]17.3
1980	1.10	[R]1.93	1.45	[R]2.54	3.72	[R]6.52	2.04	[R]3.58	[R]32.0
1981	1.18	[R]1.90	1.80	[R]2.88	5.48	[R]8.78	2.74	[R]4.40	[R]22.9
1982	1.22	[R]1.85	2.22	[R]3.35	4.92	[R]7.42	2.76	[R]4.16	[R]-5.4
1983	1.18	[R]1.71	2.32	[R]3.37	4.52	[R]6.56	2.70	[R]3.92	[R]-5.8
1984	1.16	[R]1.63	2.40	[R]3.36	4.46	[R]6.25	2.65	[R]3.70	-5.5
1985	1.15	[R]1.56	2.26	[R]3.06	4.15	[R]5.64	2.51	[R]3.41	[R]-8.0
1986	1.09	[R]1.44	1.75	[R]2.32	2.16	[R]2.86	1.65	[R]2.20	[R]-35.6
1987	1.05	[R]1.36	1.50	[R]1.94	2.66	[R]3.42	1.70	[R]2.19	-0.2
1988	1.01	[R]1.26	1.52	[R]1.90	2.17	[R]2.70	1.53	[R]1.91	[R]-12.8
1989	1.00	[R]1.20	1.53	[R]1.83	2.73	[R]3.28	1.67	[R]2.01	[R]5.0
1990	1.00	[R]1.15	1.55	[R]1.79	3.45	[R]3.99	1.84	[R]2.13	[R]6.1
1991	0.99	[R]1.10	1.48	[R]1.65	2.85	[R]3.18	1.67	[R]1.86	[R]-12.5
1992	0.97	[R]1.06	1.57	[R]1.71	2.76	[R]3.00	1.66	[R]1.80	[R]-3.1
1993	0.93	[R]0.99	1.84	[R]1.96	2.46	[R]2.61	1.67	[R]1.78	[R]-1.6
1994	0.91	[R]0.94	1.67	[R]1.74	2.27	[R]2.37	1.53	[R]1.59	[R]-10.5
1995	0.88	[R]0.90	1.40	[R]1.43	2.52	[R]2.57	1.47	[R]1.50	[R]-5.5
1996	0.87	[R]0.87	1.96	[R]1.96	3.18	[R]3.18	1.82	[R]1.82	[R]21.3
1997	0.85	[R]0.84	2.10	[R]2.06	2.97	[R]2.92	1.81	[R]1.77	-2.6
1998	0.83	[R]0.81	[R]1.75	[R]1.70	[R]1.87	[R]1.82	1.41	[R]1.36	[R]-23.1
1999	[7]0.83	[7]0.80	1.86	1.78	2.68	2.56	[7]1.63	[7]1.55	[7]13.8

[1] Bituminous coal, subbituminous coal, and lignite prices are based on the value of coal produced at free-on-board (f.o.b.) mines; anthracite prices through 1978 are f.o.b. preparation plants and for 1979 forward are f.o.b. mines.
[2] Wellhead prices.
[3] Domestic first purchase prices.
[4] Derived by multiplying the price per Btu of each fossil fuel by the total Btu content of the production of each fossil fuel and dividing this accumulated value of total fossil fuel production by the accumulated Btu content of total fossil fuel production.
[5] In chained (1996) dollars, calculated by using gross domestic product implicit price deflators.
[6] Based on real values.
[7] Calculated using the 1998 coal price for the 1999 value.
R=Revised. P=Preliminary. — = Not applicable.

SOURCE: *Annual Energy Review 1999,* Energy Information Administration, Washington, D.C., 2000

In 1998 the United States, Russia, and China were, by far, the leading producers of energy, followed by Saudi Arabia, Canada, and the United Kingdom. (See Figure 1.8.) Almost all the energy from the Middle East is in the form of oil or natural gas, while coal is a major source in China. Canada is the leading producer of hydroelectric power and, alone, accounts for 14 percent of world production. France produces the highest percentage of its energy from nuclear power.

World Consumption

Table 1.6 shows the world consumption of energy, by countries and regions, from 1989 to 1998. Five countries—the United States, China, Russia, Japan, and Germany—together consumed 50 percent of the world's total energy in 1998. The United States, by far the world's largest consumer of energy, used 94.8 quadrillion Btu, nearly triple that of China's 33.9 quadrillion Btu, while Russia consumed 26.0 quadrillion Btu.

FUTURE TRENDS IN ENERGY USAGE

The Energy Information Administration (EIA) of the U.S. Department of Energy annually forecasts energy supply, demand, and prices in its *Annual Energy Outlook,* which is used by decision-makers in the public and private sectors. The EIA's latest projections, from 1998 through 2020, are based on current U.S. laws, regulations, and economic conditions.

Worldwide oil demand is projected to increase from 75 million barrels of oil per day in 1998 to 112 million barrels per day in 2020. World oil production per person has been declining since 1979, as populations are growing faster than oil production, and this trend is expected to continue. Global oil production has been estimated to peak between 2011 and 2025. OPEC production is expected to nearly double by 2020, while non-OPEC oil production will increase as well, helped by technology improvements. Offshore production is expected to increase, particularly in Nigeria, Algeria, and Venezuela. Increased energy demand will come from developing countries as they compete with industrial nations. Meeting increasing demand will require more storage and transportation infrastructure.

In the next two decades, electricity prices in the U.S. are projected to decline slightly, due to restructuring laws designed to increase competition in the industry. (See Figure 1.9.) Coal prices are also expected to decline, while crude oil and natural gas increase in price through 2020. The jump in the price for oil using the 2000 forecast model shows the effects of OPEC's reduction in production in March 1999. Oil is projected to increase from the $12 per barrel price in 1998 to about $22 per barrel in 2020.

Energy consumed from nuclear power will decline significantly through 2020, with no new plants being

FIGURE 1.7

Energy consumption by end-use sector, 1949–99

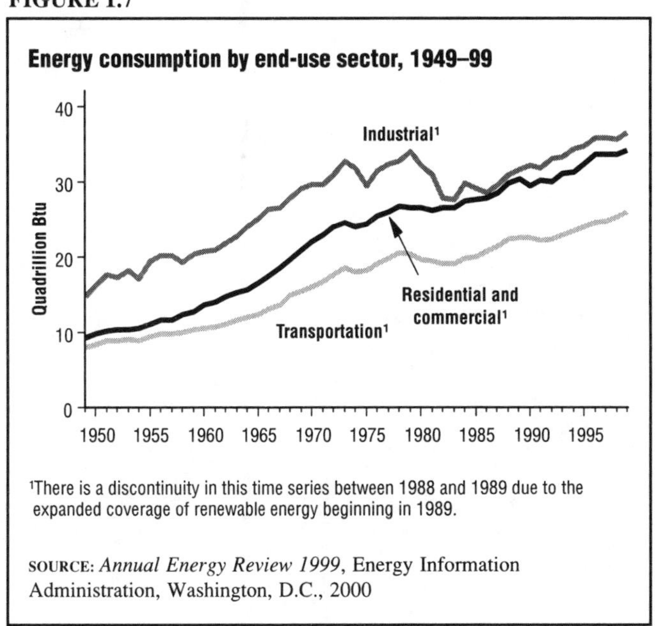

[1]There is a discontinuity in this time series between 1988 and 1989 due to the expanded coverage of renewable energy beginning in 1989.

SOURCE: *Annual Energy Review 1999*, Energy Information Administration, Washington, D.C., 2000

FIGURE 1.8

Top energy-producing countries, 1998

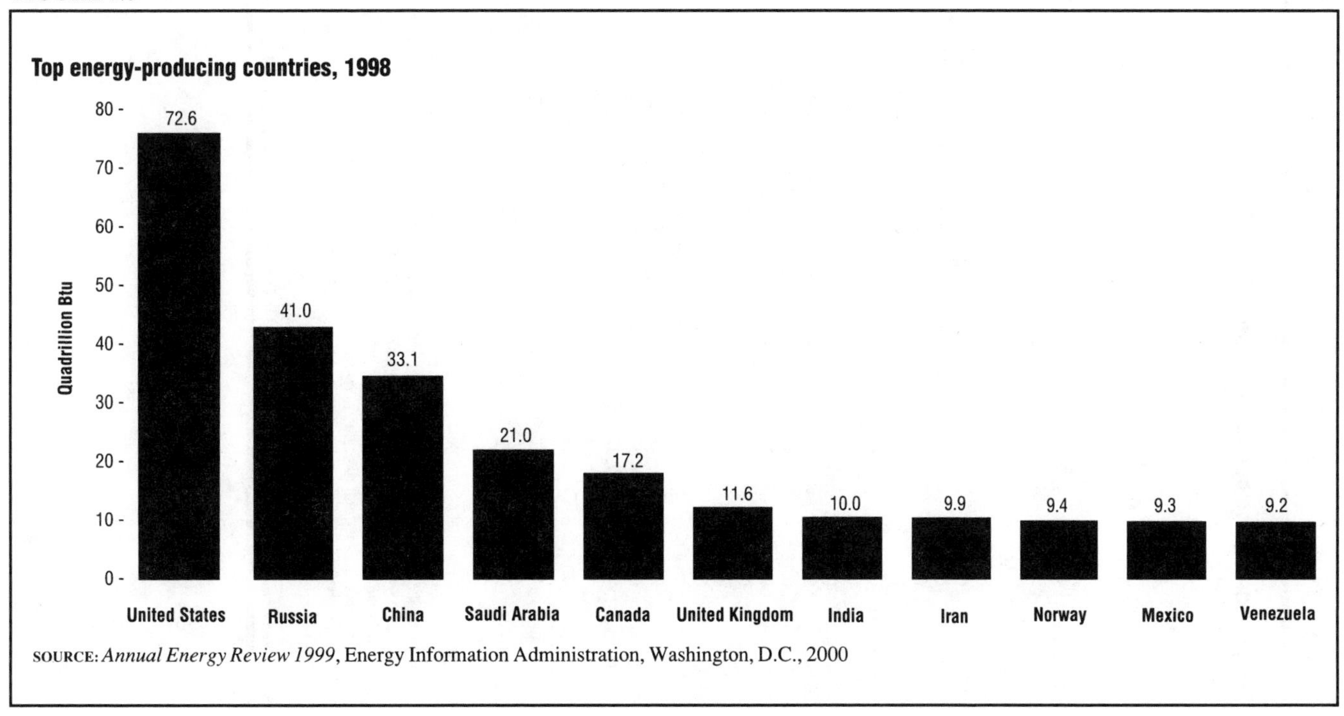

SOURCE: *Annual Energy Review 1999*, Energy Information Administration, Washington, D.C., 2000

TABLE 1.4

Energy consumption by end-use sector, 1949–99
(quadrillion Btu)

	Residential and commercial						Industrial						Transportation		
Year	Coal	Natural gas [1]	Petroleum	Electricity	Losses [2]	Total [3]	Coal	Natural gas [1]	Petroleum	Electricity	Losses [2]	Total [3,4]	Petroleum	Total [5]	Total
1949	2.83	1.39	1.85	0.43	1.72	9.28	5.43	3.19	3.47	0.42	1.68	14.73	6.15	7.99	32.00
1950	2.80	1.64	2.20	0.47	1.76	9.90	5.78	3.55	3.95	0.50	1.86	16.24	6.69	8.49	34.63
1960	0.99	4.27	3.49	1.23	3.06	13.68	4.54	5.97	5.75	1.11	2.76	20.84	10.13	10.60	45.12
1970	0.37	7.46	4.31	2.79	[R]6.77	[R]22.11	4.66	9.54	7.79	1.95	4.72	29.65	15.31	[R]16.10	67.86
1975	0.21	7.58	3.80	3.60	8.70	24.33	3.67	8.53	8.15	2.35	5.66	29.46	17.62	18.25	72.04
1980	0.15	7.54	3.04	4.35	10.58	26.53	3.16	8.39	9.53	2.78	6.76	32.21	19.01	19.69	[R]78.43
1981	0.17	7.24	2.63	4.50	10.70	26.13	3.16	8.26	8.29	2.82	6.70	30.93	18.81	[R]19.50	76.57
1982	0.19	7.43	2.45	4.57	11.00	26.59	2.55	7.12	7.80	2.54	6.12	27.78	18.42	19.07	73.44
1983	0.19	7.02	2.50	4.68	[R]11.23	[R]26.57	2.49	6.83	7.42	2.65	6.36	27.60	18.59	[R]19.14	73.32
1984	0.21	7.29	2.54	4.93	11.51	27.42	2.84	7.45	8.01	2.86	6.68	29.75	19.22	[R]19.81	76.97
1985	0.18	7.08	2.52	5.06	[R]11.86	[R]27.62	2.76	7.08	7.81	2.86	6.69	29.09	19.50	20.07	[R]76.78
1986	0.18	6.82	2.56	[R]5.23	12.06	[R]27.75	2.64	6.69	7.92	2.83	6.53	28.50	20.27	[R]20.82	[R]77.06
1987	0.16	6.95	2.59	5.44	[R]12.47	[R]28.49	2.67	7.32	8.15	2.93	6.71	29.68	20.87	[R]21.46	[R]79.63
1988	0.17	7.51	2.60	5.72	[R]12.91	[R]29.83	2.83	7.70	8.43	3.06	6.90	30.92	21.63	22.31	[R]83.07
1989	0.15	7.73	2.53	5.86	[R]13.16	[R,6]30.43	2.79	8.13	8.13	3.16	7.10	[R,6]31.58	21.87	[R]22.57	[R,6]84.59
1990	0.16	7.22	2.17	[R]6.01	13.24	[R]29.48	2.76	8.50	8.32	3.23	7.10	[R]32.15	21.81	[R]22.54	[R]84.19
1991	0.14	7.51	2.15	6.18	[R]13.44	[R]30.14	2.60	8.62	8.06	3.23	[R]7.02	[R]31.80	21.46	[R]22.13	[R]84.06
1992	0.14	7.73	2.13	[R]6.09	[R]13.18	[R]30.03	2.51	8.97	8.64	3.32	[R]7.18	33.01	21.81	[R]22.47	[R]85.51
1993	0.14	8.04	2.14	[R]6.41	[R]13.72	[R]31.12	2.50	9.41	8.45	3.33	[R]7.13	[R]33.30	22.20	[R]22.89	87.31
1994	0.14	7.97	2.09	6.56	[R]13.95	[R]31.37	2.51	9.56	8.85	3.44	7.32	[R]34.35	[R]22.76	[R]23.52	[R]89.23
1995	0.13	8.09	2.08	6.81	14.43	32.26	2.49	10.06	8.62	3.46	7.32	[R]34.70	[R]23.20	[R]23.97	[R]90.94
1996	0.14	8.63	2.20	7.04	[R]14.95	[R]33.67	2.42	10.39	9.10	3.52	[R]7.47	[R]35.71	[R]23.73	[R]24.52	[R]93.91
1997	0.15	8.42	2.14	[R]7.17	[R]15.21	[R]33.64	2.37	[R]10.31	9.31	3.52	[R]7.47	[R]35.85	[R]23.99	[R]24.82	[R]94.32
1998	[R]0.11	[R]7.77	[R]1.97	[R]7.49	[R]15.83	[R]33.68	[R]2.26	[R]10.17	[R]9.15	[R]3.55	[R]7.50	[R]35.54	[R]24.64	[R]25.36	[R]94.57
1999[P]	0.11	8.02	2.07	7.54	15.89	34.17	2.25	10.23	9.46	3.58	7.55	36.50	25.21	25.92	96.60

[1] Includes supplemental natural gas.
[2] Electrical system energy losses. Total losses are calculated as the sum of energy consumed at electric utilities to generate electricity, utility purchases of electricity from nonutility power producers, and imported electricity, minus exported electricity and electricity consumed by end users. Total losses are allocated to the end-use sectors in proportion to each sector's share of total electricity use.
[3] "Total" also includes renewable energy, which is not shown separately on this table.
[4] Also includes hydroelectric power and net imports of coal coke.
[5] Also includes coal, natural gas, electricity, and electrical system energy losses.
[6] There is a discontinuity in this time series between 1988 and 1989 due to expanded coverage of renewable energy beginning in 1989.

R=Revised. P=Preliminary.

SOURCE: *Annual Energy Review 1999*, Energy Information Administration, Washington, D.C., 2000

TABLE 1.5

World primary energy production by source, 1970–98
(quadrillion btu)

Year	Coal	Natural Gas [1]	Crude Oil [2]	Natural Gas Plant Liquids	Nuclear Electric Power [3]	Hydroelectric Power [3]	Geothermal [3] and Other [4]	Total
1970	62.96	37.09	97.09	3.61	0.90	12.15	1.59	215.39
1971	61.72	39.80	102.70	3.85	1.23	12.74	1.61	223.64
1972	63.65	42.08	108.52	4.09	1.66	13.31	1.68	234.99
1973	63.87	44.44	117.88	4.23	2.15	13.52	1.73	[R]247.82
1974	63.79	45.35	117.82	4.22	2.86	14.84	[R]1.75	250.64
1975	66.20	45.67	113.08	4.12	3.85	15.03	1.74	249.69
1976	67.32	47.62	122.92	4.24	4.52	15.08	1.97	263.67
1977	68.46	48.85	127.75	4.40	5.41	15.56	[R]2.10	272.54
1978	69.56	50.26	128.51	4.55	6.42	16.80	2.32	278.41
1979	73.83	53.93	133.87	4.87	6.69	17.69	2.48	293.36
1980	[R]72.94	[R]54.73	128.12	5.10	7.58	[R]18.06	[R]2.95	[R]289.49
1981	[R]73.06	[R]55.56	120.16	5.36	8.53	[R]18.35	[R]3.09	[R]284.12
1982	[R]75.67	[R]55.49	114.51	5.34	9.51	18.83	[R]3.24	[R]282.59
1983	[R]75.91	[R]56.13	113.97	5.34	10.72	19.73	[R]3.51	[R]285.31
1984	[R]80.12	[R]61.78	116.86	5.71	12.99	20.35	[R]3.64	[R]301.45
1985	[R]83.93	[R]64.22	115.40	5.82	15.37	20.57	[R]3.67	[R]308.98
1986	[R]86.07	[R]65.32	120.24	6.12	16.34	[R]21.03	[R]3.74	[R]318.86
1987	[R]87.89	[R]68.49	121.16	6.32	17.80	[R]21.10	[R]3.80	[R]326.55
1988	[R]89.61	[R]71.81	125.93	6.63	19.30	[R]21.90	[R]3.94	[R]339.12
1989	[R]91.05	[R]74.24	127.98	6.67	[R]19.82	21.76	[R,5]4.29	[R]345.82
1990	[R]92.28	[R]75.91	129.50	6.85	20.37	[R]22.57	[R]3.97	[R]351.46
1991	[R]87.65	[R]76.68	128.77	7.13	21.29	[R]23.00	[R]4.02	[R]348.54
1992	[R]88.35	[R]76.84	129.13	7.38	21.36	[R]22.96	[R]4.32	[R]350.33
1993	[R]85.72	78.35	128.86	7.67	22.07	[R]24.31	[R]4.35	[R]351.33
1994	[R]87.53	[R]79.16	130.46	7.84	22.50	[R]24.49	[R]4.57	[R]356.54
1995	[R]89.67	[R]80.23	133.32	8.14	23.35	[R]25.73	[R]4.72	[R]365.17
1996	[R]90.78	[R]84.03	136.64	8.30	24.17	[R]26.12	[R]4.89	[R]374.93
1997	[R]90.64	[R]84.00	[R]140.52	[R]8.49	[R]23.95	[R]26.76	[R]4.89	[R]379.25
1998[P]	88.50	85.42	143.23	8.73	24.48	26.59	4.98	381.93

[1] Dry production.
[2] Includes lease condensate.
[3] Net generation, i.e., gross generation less plant use.
[4] Includes net electricity generation from wood, other biomass, waste, solar, and wind. Data for the United States also include other renewable energy.
[5] There is a discontinuity in the series between 1988 and 1989 due to the expanded coverage of U.S. renewable energy beginning in 1989.
R=Revised. P=Preliminary.

SOURCE: *Annual Energy Review 1999,* Energy Information Administration, Washington, D.C., 2000

built and old plants being retired. (See Figure 1.10.) Coal and natural gas consumption will grow to meet the increased need for electricity. Hydroelectric power will remain steady, since no new dams are projected, while other renewable energy sources will increase slightly. Rising consumption of petroleum will lead to increasing petroleum imports by the U.S. through 2020. (See Figure 1.11.)

The EIA forecasts that total U.S. energy consumption will increase from 94.9 quadrillion Btu in 1998 to 121 quadrillion Btu by 2020, which could vary depending on economic growth and world oil prices. (See Table 1.7.) Energy use will increase in all sectors to 2020, with transportation, residential, and industrial use growing faster than commercial use. (See Figure 1.12.) Transportation demand will grow at an average annual rate of approximately 1.6 percent through 2020, with consumption increasing from 25.9 to 37.5 quadrillion Btu. In 1997 light vehicles, which include cars, light trucks, sport utility vehicles (SUVs), and vans, accounted for 63 percent of energy use in the transportation sector. (See Figure 1.13.) The agency expects increased light-vehicle travel and slower growth in efficiency of such vehicles because of continuing consumer preference for larger, more powerful vehicles over more efficient ones, and relatively low gas prices. Jet fuel use will increase due to the trend toward more jet travel, which is highly energy-intensive, combined with slower sales of more efficient wide-body aircraft.

Energy and the Environment

While historically, the production, consumption, and price of energy has largely been driven by the supply and demand of petroleum, environmental concerns may play a greater role in the future.

Whether the American public will adjust its energy use out of concern for the environment remains unclear. As of late 2000, the United States had yet to ratify the Kyoto Protocol to the United Nations Framework Convention on Climate Change (UNFCCC), an international treaty designed to curb global warming and air pollution. If the U.S. agrees to international environmental regula-

TABLE 1.6

World primary energy consumption, 1989–98

(quadrillion btu)

Region Country	1989	1990	1991	1992	1993	1994	1995	1996	1997	1998[1]
North America										
Canada	11.08	10.91	10.89	10.92	11.44	11.74	11.75	12.18	12.01	11.85
Mexico	4.80	4.98	5.02	5.13	5.14	5.29	5.36	5.59	5.68	5.92
United States	84.53	84.12	84.03	85.50	87.31	89.27	91.00	93.97	94.38	94.79
Other	0.02	0.01	0.01	0.02	0.01	0.01	0.01	0.01	0.01	0.01
Total	**100.43**	**100.02**	**99.95**	**101.56**	**103.90**	**106.31**	**108.12**	**111.74**	**112.07**	**112.57**
Central & South America										
Argentina	1.97	1.92	1.99	2.15	2.31	2.35	2.43	2.51	2.61	2.72
Brazil	5.63	5.65	5.90	5.95	6.13	6.49	6.81	7.31	7.69	8.08
Chile	0.51	0.57	0.57	0.60	0.65	0.69	0.75	0.81	0.89	0.91
Colombia	0.95	0.89	1.06	0.99	1.06	1.10	1.11	1.19	1.21	1.27
Cuba	0.51	0.50	0.46	0.41	0.41	0.41	0.42	0.43	0.40	0.42
Venezuela	2.06	2.08	2.21	2.22	2.29	2.42	2.47	2.57	2.66	2.63
Other	2.48	2.54	2.65	2.73	2.85	3.04	3.24	3.32	3.49	3.70
Total	**14.11**	**14.16**	**14.83**	**15.05**	**15.70**	**16.50**	**17.24**	**18.15**	**18.95**	**19.73**
Western Europe										
Austria	1.13	1.17	1.21	1.20	1.23	1.23	1.27	1.29	1.32	1.34
Belgium	2.08	2.14	2.27	2.25	2.25	2.31	2.37	2.54	2.60	2.66
Denmark	0.82	0.80	0.83	0.82	0.85	0.84	0.88	0.88	0.89	0.80
Finland	1.13	1.13	1.14	1.18	1.20	1.22	1.12	1.15	1.26	1.26
France	8.61	8.84	9.41	9.43	9.34	9.24	9.45	9.87	9.78	10.00
Germany	—	—	14.29	13.98	14.06	14.01	14.11	14.31	14.10	13.83
Germany, East	3.72	3.33	—	—	—	—	—	—	—	—
Germany, West	11.21	11.48	—	—	—	—	—	—	—	—
Greece	0.99	1.05	1.07	1.04	1.09	1.11	1.12	1.15	1.20	1.26
Ireland	0.35	0.37	0.39	0.40	0.40	0.42	0.45	0.46	0.48	0.52
Italy	6.99	6.98	7.15	7.19	7.01	6.93	7.52	7.64	7.70	7.98
Netherlands	3.26	3.37	3.56	3.53	3.60	3.58	3.69	3.83	3.81	3.82
Norway	1.56	1.58	1.59	1.65	1.65	1.66	1.73	1.75	1.80	1.86
Portugal	0.71	0.75	0.75	0.76	0.78	0.82	0.85	0.88	0.93	0.98
Spain	4.09	3.94	4.15	4.12	4.04	4.22	4.48	4.38	4.70	5.04
Sweden	2.20	2.17	2.17	2.18	2.18	2.16	2.35	2.27	2.17	2.27
Switzerland	1.13	1.17	1.21	1.21	1.19	1.20	1.18	1.22	1.23	1.21
Turkey	1.78	1.97	2.08	2.10	2.33	2.23	2.47	2.70	2.85	2.89
United Kingdom	9.39	9.32	9.58	9.39	9.65	9.64	9.60	10.16	9.74	9.75
Former Yugoslavia	2.23	2.13	1.87	—	—	—	—	—	—	—
Croatia	—	—	—	0.33	0.33	0.36	0.37	0.36	0.38	0.39
Serbia and Montenegro	—	—	—	0.69	0.55	0.60	0.46	0.72	0.78	0.81
Other	0.29	0.30	0.31	0.81	0.80	0.79	0.86	0.84	0.87	0.85
Total	**63.67**	**63.99**	**65.01**	**64.24**	**64.54**	**64.58**	**66.31**	**68.39**	**68.60**	**69.52**
Eastern Europe & former U.S.S.R.										
Bulgaria	1.47	1.30	1.01	1.00	0.93	0.92	0.99	1.02	0.95	0.89
Former Czechoslovakia	4.19	3.98	3.61	3.23	—	—	—	—	—	—
Czech Republic	—	—	—	—	1.57	1.56	1.60	1.66	1.69	1.75
Slovakia	—	—	—	—	0.74	0.75	0.82	0.82	0.81	0.78
Hungary	1.34	1.28	1.20	1.10	1.09	1.10	1.07	1.08	1.04	1.05
Poland	4.83	3.87	3.84	3.81	3.95	3.78	3.58	4.08	3.86	3.51
Romania	3.16	2.88	2.24	2.06	2.00	1.89	2.03	2.06	2.05	1.76
Former U.S.S.R	59.75	60.69	57.46	—	—	—	—	—	—	—
Azerbaijan	—	—	—	0.97	0.83	0.75	0.72	0.64	0.63	0.49
Belarus	—	—	—	1.57	1.34	1.10	1.06	1.06	1.06	1.06
Kazakhstan	—	—	—	3.35	2.77	2.23	2.04	1.97	1.86	1.88
Lithuania	—	—	—	0.44	0.37	0.37	0.37	0.33	0.34	0.34
Russia	—	—	—	34.87	32.66	29.61	28.19	27.82	26.11	25.99
Turkmenistan	—	—	—	0.29	0.27	0.26	0.28	0.27	0.29	0.27
Ukraine	—	—	—	8.88	8.58	7.31	7.12	6.60	6.52	6.14
Uzbekistan	—	—	—	1.66	2.04	1.76	1.85	1.91	1.89	1.84
Other	0.15	0.11	0.09	1.81	1.38	1.19	1.16	1.27	1.23	1.22
Total	**74.88**	**74.11**	**69.45**	**65.03**	**60.53**	**54.58**	**52.87**	**52.58**	**50.34**	**48.96**

See footnotes at end of table.

tions, the energy industry may face many challenges and changes. For instance, under proposed regulations, energy producers and industrialized countries may be required to pay or trade for the right to emit carbon-based gases into the atmosphere, just as they must pay for other raw materials. (See Chapter 10 for further discussion.)

TABLE 1.6

World primary energy consumption, 1989–98 [CONTINUED]

(quadrillion btu)

Region Country	1989	1990	1991	1992	1993	1994	1995	1996	1997	1998[1]
Middle East										
Bahrain	0.24	0.26	0.29	0.24	0.29	0.28	0.28	0.28	0.34	0.36
Iran	2.97	3.10	3.24	3.35	3.47	3.66	3.81	3.95	4.44	4.53
Iraq	0.82	0.92	0.60	0.84	0.96	1.08	1.13	1.12	1.03	1.06
Israel	0.44	0.45	0.48	0.54	0.60	0.62	0.61	0.65	0.72	0.76
Kuwait	0.70	0.45	0.11	0.26	0.43	0.50	0.52	0.64	0.67	0.70
Oman	0.17	0.18	0.21	0.20	0.23	0.24	0.22	0.23	0.27	0.35
Qatar	0.29	0.36	0.42	0.49	0.57	0.57	0.57	0.58	0.61	0.63
Saudi Arabia	3.07	3.15	3.28	3.39	3.52	3.64	3.85	4.05	4.08	4.19
Syria	0.50	0.59	0.56	0.59	0.64	0.68	0.65	0.70	0.77	0.83
United Arab Emirates	1.26	1.23	1.49	1.55	1.48	1.49	1.60	1.67	1.79	1.82
Yemen	0.15	0.16	0.17	0.17	0.14	0.14	0.14	0.14	0.15	0.15
Other	0.28	0.29	0.33	0.36	0.39	0.45	0.47	0.50	0.52	0.53
Total	**10.90**	**11.13**	**11.17**	**11.99**	**12.72**	**13.33**	**13.88**	**14.51**	**15.38**	**15.90**
Africa										
Algeria	1.17	1.22	1.35	1.30	1.22	1.28	1.34	1.32	1.26	1.32
Angola	0.08	0.09	0.09	0.09	0.09	0.09	0.09	0.09	0.10	0.10
Egypt	1.35	1.44	1.43	1.43	1.51	1.55	1.58	1.73	1.80	1.86
Gabon	0.04	0.04	0.04	0.05	0.04	0.05	0.05	0.06	0.06	0.06
Libya	0.48	0.51	0.53	0.49	0.51	0.53	0.54	0.57	0.59	0.62
Morocco	0.30	0.31	0.32	0.33	0.36	0.40	0.37	0.40	0.41	0.41
Nigeria	0.71	0.70	0.77	0.79	0.80	0.68	0.83	0.85	0.97	0.96
South Africa	3.40	3.36	3.52	3.79	3.87	4.07	4.10	4.15	4.31	4.43
Zimbabwe	0.22	0.24	0.23	0.24	0.22	0.23	0.24	0.22	0.25	0.23
Other	1.40	1.44	1.44	1.51	1.55	1.63	1.62	1.69	1.75	1.78
Total	**9.15**	**9.34**	**9.74**	**10.01**	**10.18**	**10.51**	**10.77**	**11.07**	**11.51**	**11.77**
Far East & Oceania										
Australia	3.55	3.67	3.70	3.83	3.95	3.98	4.13	4.14	4.07	4.30
Bangladesh	0.25	0.25	0.26	0.29	0.30	0.33	0.36	0.37	0.38	0.40
Brunei	0.08	0.07	0.04	0.05	0.05	0.05	0.06	0.06	0.07	0.07
China	26.96	27.01	28.26	29.31	31.36	34.04	35.18	35.48	35.47	33.93
Hong Kong	0.50	0.48	0.46	0.50	0.55	0.59	0.64	0.67	0.51	0.67
India	7.27	7.78	8.06	8.71	9.10	9.81	11.11	11.60	12.10	12.51
Indonesia	2.07	2.19	2.31	2.50	2.87	3.06	3.29	3.58	3.83	3.62
Japan	17.69	18.28	19.00	19.26	19.53	20.30	20.95	21.63	21.57	21.28
Korea, North	2.04	2.07	2.13	2.02	2.08	2.07	2.06	2.10	1.82	1.81
Korea, South	3.28	3.66	4.17	4.65	5.38	5.89	6.44	7.03	7.62	6.93
Malaysia	0.86	0.98	1.09	1.14	1.28	1.42	1.46	1.63	1.65	1.74
New Zealand	0.71	0.74	0.74	0.75	0.78	0.81	0.87	0.83	0.80	0.79
Pakistan	1.13	1.18	1.25	1.29	1.41	1.50	1.58	1.70	1.68	1.74
Philippines	0.71	0.73	0.73	0.77	0.84	0.90	0.96	1.02	1.09	1.08
Singapore	0.73	0.80	0.86	0.97	1.08	1.16	1.18	1.25	1.42	1.33
Taiwan	1.93	2.04	2.09	2.21	2.43	2.61	2.86	3.06	3.24	3.31
Thailand	1.07	1.25	1.37	1.47	1.68	1.87	2.24	2.44	2.52	2.34
Vietnam	0.23	0.28	0.28	0.31	0.39	0.42	0.52	0.56	0.68	0.74
Other	0.61	0.62	0.57	0.55	0.58	0.63	0.63	0.64	0.66	0.68
Total	**71.68**	**74.09**	**77.35**	**80.57**	**85.63**	**91.44**	**96.53**	**99.80**	**101.19**	**99.27**
World Total	**344.83**	**346.83**	**347.51**	**348.46**	**353.21**	**357.25**	**365.72**	**376.25**	**378.04**	**377.72**

[1] Preliminary

—= Not applicable.

(s) = Value less than 5 trillion Btu.

Notes: Sum of components may not equal total due to independent rounding.

Primary energy consumption reported in this table includes petroleum, dry natural gas, coal, net hydroelectric, nuclear, biomass, geothermal, solar, and wind electric power. Primary energy consumption for the United States also includes:

(1) the consumption of biomass, geothermal energy, and solar energy not used for electricity generation; (2) electricity imports from Mexico that are derived from geothermal energy; and (3) net imports of electricity derived from nonrenewable sources. Primary energy consumption for all countries, except the United States, has been adjusted to include total electricity imports and to exclude total electricity exports.

SOURCE: *International Energy Annual 1998,* Energy Information Administration, Washington, D.C., 2000

FIGURE 1.9

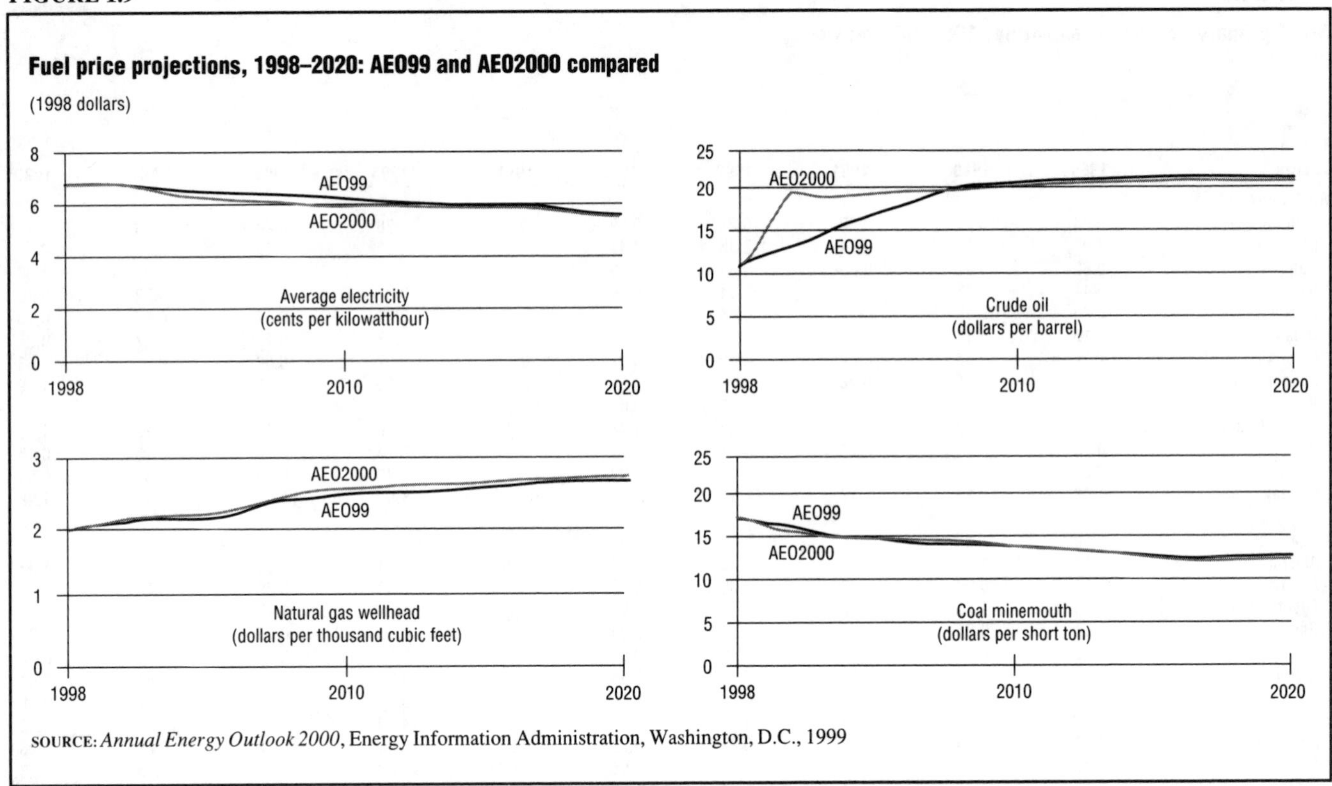

SOURCE: *Annual Energy Outlook 2000*, Energy Information Administration, Washington, D.C., 1999

FIGURE 1.10

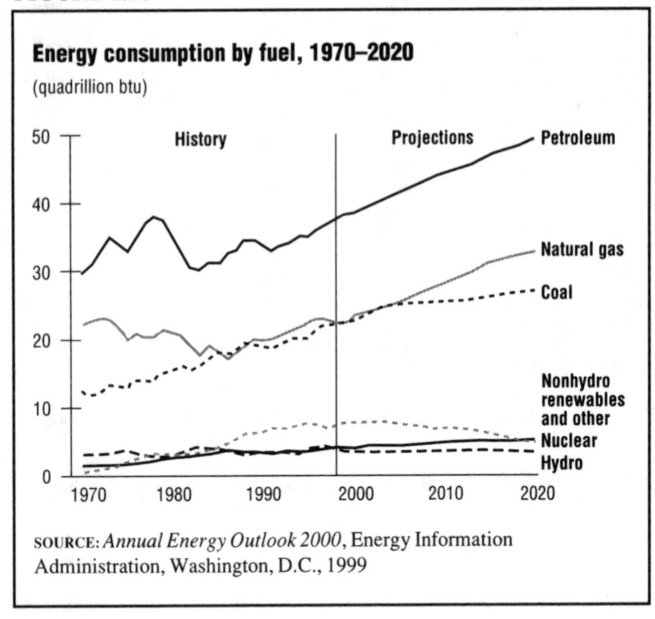

SOURCE: *Annual Energy Outlook 2000*, Energy Information Administration, Washington, D.C., 1999

FIGURE 1.11

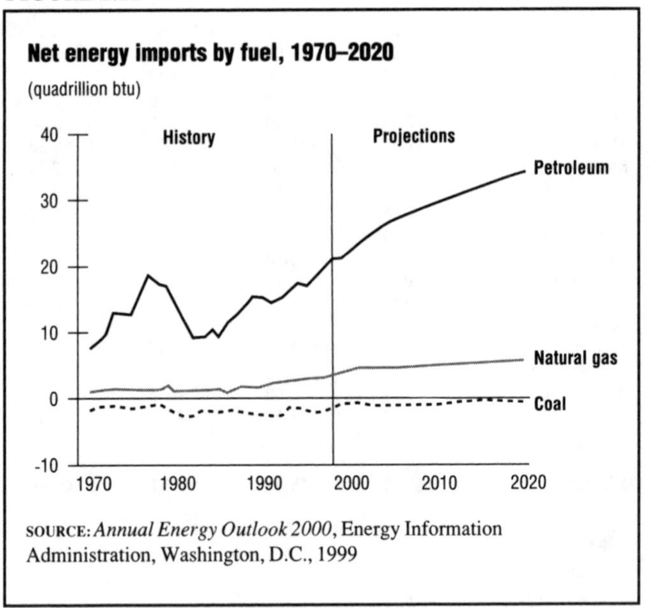

SOURCE: *Annual Energy Outlook 2000*, Energy Information Administration, Washington, D.C., 1999

TABLE 1.7

Summary of results for five cases

			2020				
Sensitivity factors	1997	1998	Reference	Low economic growth	High economic growth	Low world oil price	High world oil price
Primary production (quadrillion Btu)							
Petroleum	16.23	15.73	14.49	14.14	15.44	13.18	16.12
Natural gas	19.43	19.40	27.13	25.70	27.98	26.97	27.26
Coal	23.28	23.89	27.36	26.14	29.62	26.90	27.43
Nuclear power	6.71	7.19	4.56	4.56	4.70	4.51	4.63
Renewable energy.	7.00	6.67	7.98	7.77	8.40	7.90	8.06
Other	0.66	0.57	0.66	0.66	0.68	0.59	0.73
Total primary production	**73.30**	**73.46**	**82.18**	**78.98**	**86.82**	**80.06**	**84.23**
Net imports (quadrillion Btu)							
Petroleum (including SPR)	19.65	20.95	34.15	30.30	37.36	38.35	30.87
Natural gas	2.90	3.20	5.25	4.59	5.62	5.15	5.18
Coal/other (- indicates export)	-1.66	-1.46	-0.50	-0.55	-0.38	-0.50	-0.49
Total net imports	**20.89**	**22.69**	**38.91**	**34.35**	**42.60**	**43.00**	**35.56**
Discrepancy	-0.22	1.27	0.14	0.04	0.06	0.26	-0.09
Consumption (quadrillion Btu)							
Petroleum products	36.43	37.21	49.05	44.99	53.27	51.73	47.71
Natural gas	22.60	21.99	32.38	30.28	33.61	32.11	32.44
Coal	21.34	21.50	26.60	25.32	28.98	26.15	26.68
Nuclear power	6.71	7.19	4.56	4.56	4.70	4.51	4.63
Renewable energy	7.00	6.67	7.99	7.78	8.42	7.92	8.08
Other	0.33	0.32	0.36	0.34	0.38	0.37	0.34
Total consumption	**94.41**	**94.88**	**120.95**	**113.28**	**129.36**	**122.79**	**119.88**
Prices (1998 dollars)							
World oil price (dollars per barrel)	18.71	12.10	22.04	20.99	23.11	14.90	28.04
Domestic natural gas at wellhead (dollars per thousand cubic feet)	2.39	1.96	2.81	2.40	3.27	2.68	2.87
Domestic coal at minemouth (dollars per short ton)	18.32	17.51	12.54	12.40	12.58	12.38	12.53
Average electricity price (cents per kilowatthour)	6.9	6.7	5.8	5.5	6.1	5.8	5.9
Economic Indicators							
Real gross domestic product (billion 1992 dollars)	7,270	7,552	12,179	10,870	13,413	12,205	12,151
(annual change, 1998-2020)	—	—	2.2%	1.7%	2.6%	2.2%	2.2%
GDP implicit price deflator (index, 1992=1.00)	1.12	1.13	1.86	2.11	1.63	1.86	1.86
(annual change, 1998-2020)	—	—	2.3%	2.9%	1.7%	2.3%	2.3%
Real disposable personal income (billion 1992 dollars)	5,183	5,348	9,008	8,281	9,679	9,037	8,974
(annual change, 1998-2020)	—	—	2.4%	2.0%	2.7%	2.4%	2.4%
Index of manufacturing gross output (index, 1987=1.00)	1.365	1.411	2.160	1.972	2.483	2.166	2.158
(annual change, 1998-2020)	—	—	2.0%	1.5%	2.6%	2.0%	2.0%
Energy intensity							
(thousand Btu per 1992 dollar of GDP)	12.99	12.57	9.94	10.43	9.65	10.07	9.87
(annual change, 1998-2020)	—	—	-1.1%	-0.8%	-1.2%	-1.0%	-1.1%
Carbon emissions							
(million metric tons)	1,479	1,485	1,979	1,851	2,126	2,019	1,956
(annual change, 1998-2020)	—	—	1.3%	1.0%	1.6%	1.4%	1.3%

Notes: Quantities are derived from historical volumes and assumed thermal conversion factors. Other production includes liquid hydrogen, methanol, supplemental natural gas, and some inputs to refineries. Net imports of petroleum include crude oil, petroleum products, unfinished oils, alcohols, ethers, and blending components. Other net imports include coal coke and electricity. Some refinery inputs appear as petroleum product consumption. Other consumption includes net electricity imports, liquid hydrogen, and methanol.

SOURCE: *Annual Energy Outlook 2000*, Energy Information Administration, Washington, D.C., 1999

FIGURE 1.12

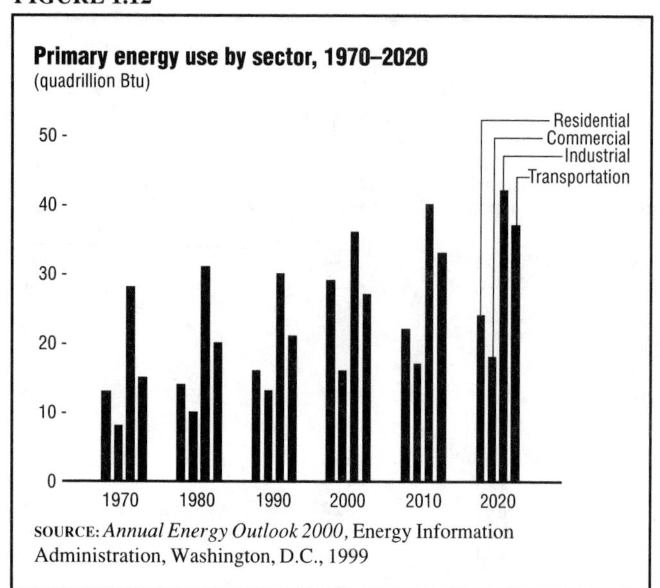

SOURCE: *Annual Energy Outlook 2000,* Energy Information Administration, Washington, D.C., 1999

FIGURE 1.13

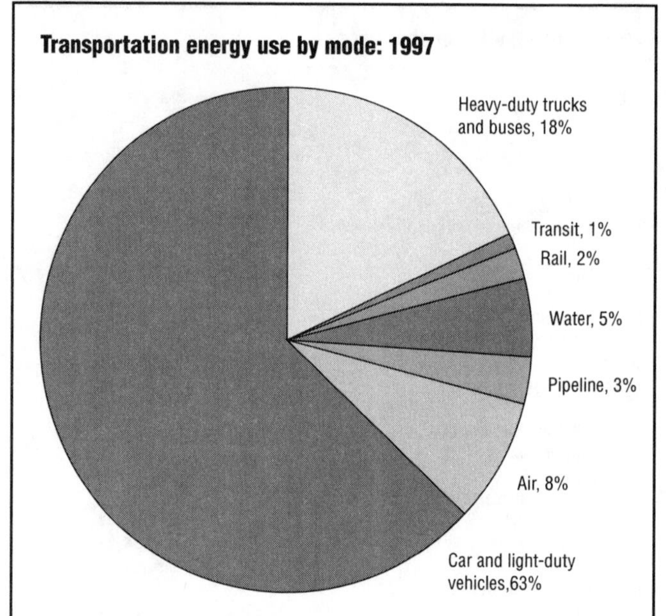

SOURCE: *Transportation Statistics Annual Report 1999*, U.S. Department of Transportation, Bureau of Transportation Statistics, Washington, D.C., 1999

CHAPTER 2

OIL

THE QUEST FOR OIL

On August 27, 1859, Edwin Drake "struck oil" 69 feet below the surface near Titusville, Pennsylvania, the first successful modern oil well, which ushered in the "Age of Petroleum." Companies were seeking to meet the growing demand for new and better fuels for heating and lighting. The development of the internal combustion engine in the late 1800s opened a huge market for oil products.

Sources of Oil

Oil is a limited resource. Almost all oil comes from underground reservoirs. Most scientists believe that oil and natural gas are the products of intense heat and pressure applied over millions of years to organic (formerly alive) sediments buried in geological formations. A few scientists, however, believe that at least some of the deposits are accumulations of methane gas seeping out from the molten core of the earth.

At one time, it was believed that crude oil flowed in underground streams and accumulated in lakes or caverns in the earth. Today, scientists know that a petroleum reservoir is usually a solid sandstone or limestone formation overlaid with a layer of impermeable rock or shale that creates a shield. The petroleum accumulates within the pores and fractures of the rock and is trapped beneath the seal. Anticlines (archlike folds in a bed of rock), faults, and salt domes are common trapping formations. (See Figure 2.1.) Oil and natural gas deposits can be at varying depths. Wells are drilled to reach the reservoirs and extract the oil. Deep wells are more expensive to drill and are usually attempted only to reach large reservoirs or when the price of oil is high.

The Cost of Exploration

Oil is crucial to the economies of developed nations, and oil production is strongly influenced by market prices. Exploring for new deposits and drilling exploratory wells is expensive. It is also costly to drill, maintain, and operate production wells. If the price of a barrel of oil falls too low, operators will shut off expensive wells because they cannot recover their higher operating costs. In general, when oil prices are high, oil companies drill; when prices are low, drilling drops off. Figure 2.2 shows a diagram of a rotary drilling system.

Oil deposits are not distributed evenly over the world, and some that have been exploited for decades are being exhausted. As an oil reservoir is depleted, various tech-

FIGURE 2.1

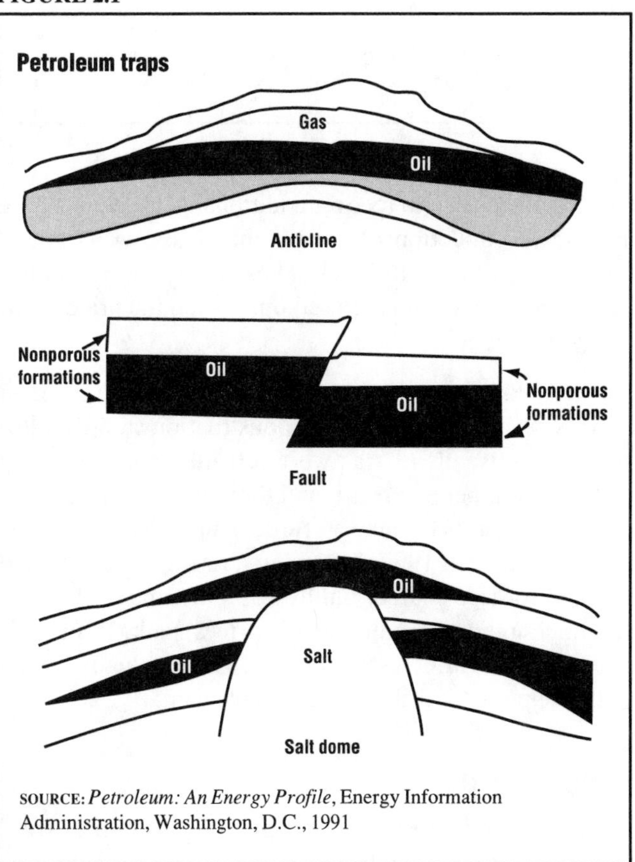

SOURCE: *Petroleum: An Energy Profile*, Energy Information Administration, Washington, D.C., 1991

FIGURE 2.2

A rotary drilling system

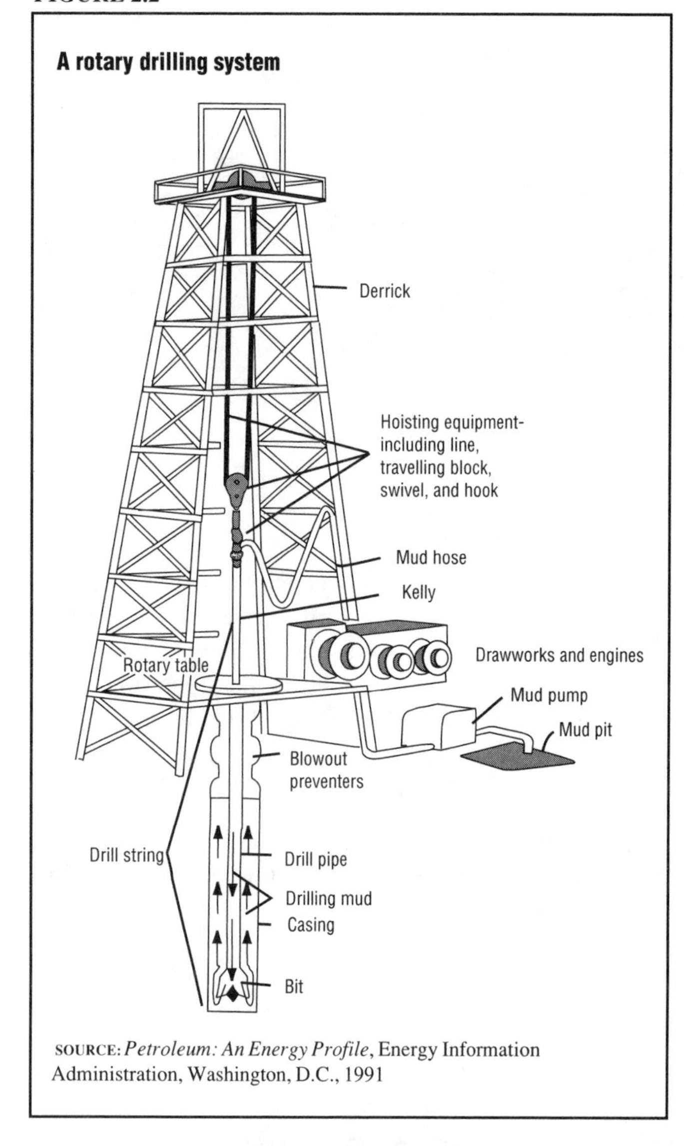

SOURCE: *Petroleum: An Energy Profile*, Energy Information Administration, Washington, D.C., 1991

niques can be used to recover additional petroleum. These include the injection of water, chemicals, or steam to force more oil from the rock. These recovery techniques can be expensive, and are used only when the price of oil is relatively high.

In 1996, after a decade of low oil prices, the demand for rigs collapsed, and new rig construction stopped altogether. Thousands of rigs were left idle, sold for scrap metal, or shipped overseas, and their crews were put out of work. Idle rigs became a source of spare parts for ones still operating. In 1997, following a rise in oil prices, the demand for rigs soared, but by 1998 the market for rigs had once again dwindled as oil prices sank. Rising oil prices due to OPEC restrictions in 1999 eventually boosted the demand for drilling equipment again.

TYPES OF OIL

While crude oil is usually dark when it comes from the ground, it may also be green, red, yellow, or colorless, depending on its chemical composition and the amount of sulfur, oxygen, nitrogen, and trace minerals present. Its viscosity (thickness, or resistance to flow) can range from being thin as water to thick as tar. Crude oil has to be refined in order to be made into thousands of useful products, as it has limited uses in its natural form.

Crude oils vary in quality. "Sweet" crudes have little sulfur, refine easily, and are worth more than "sour" crudes, which contain more impurities. "Light" crudes have more short molecules, which yield more gasoline and are more profitable than "heavy" crudes, which bring a lower price in the market.

In addition to crude oil, there are two other sources of primary petroleum: lease condensate and natural gas plant liquids. Lease condensate is the liquid condensed from natural gas at or near the wellhead during production operations. It consists primarily of chemical compounds called penthanes and heavier hydrocarbons, and is generally blended with crude oil for refining. Natural gas plant liquids are collected when natural gas is refined. (See Chapter 3.)

HOW OIL IS REFINED

Before oil can be used by consumers, the crude oil, lease condensate, and natural gas plant liquids must be processed into finished products. The first step in refining is distillation, in which crude oil molecules are separated according to size and weight. After the crude oil is heated and turned to vapor, it enters the bottom of a distillation tower where the vaporized crude rises and condenses on trays—gasoline on the top, distillates in the middle, and the heavy residual fuels at the bottom. Each separate portion can then be further refined through the processes of cracking and reforming, which transform the condensed molecules into other forms of petroleum. Cracking converts the heaviest fractions of separated petroleum into lighter fractions to produce jet fuel, motor gasoline, home heating oil, and less-residual fuel oils, which are heavier and used for naval ships, commercial and industrial heating, and some power generation. Reforming is used to increase the octane rating of gasoline. (See Figure 2.3.)

Refining is a continuous process, with crude oil entering the refinery while finished products leave by pipeline, truck, and train. Although refineries are surrounded by storage tanks, they have limited storage capacity. If there is a malfunction and products cannot be processed, they may be burned off (flared) if no storage facility is available. While a small flare is normal at a refinery or a chemical plant, a large flare, or many flares, likely indicates a processing problem.

REFINERIES

In 1999, 159 refineries were operating in the United States, a continuing drop from 336 in 1949 and 324 in

FIGURE 2.3

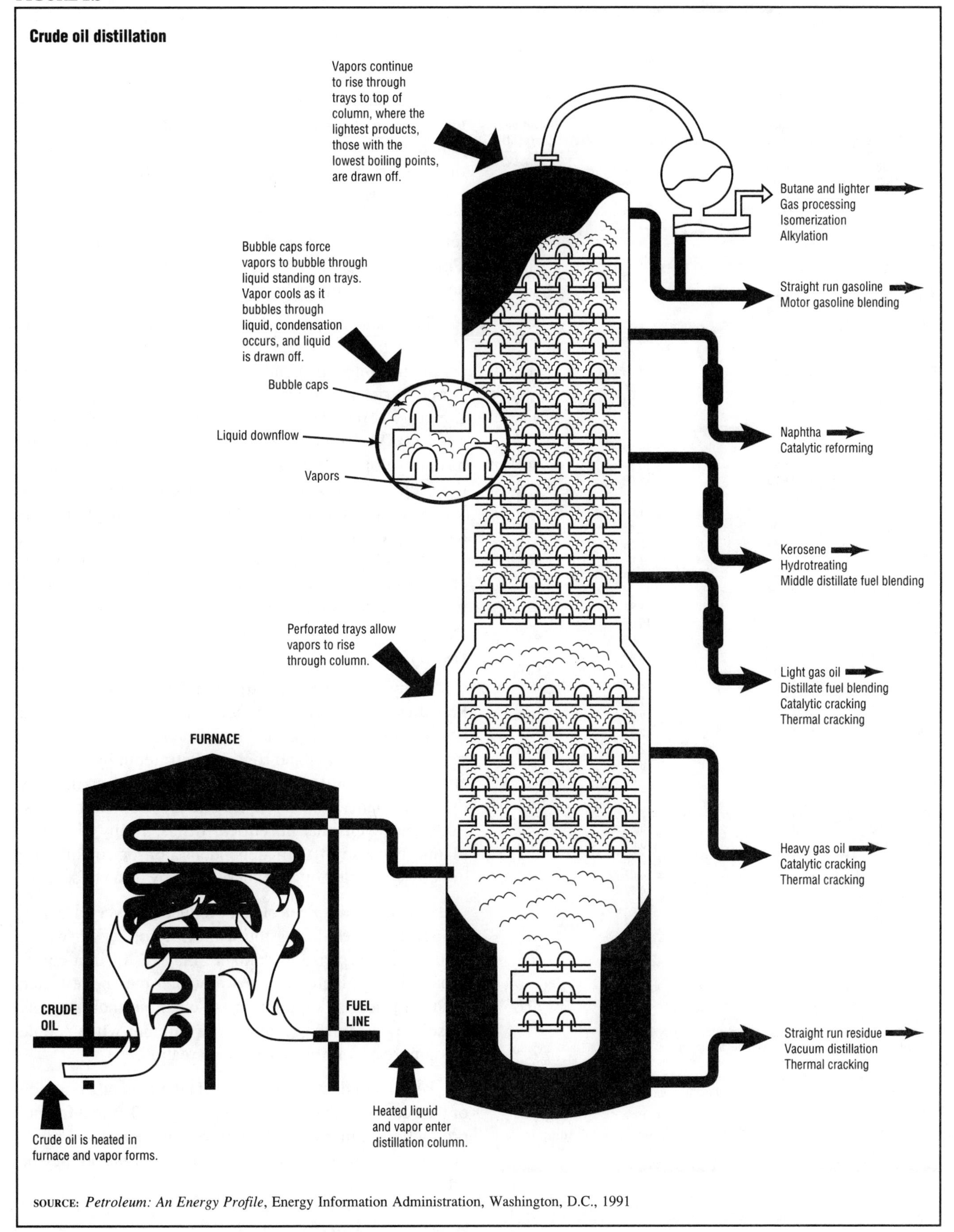

SOURCE: *Petroleum: An Energy Profile*, Energy Information Administration, Washington, D.C., 1991

TABLE 2.1

Refinery capacity and utilization, 1949–99

	Operable refineries			
Year	Number[1]	Capacity[2] (million barrels per day)	Gross input to distillation units (million barrels per day)	Utilization[3] (percent)
1949	336	6.23	5.56	89.2
1950	320	6.22	5.98	92.5
1951	325	6.70	6.76	97.5
1952	327	7.16	6.93	93.8
1953	315	7.62	7.26	93.1
1954	308	7.98	7.27	88.8
1955	296	8.39	7.82	92.2
1956	317	8.58	8.25	93.5
1957	317	9.07	8.22	89.2
1958	315	9.36	8.02	83.9
1959	313	9.76	8.36	85.2
1960	309	9.84	8.44	85.1
1961	309	10.00	8.57	85.7
1962	309	10.01	8.83	88.2
1963	304	10.01	9.14	90.0
1964	298	10.31	9.28	89.6
1965	293	10.42	9.56	91.8
1966	280	10.39	9.99	94.9
1967	276	10.66	10.39	94.4
1968	282	11.35	10.89	94.5
1969	279	11.70	11.25	94.8
1970	276	12.02	11.52	92.6
1971	272	12.86	11.88	90.9
1972	274	13.29	12.43	92.3
1973	268	13.64	13.15	93.9
1974	273	14.36	12.69	86.6
1975	279	14.96	12.90	85.5
1976	276	15.24	13.88	87.8
1977	282	16.40	14.98	89.6
1978	296	17.05	15.07	87.4
1979	308	17.44	14.96	84.4
1980	319	17.99	13.80	75.4
1981	324	18.62	12.75	68.6
1982	301	17.89	12.17	69.9
1983	258	16.86	11.95	71.7
1984	247	16.14	12.22	76.2
1985	223	15.66	12.17	77.6
1986	216	15.46	12.83	82.9
1987	219	15.57	13.00	83.1
1988	213	15.92	13.45	84.7
1989	204	15.65	13.55	86.6
1990	205	15.57	13.61	87.1
1991	202	15.68	13.51	86.0
1992	199	15.70	13.60	87.9
1993	187	15.12	13.85	91.5
1994	179	15.03	14.03	92.6
1995	175	15.43	14.12	92.0
1996	170	15.33	14.34	94.1
1997	164	15.45	14.84	95.2
1998	163	15.71	R15.11	R95.6
1999P	159	16.26	15.09	92.7

1 Prior to 1956, the number of refineries included only those in operation on January 1. For 1957 forward, the number of refineries has included all operable refineries on January 1.
2 Capacity in million barrels per calendar day on January 1.
3 For 1949-1980, utilization is derived by dividing gross input to distillation units by one-half of the current year January 1 capacity and the following year January 1 capacity. Percentages were derived from unrounded numbers. For 1981 forward, utilization is derived by averaging reported monthly utilization.
R=Revised. P=Preliminary.

SOURCE: *Annual Energy Review 1999,* Energy Information Administration, Washington, D.C., 2000

1981. Refinery capacity was about 16.3 million barrels per day, below the 1981 peak of 18.6 million barrels. (See Table 2.1.) Currently, U.S. refineries are operating near full capacity. Utilization rates have steadily increased from a low of 69 percent in 1981, which was a period of low demand due to economic recession, to a very busy utilization rate of 93 percent in 1999.

The petroleum industry has been going through "rationalization," shutting down older, inefficient refineries and concentrating production in more efficient plants, which are usually newer and larger. Consolidation within the industry has also played a role in refinery operation. For example, the merger of Gulf Oil Corporation into Chevron Corporation led to the closing of two large refineries, one in Bakersfield, California, and the other in Cincinnati, Ohio. In 1998 Exxon merged with Mobile Oil, and large merger possibilities continue in the industry.

The cheap oil prices of the 1990s caused some OPEC (Organization of Petroleum Exporting Countries) producers to encounter financial problems. (Note: OPEC member countries are Algeria, Indonesia, Iran, Iraq, Kuwait, Libya, Nigeria, Qatar, Saudi Arabia, United Arab Emirates, and Venezuela.) In response, some OPEC members have developed "downstream" marketing, which means they refine their own oil products to make extra profit. This strategy, particularly employed by Saudi Arabia, has led to a drop in demand for U.S. refineries. The National Petroleum Refiners Association notes that no major refinery manufacturer has plans to begin construction, a process that takes three to five years to complete.

USES FOR OIL

Many of the uses of petroleum are well known: gasoline, diesel fuel, jet fuel, and lubricants for transportation; heating oil, residual oil, and kerosene for heat; and heavy residuals for paving and roofing. Petroleum by-products are also vital to the chemical industry, ending up in many different foams, plastics, synthetic fabrics, paints, dyes, inks, and even pharmaceutical drugs. Many chemical plants, because of their dependence on petroleum, are directly connected by pipelines to nearby refineries.

The demand for petroleum products can vary. Heating oil demand rises during the winter. A cold spell, which leads to a sharp rise in demand, may result in a corresponding price increase. A warm winter may be reflected in lower prices as suppliers try to clear out their storage. Gasoline demand rises during the summer when people drive more on vacations and for recreation.

Petroleum demand also reflects the general condition of the economy. During a recession, when demand for and production of many items incorporating petroleum products drops, demand for oil products also falls and may be reflected in lower prices.

DOMESTIC PRODUCTION

Figure 2.4 shows the overall flow of oil in the United States for 1999. U.S. production of petroleum reached its

FIGURE 2.4

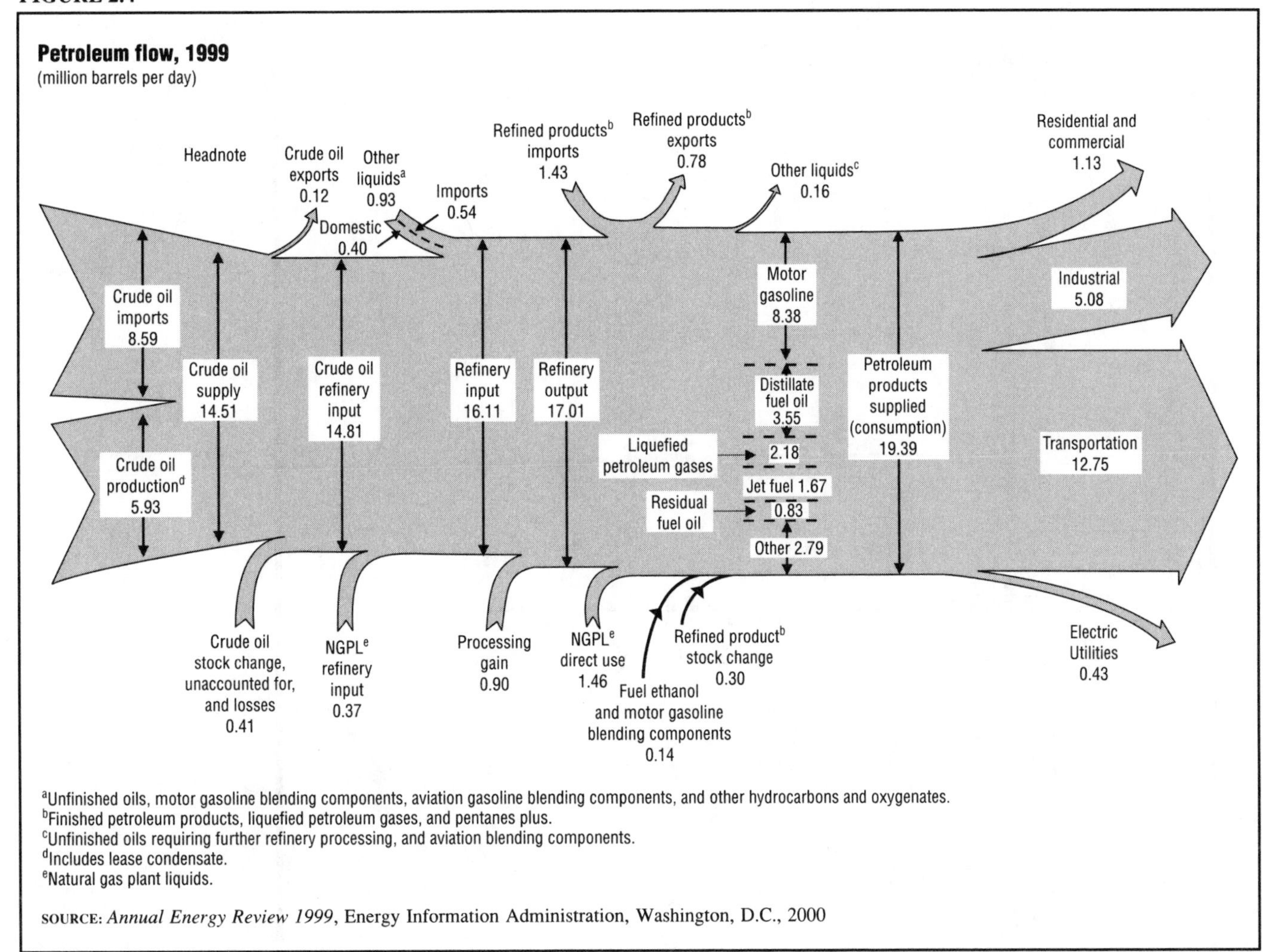

SOURCE: *Annual Energy Review 1999*, Energy Information Administration, Washington, D.C., 2000

highest level in 1970 at 11.3 million barrels per day total (9.6 of that was crude oil) and then turned downward. By 1999 U.S. domestic production of petroleum averaged about 7.8 million barrels per day; of that amount, 5.9 million barrels per day was crude oil. (See Table 2.2 and Figure 2.5.) Texas, Alaska, Louisiana, California, and the offshore areas around these states produce about 75 percent of the nation's oil. Supplies from Alaska, which increased with the construction of a direct pipeline in the late 1970s, have begun to decline as supplies are used up. (See Figure 2.6. For further discussion of offshore drilling, see Chapter 5.)

Any new oil discoveries are unlikely to lead to a significant increase in domestic production in the near future because of the long lead-time needed to prepare for production. Most oil (75 percent, or nearly 4.5 million barrels per day) comes from onshore drilling, while the remaining 1.4 million barrels come from offshore sources. (See Figure 2.7.) In 1999 the 554,000 producing wells in the United States produced an average of 10.7 barrels per day per well, significantly below peak levels of over 18 barrels per day in the early 1970s.

FIGURE 2.5

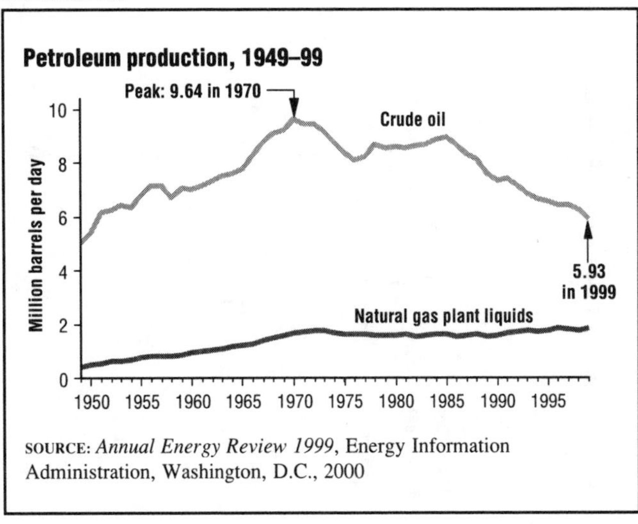

SOURCE: *Annual Energy Review 1999*, Energy Information Administration, Washington, D.C., 2000

Even with oil prices expected to increase in the future, U.S. oil production will likely continue to drop. The Alaskan oil boom appears to be near an end, unless protected wildlife refuges are opened up for drilling. The United States is considered to be in a "mature" oil development

TABLE 2.2

Petroleum overview, 1950–99
(million barrels per day)

	Production				Trade							
Year	Crude oil [1]	Natural gas plant liquids	Total petroleum	Other domestic supply [2]	Crude oil imports [3]	Petroleum product imports [4]	Total imports	Total exports	Net imports [5]	Crude oil losses	Stock change [6]	Petroleum products supplied
1950	5.41	0.50	5.91	(s)	0.49	0.36	0.85	0.30	0.55	0.05	-0.06	6.46
1960	7.04	0.93	7.96	0.15	1.02	0.80	1.81	0.20	1.61	0.01	-0.08	9.80
1965	7.80	1.21	9.01	0.22	1.24	1.23	2.47	0.19	2.28	0.01	-0.01	11.51
1970	9.64	1.66	11.30	0.35	1.32	2.10	3.42	0.26	3.16	0.01	0.10	14.70
1975	8.37	1.63	10.01	0.51	4.10	1.95	6.06	0.21	5.85	0.01	0.03	16.32
1980	8.60	1.57	10.17	0.68	5.26	1.65	6.91	0.54	6.36	0.01	0.14	17.06
1981	8.57	1.61	10.18	0.64	4.40	1.60	6.00	0.59	5.40	(s)	0.16	16.06
1982	8.65	1.55	10.20	0.65	3.49	1.63	5.11	0.82	4.30	(s)	-0.15	15.30
1983	8.69	1.56	10.25	0.65	3.33	1.72	5.05	0.74	4.31	(s)	-0.02	15.23
1984	8.88	1.63	10.51	0.78	3.43	2.01	5.44	0.72	4.72	(s)	0.28	15.73
1985	8.97	1.61	10.58	0.76	3.20	1.87	5.07	0.78	4.29	(s)	-0.10	15.73
1986	8.68	1.55	10.23	0.81	4.18	2.05	6.22	0.78	5.44	(s)	0.20	16.28
1987	8.35	1.60	9.94	0.85	4.67	2.00	6.68	0.76	5.91	(s)	0.04	16.67
1988	8.14	1.62	9.76	0.90	5.11	2.30	7.40	0.82	6.59	(s)	-0.03	17.28
1989	7.61	1.55	9.16	0.92	5.84	2.22	8.06	0.86	7.20	(s)	-0.04	17.33
1990	7.36	1.56	8.91	1.02	5.89	2.12	8.02	0.86	7.16	(s)	0.11	16.99
1991	7.42	1.66	9.08	1.00	5.78	1.84	7.63	1.00	6.63	(s)	-0.01	16.71
1992	7.17	1.70	8.87	1.16	6.08	1.80	7.89	0.95	6.94	(s)	-0.07	17.03
1993	6.85	1.74	8.58	1.19	6.79	1.83	8.62	1.00	7.62	(s)	0.15	17.24
1994	6.66	1.73	8.39	1.29	7.06	1.93	9.00	0.94	8.05	(s)	0.02	17.72
1995	6.56	1.76	8.32	1.27	7.23	1.61	8.83	0.95	7.89	(s)	-0.25	17.72
1996	6.46	1.83	8.29	1.36	7.51	1.97	9.48	0.98	8.50	(s)	-0.15	18.31
1997	6.45	1.82	8.27	1.34	8.23	1.94	10.16	1.00	9.16	0.00	0.14	18.62
1998	R6.25	R1.76	R8.01	R1.38	R8.71	R2.00	R10.71	R0.94	R9.76	(s)	R0.24	R18.92
1999P	5.93	1.83	7.76	1.59	8.59	1.96	10.55	0.94	9.61	(s)	-0.44	19.39

[1] Includes lease condensate.
[2] Other hydrocarbons, hydrogen, oxygenates (ethers and alcohols), gasoline blending components, finished petroleum products, processing gains, and unaccounted-for crude oil.
[3] Includes any imports for the Strategic Petroleum Reserve, which began in 1977.
[4] For 1981 forward, includes motor gasoline blending components and aviation gasoline blending components.
[5] Net imports = imports minus exports.
[6] A negative value indicates a net decrease in stocks; a positive value indicates a net increase in stocks.
R=Revised. P=Preliminary. (s)=Less than 0.005 million barrels per day and greater than -0.005 million barrels per day.

SOURCE: *Annual Energy Review 1999*, Energy Information Administration, Washington, D.C., 2000

FIGURE 2.6

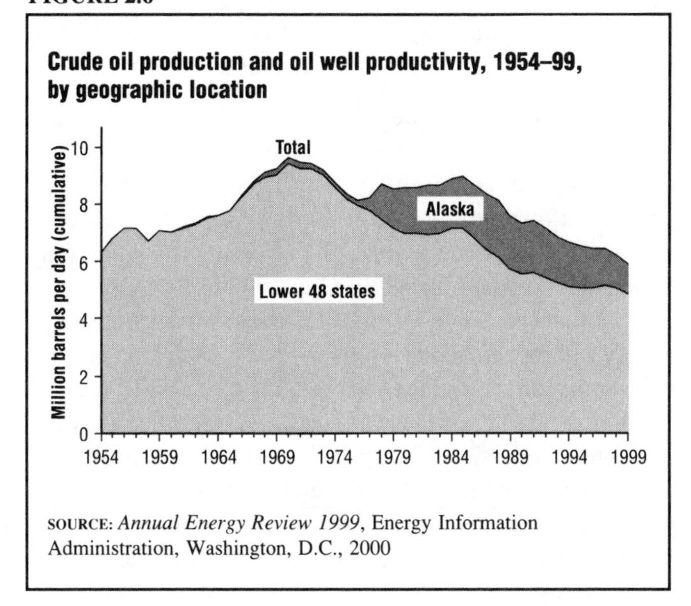

Crude oil production and oil well productivity, 1954–99, by geographic location

SOURCE: *Annual Energy Review 1999*, Energy Information Administration, Washington, D.C., 2000

FIGURE 2.7

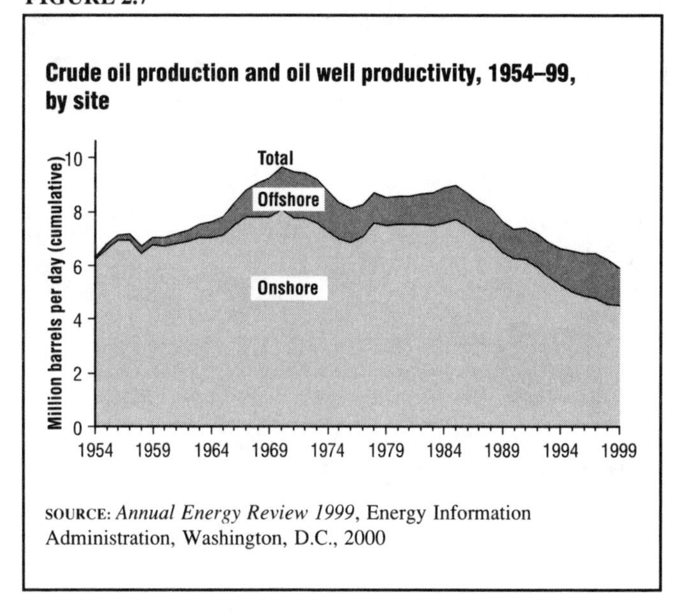

Crude oil production and oil well productivity, 1954–99, by site

SOURCE: *Annual Energy Review 1999*, Energy Information Administration, Washington, D.C., 2000

phase, meaning that much of the oil has been found and the nation is producing what it can. The amount of oil discovered per foot of exploratory well in the United States has fallen to less than half the rate of the early 1970s. Of the country's 13 largest oil fields, 7 are at least 80 percent depleted. Geological studies have estimated that 34 percent of U.S. undiscovered recoverable resources are in Alaska, but it is uncertain whether the oil will ever be recovered.

Middle Eastern producers can drill and bring out crude oil from enormous, easily accessible reservoirs for around $2 a barrel. In contrast, the U.S. Department of Energy estimates it costs an American oil producer about $14 to produce a barrel of oil, not counting royalty payments and taxes, which add to the cost. Of all the successful domestic oil wells drilled, only approximately 1 percent have been "wildcat" wells that have led to the discovery of new large fields, and these discoveries have provided only minor additions to the total proved reserves. (See Chapter 5 for a more complete discussion of reserves.)

FIGURE 2.8

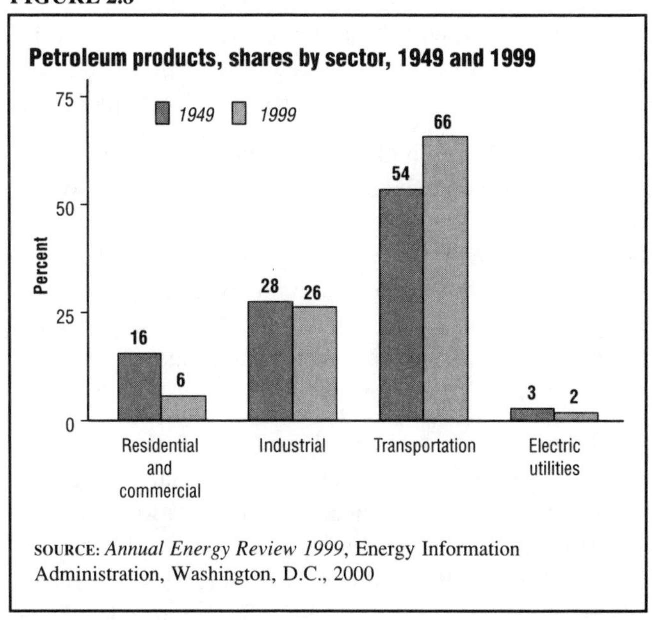

Petroleum products, shares by sector, 1949 and 1999

SOURCE: *Annual Energy Review 1999*, Energy Information Administration, Washington, D.C., 2000

DOMESTIC CONSUMPTION

In 1999 most petroleum, by far, was used for transportation (66 percent), followed by industrial use (26 percent), residential and commercial use (6 percent), and electric utilities (2 percent). (See Figure 2.8.)

Most petroleum used for transportation is for motor gasoline. In the residential and commercial sectors, distillate fuel oil (refined fuels used for space heaters, diesel engines, and electric power generation) accounts for most petroleum use. LPG (Liquid Petroleum Gas) is the primary oil used in the industrial sector. In electric utilities, residual fuel oils are used most.

A modest decline in residual fuel oil consumption has been caused by the conversion of electric utilities and plants from heavy oil to coal or natural gas energy. An initial decline in the amount of motor gasoline used, beginning in 1978, was attributed to the federal CAFE (Corporate Average Fuel Economy) regulations, which required increased miles-per-gallon efficiency in new automobiles. However, motor gasoline use has increased steadily since then, partly from an increase in users and partly from a leveling off in vehicle efficiency as consumers once again prefer less efficient vehicles, such as larger automobiles and sport utility vehicles (SUVs).

OIL IMPORTS AND EXPORTS

Around the world, there are inequities between those countries which use petroleum and those which possess it. Countries with surpluses (Saudi Arabia, for example) sell

FIGURE 2.9

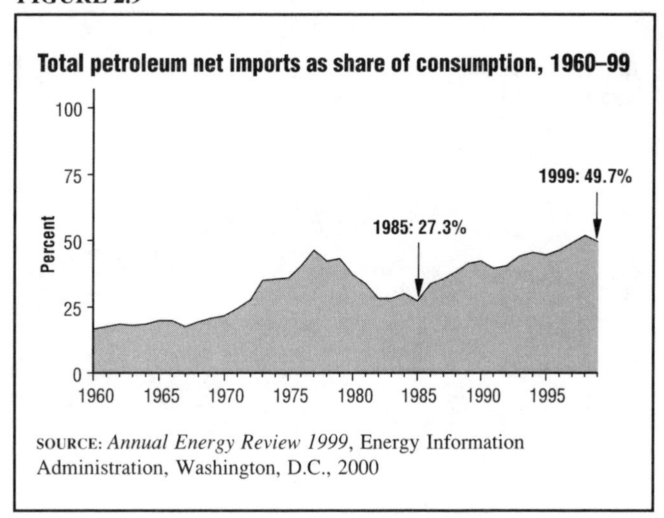

SOURCE: *Annual Energy Review 1999*, Energy Information Administration, Washington, D.C., 2000

their excess to others that need more than they can produce (the United States, Japan, and European countries). Petroleum is sold as crude oil or as refined products. World trade has been moving towards refined products, as the petroleum exporting countries realize that they can make more profit from refined oil products than from crude oil.

Though the United States produces a significant amount of petroleum, it has been importing oil since World War II. This reflects the gradual exhaustion of reserves in the United States and the growing energy demand caused by population growth and economic expansion. The relatively low price of foreign oil has encouraged dependence on imported oil. American industry and economic life have been built on oil's cheap availability.

Relatively low crude oil prices and the resulting reduced domestic oil production are the major causes of an increase in imports since 1985. From a low total net import (imports minus exports) of 4.3 million barrels per day in 1985, oil net imports in 1999 were 9.6 million barrels per day. (See Table 2.2.) In 1985, imported oil supplied only 27 percent of American oil consumption. Just five years later, in 1990, the proportion had risen to 42 percent and by 1999 to 50 percent as demand continued to grow. (See Figure 2.9.) The Persian Gulf nations, Canada, Saudi Arabia, Venezuela, and Mexico, respectively, were the leading suppliers of petroleum to the United States in 1999.

OIL PRICES

The law of supply and demand, which means that the price of goods reflects a relationship between the supply (availability) and the demand (need), usually explains oil price changes. Changes in the price of oil also eventually affect both supply and demand. Higher prices will increase production as it becomes profitable to operate more expensive wells, as well as reduce demand as consumers lower use and increase conservation. The factors also work the other way: reduced demand or increased supply will generally cause the price of oil to drop.

While consumers prefer low prices that allow them to save money or get more of the commodity for the same price, producers naturally prefer to keep prices high. Oil producers formed the cartel (a group of businesses who agree to control production and marketing to avoid competing with one another) OPEC in 1960 to try to manipulate the market. Since 1973 OPEC has tried to control the supply in order to achieve higher prices. Over this period, each individual producer has restricted exports for the good of the cartel, because an increased supply causes the price to drop. OPEC has faced long-term problems because earlier higher prices encouraged conservation, reducing demand for oil and leading to a sharp decline in oil prices. As a result of the decreased demand for oil and lower prices, OPEC lost some of the ability to control its members and the power it once had over prices. In 1999, 46 percent of the oil that the United States imported came from OPEC nations. OPEC actions can still effectively influence the petroleum market. In 1998 gas was cheaper than ever, but an OPEC restriction in supply in 1999 caused oil and gasoline prices to surge in 2000.

Gasoline Prices

Not only senior citizens can reminisce about the "good old days" of cheap gas. A middle-aged person can remember when gas cost $.50 a gallon in the early 1970s. From 1973 to 1981, the price of a gallon of gas (in current dollars that do not consider inflation) more than tripled, while the price in real dollars (adjusted for inflation) rose 77 percent. However, after 1981, as a result of the international oil glut, real prices tumbled. In real dollars, the price of a gallon of regular unleaded gasoline was $2.21 in 1981; by 1999, the price was only $1.11. (See Table 2.3.) According to the Energy Information Administration, the price paid by refiners for crude oil in 1999 averaged $16.69 per barrel in real dollars, compared to a high of $56.50 per barrel in 1981.

A limited supply of oil was an issue during the Persian Gulf crisis in 1990–91. The conflict caused a sharp, momentary spike in prices as Iraqi and Kuwaiti oil was withdrawn from the market, which was coupled with increased consumer demand. The price then dropped as other oil producers, particularly Saudi Arabia, increased production, and an economic recession dampened demand.

On the other hand, low prices can cause economic disasters for oil-producing nations. This can lead many of these countries to cut oil production. Saudi Arabia, Mexico, and Venezuela agreed to cut production by 1.6 to 2 million barrels a day beginning in 1999 in an attempt to halt the downward slide of prices. Many other oil-producing nations also limited their production. The limitations

TABLE 2.3

Retail motor gasoline and on-highway diesel fuel prices, 1950–1999

(dollars per gallon)

	Motor gasoline by grade [1]								Regular motor gasoline by area type [2,3]					On-highway diesel fuel [3]
	Leaded regular		Unleaded regular		Unleaded premium		All types							
Year	Nominal	Real [4]	Nominal	Real [4]	Nominal	Real [4]	Nominal	Real [4]	Conventional	Oxygenated	Oxygenated and reformulated	reformulated	All area types	
1950	0.27	[R]1.54	NA	NA	NA	NA	NA	NA	NA	NA	NA	NA	NA	NA
1960	0.31	[R]1.40	NA	NA	NA	NA	NA	NA	NA	NA	NA	NA	NA	NA
1970	0.36	[R]1.23	NA	NA	NA	NA	NA	NA	NA	NA	NA	NA	NA	NA
1971	0.36	[R]1.19	NA	NA	NA	NA	NA	NA	NA	NA	NA	NA	NA	NA
1972	0.36	[R]1.14	NA	NA	NA	NA	NA	NA	NA	NA	NA	NA	NA	NA
1973	0.39	[R]1.16	NA	NA	NA	NA	NA	NA	NA	NA	NA	NA	NA	NA
1974	0.53	[R]1.45	NA	NA	NA	NA	NA	NA	NA	NA	NA	NA	NA	NA
1975	0.57	[R]1.42	NA	NA	NA	NA	NA	NA	NA	NA	NA	NA	NA	NA
1976	0.59	[R]1.40	0.61	[R]1.45	NA	NA	NA	NA	NA	NA	NA	NA	NA	NA
1977	0.62	[R]1.38	0.66	[R]1.46	NA	NA	NA	NA	NA	NA	NA	NA	NA	NA
1978	0.63	[R]1.30	0.67	[R]1.39	NA	NA	0.65	[R]1.35	NA	NA	NA	NA	NA	NA
1979	0.86	[R]1.64	0.90	[R]1.73	NA	NA	0.88	[R]1.69	NA	NA	NA	NA	NA	NA
1980	1.19	[R]2.09	1.25	[R]2.18	NA	NA	1.22	[R]2.14	NA	NA	NA	NA	NA	NA
1981	1.31	[R]2.10	1.38	[R]2.21	1.47	[R]2.36	1.35	[R]2.17	NA	NA	NA	NA	NA	NA
1982	1.22	[R]1.85	1.30	[R]1.96	1.42	[R]2.14	1.28	[R]1.93	NA	NA	NA	NA	NA	NA
1983	1.16	[R]1.68	1.24	[R]1.80	1.38	[R]2.01	1.23	[R]1.78	NA	NA	NA	NA	NA	NA
1984	1.13	[R]1.58	1.21	[R]1.70	1.37	[R]1.91	1.20	[R]1.68	NA	NA	NA	NA	NA	NA
1985	1.12	[R]1.51	1.20	[R]1.63	1.34	[R]1.82	1.20	[R]1.62	NA	NA	NA	NA	NA	NA
1986	0.86	[R]1.14	0.93	[R]1.23	1.09	[R]1.44	0.93	[R]1.24	NA	NA	NA	NA	NA	NA
1987	0.90	[R]1.16	0.95	[R]1.22	1.09	[R]1.41	0.96	[R]1.23	NA	NA	NA	NA	NA	NA
1988	0.90	[R]1.12	0.95	[R]1.18	1.11	[R]1.38	0.96	[R]1.20	NA	NA	NA	NA	NA	NA
1989	1.00	[R]1.20	1.02	[R]1.23	1.20	[R]1.44	1.06	[R]1.27	NA	NA	NA	NA	NA	NA
1990	1.15	[R]1.33	1.16	[R]1.35	1.35	[R]1.56	1.22	[R]1.41	NA	NA	NA	NA	NA	NA
1991	NA	NA	1.14	[R]1.27	1.32	[R]1.47	1.20	[R]1.33	1.10	NA	NA	NA	1.10	NA
1992	NA	NA	1.13	[R]1.23	1.32	[R]1.43	1.19	[R]1.30	1.09	NA	NA	NA	1.09	NA
1993	NA	NA	1.11	[R]1.18	1.30	[R]1.38	1.17	[R]1.25	1.05	1.14	NA	NA	1.07	NA
1994	NA	NA	1.11	[R]1.16	1.31	[R]1.36	1.17	[R]1.22	1.06	1.14	NA	NA	1.08	NA
1995	NA	NA	1.15	[R]1.17	1.34	[R]1.36	1.21	[R]1.23	1.09	1.16	1.18	1.16	1.11	1.11
1996	NA	NA	1.23	[R]1.23	1.41	[R]1.41	1.29	[R]1.29	1.18	1.27	1.27	1.24	1.20	1.24
1997	NA	NA	1.23	[R]1.21	1.42	[R]1.39	1.29	[R]1.27	1.18	1.26	1.28	1.25	1.20	1.20
1998	NA	NA	1.06	[R]1.03	1.25	[R]1.21	1.12	[R]1.08	1.01	1.08	1.09	1.08	1.03	1.04
1999	NA	NA	1.17	1.11	1.36	1.30	1.22	1.17	1.11	1.20	1.19	1.20	1.14	1.12

[1] Average motor gasoline prices are calculated from a sample of service stations providing all types of service (i.e., full-, mini-, and self-serve). Geographic coverage - 1950-1973, 55 representative cities; 1974-1977, 56 urban areas; 1978 forward, 85 urban areas.

[2] "Area type" refers to the specific types of motor gasoline that are mandated by the Environmental Protection Agency to be sold in designated areas of the country. Only cash self-service prices are included.

[3] Nominal dollars.

[4] In chained (1996) dollars, calculated by using gross domestic product implicit price deflators.

R=Revised. NA=Not available.

SOURCE: Annual Energy Review 1999, Energy Information Administration, Washington, D.C., 2000

worked: during the summer of 2000, the price of gas surged to over $2.00 a gallon in parts of the U.S.

LESS CONCERN ABOUT OIL DEPENDENCY

Two decades ago the United States and its leaders were very concerned that so much of the U.S. economic structure, based heavily on oil, was dependent upon the whims of OPEC countries. Oil resources became an issue of national security, and OPEC countries, especially the Arab members, were often portrayed as potentially strangling the United States. Now, while nearly half of the nation's oil comes from outside the country and a majority of that coming from OPEC nations, there seems to be little public concern. The Reagan and Bush Administrations' decisions to permit the energy issue to be handled by the marketplace was consistent with their economic philosophy but indicated that they saw oil supply as an economic, not a political, issue. This downplayed the international political side of the energy problem. The Clinton Administration, as well, was unable to do much about America's dependence on foreign oil, as low prices throughout most of the 1990s set back energy conservation measures and public concern. In the U.S., efficiency gains in automobiles have been offset by the preference for large vehicles such as SUVs, which further shows that the public is uninterested in reducing the consumption of foreign oil.

The decline in public concern is due to other factors as well. The United States and European nations have developed substantial oil reserves to ride out oil supply fluctuations. This includes the Strategic Petroleum Reserve (SPR) in the United States (see below), and the government-required reserves in Europe. Furthermore, in emergencies, non-OPEC oil producers such as the United Kingdom and Norway can increase their supplies. Demand for oil has stabilized because of the conservation efforts of many industrialized nations, particularly in Europe.

The increased use of pipelines across Saudi Arabia, Turkey, and, until the Gulf War in 1991, through Iraq has resulted in a growing number of tankers picking up their oil in either the Red Sea or the Mediterranean Sea, before delivering it to Europe or the United States. These ships do not have to go through the potentially dangerous Persian Gulf. In addition, since many American strategists are wary of the "choke-point" of the Straits of Hormuz, where a future enemy could supposedly stop the flow of oil to the West, shipment through pipelines lessens the importance of the waterway. On the other hand, such pipelines could be destroyed in a war relatively easily.

Based on these factors, if the oil shortages that developed in 1973 or 1979 occurred again, the result would not likely be the same. Despite the United States' role as protector of Kuwait's oil in the Persian Gulf, there seems to be little real concern that America's dependence on foreign oil, especially OPEC oil, represents a threat to national security or national stability.

The 1990–91 Persian Gulf War illustrated the effectiveness of both the reserves and market mechanisms. Prices rose in reaction to the loss of Iraqi and Kuwaiti oil supplies and to speculation that hostilities in the region could restrict other Gulf oil supplies. Demand dropped in reaction to the higher prices, which in turn was one factor in the recession that followed. In addition, the price increases were somewhat moderated by the availability of reserve supplies, as the United States and other Western nations agreed to release reserves. The United States even sold some of its oil to other countries.

In 1991 the World Bank predicted that prices could go as high as $65 a barrel. Instead, prices plummeted. Saudi Arabia and other producers, taking advantage of the war-reduced supply and higher prices, increased production. Production continued to rise even as demand fell, resulting in record-high world petroleum stockpiles. As the imbalance persisted, oil prices fell to below the pre-crisis level. Gulf oil production remained high after the war.

Nonetheless, the war highlighted the dependence on oil by the United States and its Western allies. In order to guarantee its oil supply, the United States and its allies committed 500,000 troops to war. Though world leaders spoke of stopping aggression, many observers wondered if the reaction would have been the same if Kuwait did not have large oil fields.

CONSERVATION AND OIL DEMAND

The relatively slow growth in total demand for oil can be attributed to the conservation measures adopted in response to the oil price shocks of the 1970s. Oil companies have overhauled existing refineries and built more efficient new ones. Energy efficiency will become increasingly effective as oil-powered equipment continues to be replaced with new technology. Oil may be saved as well if energy users switch to other forms of fuel, including renewable sources. (See Chapter 9.)

Many quick and relatively inexpensive conservation measures have become available. Moderate investments in insulation and heater efficiency in existing buildings can cut fuel needs for these buildings by 25 percent, and appliances and engines have become more efficient. Most of the increase in oil demand has been for transportation. Much more efficient vehicles can be built with today's technology, but low gasoline prices have eliminated consumer demand for them. (For information on automobile fuel efficiency measures, see Chapter 10.)

Because the United States is a "mature" oil-producing nation, with few new sources predicted, an increased national commitment to conservation could be an important way to resolve the problem of an ever-greater dependence on foreign oil. Some observers believe it is better to develop programs promoting conservation now because the cost would be less than if steps were taken in a hurried manner during a crisis in the future. Others argue, however, that the marketplace will resolve the problem if shortages arise. Either way, the once important issue of "energy independence" is no longer generally considered a national concern.

ENVIRONMENTAL CONCERNS

Fossil fuels, including petroleum, have been responsible for much contamination of the environment. Because of public sentiment and legislation passed to slow the environmental damage, efforts are being made to shift to other means of energy production, to conserve existing oil supplies, and to find cleaner ways to produce, transport, and burn oil. Transporting oil carries significant environmental risks. According to the U.S. Department of the Interior, the cause of most transportation spills is oil tanker accidents, such as the grounding of the *Exxon Valdez* in 1989.

The *Exxon Valdez* Oil Spill

A number of events of the past decade have influenced American attitudes toward oil production and use. In March 1989, the *Exxon Valdez* oil tanker hit a reef in Alaska, spilling 11 million gallons of crude oil into the waters of Prince William Sound—the largest oil spill ever. The cleanup cost Exxon $1.28 billion, which does not include the legal costs or the cost of the lost wildlife. The spill was an environmental disaster for a formerly pristine area. Even the measures used to clean up the spill, such as washing the beaches with hot water, proved to be additionally damaging.

The *Exxon Valdez* spill also led to debate about added safety measures in the design of tankers. Tankers are bigger than ever before. In 1945 the largest tanker held 16,500 tons of oil; today, supertankers carry more than 550,000 tons. These supertankers are difficult to maneuver because of their size and are likely to spill more oil if damaged. Although there have been fewer spills since 1973, the amount of oil lost is roughly the same.

The Oil Pollution Act of 1990

The *Exxon Valdez* oil spill led Congress to pass the Oil Pollution Act of 1990 (PL 101-380), after having debated the issue for 16 years. The bill increases, but still limits, oil spillers' federal liability (financial responsibility) as long as a spill is not the result of "gross negligence." The bill also mandates compensation to those who are economically injured by oil-spill accidents. Damages that can be charged to oil companies are limited to $60 million for tanker accidents and $75 million for accidents at offshore facilities. The rest of the cleanup costs are paid from a $500 million oil-spill fund generated by a 1.3 cents-per-barrel tax on oil. The individual states still maintain the right to impose unlimited liability on spillers. Oil companies are also required to phase in double hulls on oil vessels over a 25-year period. Double hulls provide an extra container in the case of accidents.

THE SHRINKING U.S. OIL INDUSTRY

During the past two decades, the U.S. petroleum industry has experienced a severe loss of jobs. The number of seismic land crews and marine vessels searching for oil in the United States and its waters has decreased sharply since 1981. From 1982 to 1992 alone, oil-extraction companies lost 51 percent of their work force, and petroleum refiners experienced a decline of 28 percent. In Texas alone, once the center of the U.S. oil industry, jobs in the industry plummeted from 80,000 in 1981 to 25,000 in 1996. In 1998 the Texas Comptroller of Public Accounts estimated that for every dollar drop in the price of oil per barrel, 10,000 jobs are lost in the Texas economy. That translated into 100,000 jobs lost in the Texas oil industry from October 1997 to December 1998. Oil companies have blamed environmental regulations and the restriction of oil exploration on some federal lands, such as Arctic wildlife preserves and some offshore areas. However, mergers in the industry have cost thousands of jobs as companies restructure and downsize.

STRATEGIC PETROLEUM RESERVES

In 1923 the Harding Administration set up the Petroleum Reserve to ensure that the U.S. Navy would have adequate fuel in the event of war. In 1975, in response to the growing concern over America's energy dependence, Congress turned the Strategic Petroleum Reserve (SPR) over to the Interior Department under the Energy Policy and Conservation Act (PL 94-163). An additional law, PL 101-383, expanded the SPR and created a second reserve for refined products. A small amount of oil was withdrawn from the reserves at the start of the Persian Gulf War, but it became clear that large withdrawals were unnecessary.

In the SPR oil is stored in deep salt caverns located at four storage sites in Louisiana and two in Texas (oil does not dissolve salt the way water does). Each site has caverns that vary in capacity, but most can hold approximately 10 million barrels of oil. The SPR system currently has 41 such caverns. If the United States suddenly found its supplies cut off, the reserve system would be connected to existing commercial lines which would start pumping the oil.

At the end of 1999 the SPR contained about 567 million barrels (see Figure 2.10), enough to equal 59 days of imports should the supply be cut off. (See Figure 2.11.) This is over one-third of the total stock available should

FIGURE 2.10

End-of-year stocks in Strategic Petroleum Reserve (SPR), 1977–99

567 in 1999

SOURCE: *Annual Energy Review 1999*, Energy Information Administration, Washington, D.C., 2000

FIGURE 2.11

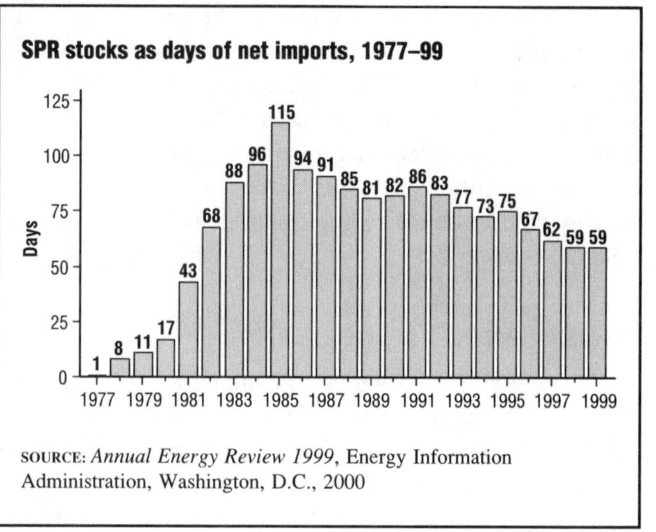

SPR stocks as days of net imports, 1977–99

SOURCE: *Annual Energy Review 1999*, Energy Information Administration, Washington, D.C., 2000

an emergency develop. The remaining two-thirds is in the hands of private oil companies. Though Figure 2.11 appears to show a decline in the reserves, from a high of 115 days in 1985 to the current 59 days, it is not because of a drop in quantity. Rather, it reflects the increase in imports since 1985. As the nation has imported a greater amount of oil, the days of import replacement represented by the amount of oil in the SPR have dropped.

In 1999 the Clinton Administration announced that it would add oil to the reserves for the first time since 1994. The government will replace the 28 million barrels that have been withdrawn and sold to the public and private sector over the past few years at a rate of about 150,000 barrels a day.

SYNTHETIC FUELS

Fuel can be produced from other energy sources besides oil, including coal, natural gas, grain, garbage, and oil shale. Some common examples of synthetic fuels are alcohol, methane gas, methanol, and gasohol. Many years ago, kerosene was made from coal oil, so it was an early synthetic fuel.

In 1980 President Jimmy Carter signed legislation creating the Synthetic Fuels Corporation Act (PL 96-294) in an attempt to make the United States less dependent on foreign oil. The legislation was intended to provide funding for companies that would turn coal, shale oil, and tar sands into oil and natural gas. Congress originally financed the program with $15 billion because people feared the price of oil had no upper limit, and experts predicted oil would reach $90 a barrel. The synthetic fuels industry quickly lost support when the United States experienced a recession in the 1980s and world oil prices dropped, making synthetic fuels uneconomical. It cost much less to import oil than to manufacture synthetic fuels, and the whole idea of "energy independence" quickly faded away. In its five-year existence, the Synthetic Fuels Corporation planned six manufacturing plants. After gradual reductions in financing, the entire program was discontinued in 1985.

The United States government has continued a program to study and subsidize the use of ethanol, which is grain alcohol usually produced from corn, as a gasoline extender or substitute. Ethanol receives subsidies in the form of fuel tax exemptions that compensate for its higher cost. An ethanol-gasoline mixture is usually 10 percent ethanol and 90 percent gasoline. Gasoline with higher percentages of ethanol tends to corrode parts of automobile fuel systems, makes cold engines hard to start, and produces less energy per gallon which reduces fuel efficiency. Brazil currently pursues an aggressive program to substitute fuels from grain for gasoline, with mixed results.

INTERNATIONAL OIL USAGE

World Production

Total world petroleum production has generally increased since the early 1980s, reaching nearly 66 million barrels per day in 1999. (See Table 2.4.) The major producers have been Saudi Arabia, Russia, and the United States. Together, these three countries account for 30 percent of the world's crude oil production. Other leading oil producers include Iran, China, Norway, Mexico, Venezuela, the United Kingdom, Iraq, United Arab Emirates, and Nigeria. (See Figure 2.12.)

World Consumption

World petroleum consumption in 1998 was estimated at 73.6 million barrels per day. (See Table 2.5.) The United States was by far the leading consumer, using 18.9 million barrels per day, followed by Japan (5.5 million barrels per day), China (4.1 million barrels a day), Ger-

TABLE 2.4

World crude oil production, 1960–99

(million barrels per day)

	Selected OPEC [1] producers									Selected non-OPEC producers									
Year	Persian Gulf nations [2]	Iran	Iraq	Kuwait [3]	Nigeria	Saudi Arabia [3]	United Arab Emirates	Venezuela	Total OPEC	Canada	China	Mexico	Norway	Former U.S.S.R.	Russia	United Kingdom	United States	Total non-OPEC [4]	World
1960	5.27	1.07	0.97	1.69	0.02	1.31	0.00	2.85	8.70	0.52	0.10	0.27	0.00	2.91	—	(s)	7.04	12.29	20.99
1961	5.65	1.20	1.01	1.74	0.05	1.48	0.00	2.92	9.36	0.61	0.11	0.29	0.00	3.28	—	(s)	7.18	13.09	22.45
1962	6.19	1.33	1.01	1.96	0.07	1.64	0.01	3.20	10.51	0.67	0.12	0.31	0.00	3.67	—	(s)	7.33	13.84	24.35
1963	6.82	1.49	1.16	2.10	0.08	1.79	0.05	3.25	11.51	0.71	0.13	0.31	0.00	4.07	—	(s)	7.54	14.62	26.13
1964	7.61	1.71	1.26	2.30	0.12	1.90	0.19	3.39	12.98	0.75	0.18	0.32	0.00	4.60	—	(s)	7.61	15.20	28.18
1965	8.37	1.91	1.32	2.36	0.27	2.21	0.28	3.47	14.35	0.81	0.23	0.32	0.00	4.79	—	(s)	7.80	15.98	30.33
1966	9.32	2.13	1.39	2.48	0.42	2.60	0.36	3.37	15.77	0.88	0.29	0.33	0.00	5.23	—	(s)	8.30	17.19	32.96
1967	9.91	2.60	1.23	2.50	0.32	2.81	0.38	3.54	16.85	0.96	0.28	0.36	0.00	5.68	—	(s)	8.81	18.54	35.39
1968	10.91	2.84	1.50	2.61	0.14	3.04	0.50	3.60	18.79	1.19	0.30	0.39	0.00	6.08	—	(s)	9.10	19.84	38.63
1969	11.95	3.38	1.52	2.77	0.54	3.22	0.63	3.59	20.91	1.13	0.48	0.46	0.00	6.48	—	(s)	9.24	20.79	41.70
1970	13.39	3.83	1.55	2.99	1.08	3.80	0.78	3.71	23.30	1.26	0.60	0.49	0.00	6.99	—	(s)	9.64	22.59	45.89
1971	15.77	4.54	1.69	3.20	1.53	4.77	1.06	3.55	25.21	1.35	0.78	0.49	0.01	7.48	—	(s)	9.46	23.31	48.52
1972	17.54	5.02	1.47	3.28	1.82	6.02	1.20	3.22	26.89	1.53	0.90	0.51	0.03	7.89	—	(s)	9.44	24.25	51.14
1973	20.67	5.86	2.02	3.02	2.05	7.60	1.53	3.37	30.63	1.80	1.09	0.47	0.03	8.32	—	(s)	9.21	25.05	55.68
1974	21.28	6.02	1.97	2.55	2.26	8.48	1.68	2.98	30.35	1.55	1.32	0.57	0.04	8.91	—	(s)	8.77	25.37	55.72
1975	18.93	5.35	2.26	2.08	1.78	7.08	1.66	2.35	26.77	1.43	1.49	0.71	0.19	9.52	—	0.01	8.37	26.06	52.83
1976	21.51	5.88	2.42	2.15	2.07	8.58	1.94	2.29	30.33	1.31	1.67	0.83	0.28	10.06	—	0.25	8.13	27.01	57.34
1977	21.73	5.66	2.35	1.97	2.09	9.25	2.00	2.24	30.89	1.32	1.87	0.98	0.28	10.60	—	0.77	8.24	28.82	59.71
1978	20.61	5.24	2.56	2.13	1.90	8.30	1.83	2.17	29.46	1.32	2.08	1.21	0.36	11.11	—	1.08	8.71	30.70	60.16
1979	21.07	3.17	3.48	2.50	2.30	9.53	1.83	2.36	30.58	1.50	2.12	1.46	0.40	11.38	—	1.57	8.55	32.09	62.67
1980	17.96	1.66	2.51	1.66	2.06	9.90	1.71	2.17	26.61	1.44	2.11	1.94	0.53	11.71	—	1.62	8.60	32.99	59.60
1981	15.25	1.38	1.00	1.13	1.43	9.82	1.47	2.10	22.48	1.29	2.01	2.31	0.50	11.85	—	1.81	8.57	33.60	56.08
1982	12.16	2.21	1.01	0.82	1.30	6.48	1.25	1.90	18.78	1.27	2.05	2.75	0.52	11.91	—	2.07	8.65	34.70	53.48
1983	11.08	2.44	1.01	1.06	1.24	5.09	1.15	1.80	17.50	1.36	2.12	2.69	0.61	11.97	—	2.29	8.69	35.76	53.26
1984	10.78	2.17	1.21	1.16	1.39	4.66	1.15	1.80	17.44	1.44	2.30	2.78	0.70	11.86	—	2.48	8.88	37.05	54.49
1985	9.63	2.25	1.43	1.02	1.50	3.39	1.19	1.68	16.18	1.47	2.51	2.75	0.79	11.59	—	2.53	8.97	37.80	53.98
1986	11.70	2.04	1.69	1.42	1.47	4.87	1.33	1.79	18.28	1.47	2.62	2.44	0.87	11.90	—	2.54	8.68	37.95	56.23
1987	12.10	2.30	2.08	1.59	1.34	4.27	1.54	1.75	18.52	1.54	2.69	2.55	1.02	12.05	—	2.41	8.35	38.15	56.67
1988	13.46	2.24	2.69	1.49	1.45	5.09	1.57	1.90	20.32	1.62	2.73	2.51	1.16	12.05	—	2.23	8.14	38.42	58.74
1989	14.84	2.81	2.90	1.78	1.72	5.06	1.86	1.91	22.07	1.56	2.76	2.52	1.55	11.72	—	1.80	7.61	37.79	59.86
1990	15.28	3.09	2.04	1.18	1.81	6.41	2.12	2.14	23.20	1.55	2.77	2.55	1.70	10.98	—	1.82	7.36	37.37	60.57
1991	14.74	3.31	0.31	0.19	1.89	8.12	2.39	2.38	23.27	1.55	2.84	2.68	1.89	9.99	—	1.80	7.42	36.94	60.21
1992	15.97	3.43	0.43	1.06	1.94	8.33	2.27	2.37	24.40	1.61	2.85	2.67	2.23	—	7.63	1.83	7.17	35.81	60.21
1993	16.71	3.54	0.51	1.85	1.96	8.20	2.16	2.45	25.12	1.68	2.89	2.67	2.35	—	6.73	1.92	6.85	35.12	60.24
1994	16.96	3.62	0.55	2.03	1.93	8.12	2.19	2.59	25.51	1.75	2.94	2.69	2.52	—	6.14	2.37	6.66	35.48	60.99
1995	17.21	3.64	0.56	2.06	1.99	8.23	2.23	2.75	26.00	1.81	2.99	2.62	2.77	—	6.00	2.49	6.56	36.33	62.33
1996	17.37	3.69	0.58	2.06	[R]2.00	8.22	2.28	[R]2.94	[R]26.46	1.84	3.13	2.86	3.10	—	5.85	2.57	6.46	[R]37.25	[R]63.71
1997	[R]18.47	3.66	[R]1.16	2.08	[R]2.33	8.56	2.32	[R]3.32	[R]28.32	[R]1.92	3.20	3.02	[R]3.14	—	[R]5.92	2.52	6.45	[R]38.10	[R]66.42
1998	19.33	3.63	2.15	2.09	2.15	8.39	2.35	3.17	28.76	[R]1.98	3.20	3.07	3.02	—	5.94	2.62	[R]6.25	38.11	66.87
1999[P]	18.69	3.56	2.51	1.90	2.13	7.83	2.17	2.83	27.64	1.91	3.21	2.91	3.02	—	6.07	2.69	5.93	38.02	65.66

[1] Organization of Petroleum Exporting Countries.
[2] Persian Gulf Nations are Bahrain, Iran, Iraq, Kuwait, Qatar, Saudi Arabia, and United Arab Emirates.
[3] Includes about one-half of the production in the Neutral Zone between Kuwait and Saudi Arabia.
[4] Ecuador, which withdrew from OPEC on December 31, 1992, and Gabon, which withdrew on December 31,1994, are included in "Non-OPEC" for all years.
R=Revised. P=Preliminary.— = Not applicable. (s)=Less than 0.005 million barrels per day.

SOURCE: *Annual Energy Review 1999,* Energy Information Administration, Washington, D.C., 2000

FIGURE 2.12

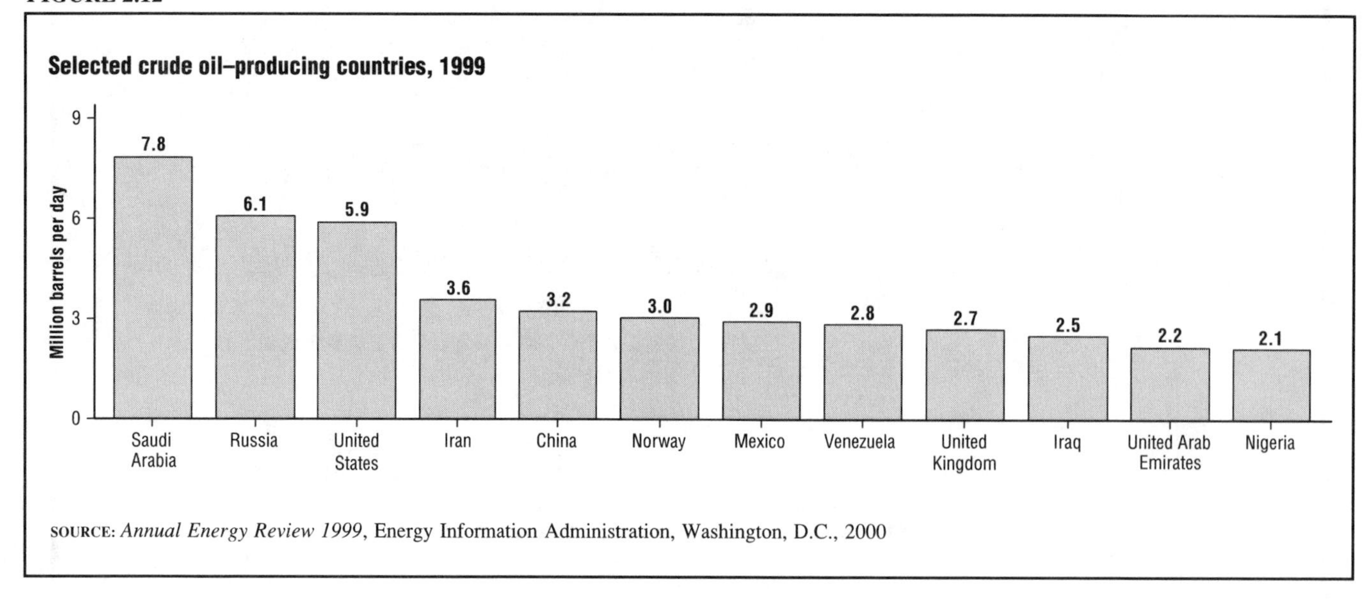

SOURCE: *Annual Energy Review 1999*, Energy Information Administration, Washington, D.C., 2000

many (2.9 million barrels a day), and Russia (2.5 million barrels per day).

FUTURE TRENDS IN THE OIL INDUSTRY

The Energy Information Administration of the U.S. Department of Energy, in its *Annual Energy Outlook 2000* (1999), has projected that domestic crude oil production will continue to decline over the next two decades, from 6.3 million barrels per day in 1998 to about 5.3 million barrels per day in 2020. Some of this drop will be compensated by improved refinery efficiency and increased natural gas plant liquids, so that total domestic supply only drops from 9.2 million barrels per day in 1998 to 9.1 million in 2020. (See Figure 2.13.)

Domestic petroleum consumption is expected to increase approximately 1.3 percent per year to 25 million barrels per day in 2020, depending on economic growth and world oil prices. Petroleum consumption increases mainly in the transportation sector, which used two-thirds of petroleum in 1998. In transportation, efficiency gains will be offset by increased travel. Continuing consumer preference for large, more powerful automobiles will raise gasoline demand, and more frequent air travel heightens the demand for jet fuel. Petroleum use by the industrial sector is projected to increase slightly through 2020, while the use of petroleum for residential, commercial, and electricity generation purposes is projected to decrease.

FIGURE 2.13

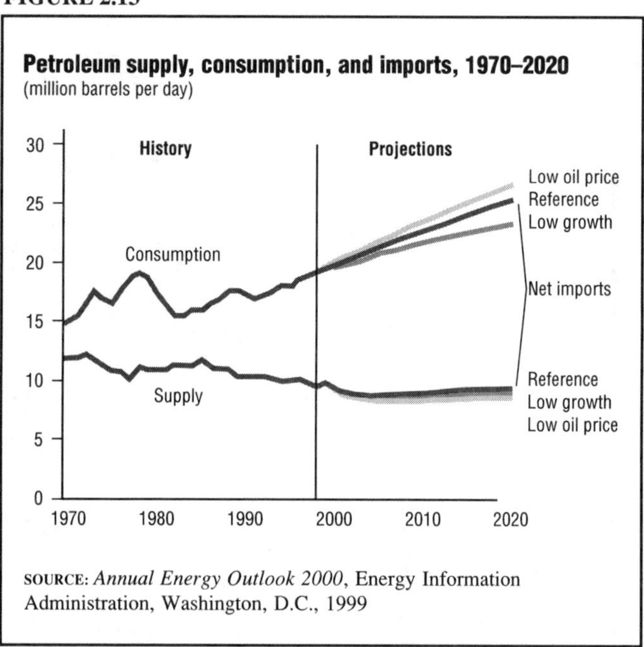

SOURCE: *Annual Energy Outlook 2000*, Energy Information Administration, Washington, D.C., 1999

The U.S. could import as many as 18.1 million barrels per day in 2020, compared to 9.8 million in 1998, depending on world economic conditions. The U.S. could pay as much as $161.3 billion for oil imports in 2020, compared to $46.6 billion in 1998. Oil is projected to increase from the $12 per barrel price in 1998 to about $22 per barrel in 2020.

TABLE 2.5

World petroleum consumption, 1960–98

(million barrels per day)

	Selected OECD [1] consumers											Selected non-OECD consumers						
Year	Canada	France	Germany [2]	Italy	Japan	Mexico [3]	South Korea [3]	Spain	United Kingdom	United States	Total OECD [4]	Brazil	China	India	Former U.S.S.R.	Russia	Total non-OECD	World
1960	0.84	0.56	0.63	0.44	0.66	0.30	0.01	0.10	0.94	9.80	15.78	0.27	0.17	0.16	2.38	—	5.56	21.34
1961	0.87	0.63	0.79	0.54	0.82	0.29	0.02	0.12	1.04	9.98	16.77	0.28	0.17	0.17	2.57	—	6.23	23.00
1962	0.92	0.73	1.00	0.67	0.93	0.30	0.02	0.12	1.12	10.40	18.06	0.31	0.14	0.18	2.87	—	6.83	24.89
1963	0.99	0.86	1.17	0.77	1.21	0.31	0.03	0.12	1.27	10.74	19.60	0.34	0.17	0.21	3.15	—	7.32	26.92
1964	1.05	0.98	1.36	0.90	1.48	0.33	0.02	0.20	1.36	11.02	21.05	0.35	0.20	0.22	3.58	—	8.03	29.08
1965	1.14	1.09	1.61	0.98	1.74	0.34	0.03	0.23	1.49	11.51	22.81	0.33	0.23	0.25	3.61	—	8.33	31.14
1966	1.21	1.19	1.80	1.08	1.98	0.36	0.04	0.31	1.58	12.08	24.60	0.38	0.30	0.28	3.87	—	8.96	33.56
1967	1.25	1.34	1.86	1.19	2.14	0.39	0.07	0.36	1.64	12.56	25.94	0.38	0.28	0.26	4.22	—	9.65	35.59
1968	1.34	1.46	1.99	1.40	2.66	0.41	0.10	0.46	1.82	13.39	28.56	0.46	0.31	0.31	4.48	—	10.40	38.96
1969	1.42	1.66	2.33	1.69	3.25	0.45	0.15	0.49	1.98	14.14	31.54	0.48	0.44	0.34	4.87	—	11.35	42.89
1970	1.52	1.94	2.83	1.71	3.82	0.50	0.20	0.58	2.10	14.70	34.49	0.53	0.62	0.40	5.31	—	12.32	46.81
1971	1.56	2.12	2.94	1.84	4.14	0.52	0.23	0.64	2.14	15.21	36.07	0.58	0.79	0.42	5.66	—	13.35	49.42
1972	1.66	2.32	3.13	1.95	4.36	0.59	0.23	0.68	2.28	16.37	38.74	0.66	0.91	0.46	6.12	—	14.35	53.09
1973	1.73	2.60	3.34	2.07	4.95	0.67	0.28	0.78	2.34	17.31	41.53	0.78	1.12	0.49	6.60	—	15.71	57.24
1974	1.78	2.45	3.06	2.00	4.86	0.71	0.29	0.86	2.21	16.65	40.12	0.86	1.19	0.47	7.28	—	16.56	56.68
1975	1.78	2.25	2.96	1.86	4.62	0.75	0.31	0.87	1.91	16.32	38.82	0.92	1.36	0.50	7.52	—	17.38	56.20
1976	1.82	2.42	3.21	1.97	4.84	0.83	0.36	0.97	1.89	17.46	41.39	1.00	1.53	0.51	7.78	—	18.28	59.67
1977	1.85	2.29	3.21	1.90	4.88	0.88	0.42	0.94	1.91	18.43	42.43	1.02	1.64	0.55	8.18	—	19.40	61.83
1978	1.90	2.41	3.29	1.95	4.95	0.99	0.48	0.98	1.94	18.85	43.62	1.11	1.79	0.62	8.48	—	20.54	64.16
1979	1.97	2.46	3.37	2.04	5.05	1.10	0.53	1.02	1.97	18.51	44.01	1.18	1.84	0.66	8.64	—	21.21	65.22
1980	1.87	2.26	3.08	1.93	4.96	1.27	0.54	0.99	1.73	17.06	41.41	1.15	1.77	0.64	9.00	—	21.66	63.07
1981	1.77	2.02	2.80	1.87	4.85	1.40	0.54	0.94	1.59	16.06	39.14	1.09	1.71	0.73	8.94	—	21.76	60.90
1982	1.58	1.88	2.74	1.78	4.58	1.48	0.53	1.00	1.59	15.30	37.45	1.06	1.66	0.74	9.08	—	22.05	59.50
1983	1.45	1.84	2.66	1.75	4.40	1.35	0.56	1.01	1.53	15.23	36.59	0.98	1.73	0.77	8.95	—	22.15	58.74
1984	1.47	1.75	2.66	1.65	4.58	1.45	0.59	0.91	1.85	15.73	37.43	1.03	1.74	0.82	8.91	—	22.41	59.84
1985	1.50	1.78	2.70	1.72	4.38	1.47	0.57	0.85	1.63	15.73	37.23	1.08	1.89	0.90	8.95	—	22.87	60.10
1986	1.51	1.77	2.86	1.74	4.44	1.49	0.61	0.88	1.65	16.28	38.28	1.24	2.00	0.95	8.98	—	23.48	61.76
1987	1.55	1.79	2.77	1.86	4.48	1.52	0.64	0.90	1.60	16.67	38.96	1.26	2.12	0.99	9.00	—	24.04	63.00
1988	1.69	1.80	2.74	1.84	4.75	1.55	0.73	0.98	1.70	17.28	40.24	1.30	2.28	1.08	8.89	—	24.58	64.82
1989	1.73	1.86	2.58	1.93	4.98	1.64	0.84	1.03	1.74	17.33	40.88	1.32	2.38	1.15	8.74	—	25.04	65.92
1990	1.69	1.82	2.66	1.87	5.14	1.68	1.03	1.01	1.75	16.99	40.92	1.34	2.30	1.17	8.39	—	25.06	65.98
1991	1.62	1.94	2.83	1.86	5.28	1.70	1.20	1.07	1.80	16.71	41.40	1.35	2.50	1.19	8.35	—	25.17	66.57
1992	1.64	1.93	2.84	1.94	5.45	1.72	1.46	1.11	1.80	17.03 R	42.42	1.37	2.66	1.28	—	4.42	R24.34	R66.76
1993	1.69	1.88	2.90	1.85	5.40	1.71	1.69	1.06	1.82	17.24	42.98	1.43	2.96	1.31	—	3.75	R24.02	R67.00
1994	1.73	1.83	2.88	1.84	5.67	1.80	1.86	1.13	1.84	17.72	44.17	1.51	R3.16	1.41	—	3.18	R24.11	R68.28
1995	1.76	1.90	2.88	2.05	5.71	1.72	2.03	1.26	1.85	17.72 R	44.96	1.60	R3.36	1.58	—	2.98	R24.91	R69.87
1996	1.80	1.94	2.91	2.06	5.87	1.76	2.18	1.18	1.85	18.31	46.07	1.72	R3.61	1.68	—	2.62	R25.33	R71.40
1997	1.86	1.96	2.90	2.05	5.71	R1.87	R2.39	1.30	1.80	18.62 R	46.83	R1.82	R3.92	R1.77	—	R2.56	R26.30	R73.13
1998P	1.87	2.03	2.92	2.07	5.51	1.95	2.00	1.39	1.78	18.92	46.98	1.88	4.11	1.84	—	2.46	26.66	73.64

[1] Organization for Economic Cooperation and Development.

[2] Through 1969, the data for Germany are for the former West Germany only. For 1970 through 1990, this is East and West Germany. Beginning in 1991, this is unified Germany.

[3] Mexico, which joined the OECD on May 18, 1994, and South Korea, which joined the OECD on December 12, 1996, are included in the OECD for all years shown in this table.

[4] Hungary and Poland, which joined the OECD on May 7, 1996, and November 22, 1996, respectively, are included in Total OECD beginning in 1970, the first year that data for these countries were available. The Czech Republic, which joined the OECD on December 21, 1995, is included in Total OECD beginning in 1993, the year that it came into existence.

R=Revised. P=Preliminary. — = Not applicable.

SOURCE: *Annual Energy Review 1999*, Energy Information Administration, Washington, D.C., 2000

CHAPTER 3

NATURAL GAS

Natural gas is an important source of energy in the United States. The natural gas industry first developed out of the growing petroleum industry. Wells drilled for oil often produced considerable amounts of natural gas, but the early oilmen had no idea what to do with it.

Natural gas is mostly a mixture of methane, ethane, and propane, with methane making up 73 to 95 percent of the total. Originally considered a waste by-product of oil production, natural gas had no market, nor were transmission lines available to deliver it even if a use could have been found. As a result, the gas was burned off (flared). Pictures of southeast Texas at the turn of the century showed thousands of wooden drilling rigs topped with flaming plumes of gas like burning candles. Even today, flaring sites are sometimes the brightest areas visible in nighttime satellite images, outshining even the brightest urban areas.

Nonetheless, researchers soon found ways to use natural gas. In 1925 the first welded pipeline was built from Louisiana to Texas and was over 200 miles long. U.S demand grew rapidly, especially after World War II. By the 1950s natural gas was providing one-quarter of the nation's energy needs. Today, natural gas is second only to petroleum in the share of U.S. energy produced, and a vast pipeline transmission system connects the production facilities in the United States, Canada, and Mexico with natural gas distributors. Figures 3.1 and 3.2 show the flow of natural gas.

THE PRODUCTION OF NATURAL GAS

Natural gas is produced from gas wells and oil wells. There is little delay between production and consumption, except for the gas that is placed in storage. Changes in demand are almost immediately reflected by changes in wellhead flows, or supply.

Total gas production in America was 18.7 trillion cubic feet in 1999, below the levels produced during the early 1970s. (See Figures 3.3 and 3.4.) Although production levels are being driven up by increasing demand and rising prices, production continues to be outpaced by consumption. Imported gas makes up the difference between supply and demand. Texas, Louisiana, and Oklahoma accounted for over half the natural gas produced in the United States in 1999.

Natural Gas Wells

In 1999, 306,000 gas wells were in operation, down slightly from 1998's all-time high. (See Figure 3.5.) Average productivity of these wells has remained relatively low since the 1980s. (See Figure 3.6.) The number of producing wells fluctuates due to drilling activity for new sources of gas, the rate of abandoning old wells, weather conditions, and the economic conditions of the nation. Wells are abandoned when they can no longer produce economically.

Offshore Production

Offshore drilling for natural gas accounted for about one-fourth of the total U.S. production in 1999. (See Figure 3.7.) Almost all natural gas produced offshore comes from the Gulf of Mexico and offshore California. U.S. offshore production is expected to increase to meet the nation's growing need for energy, although this type of production could be slowed by environmental restrictions.

Offshore drilling generally occurs on the outer continental shelf, the submerged area offshore with a depth of up to 200 meters (656 feet). Figure 3.8 shows a diagram of a continental margin. The continental shelf varies from one coastal area to another. The shelf is relatively narrow along the Pacific coast, wide along much of the Atlantic coast and the Gulf of Alaska, and widest in the Gulf of Mexico. Offshore waters contain state and federal leases for gas rights involving more than 1.5 billion acres.

The development of offshore oil and gas resources began with the drilling of the Summerland oil field in Cal-

FIGURE 3.1

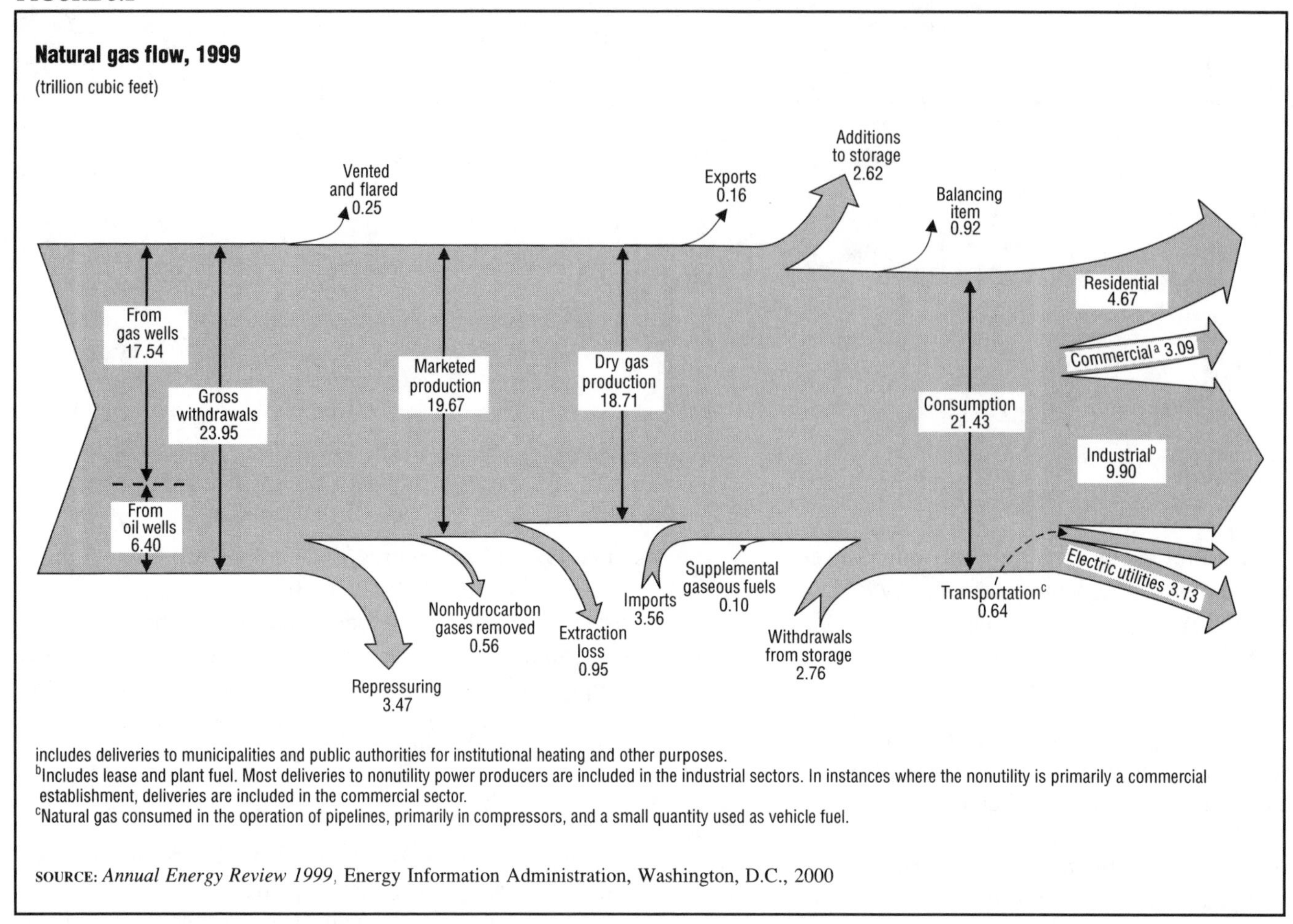

includes deliveries to municipalities and public authorities for institutional heating and other purposes.
[b]Includes lease and plant fuel. Most deliveries to nonutility power producers are included in the industrial sectors. In instances where the nonutility is primarily a commercial establishment, deliveries are included in the commercial sector.
[c]Natural gas consumed in the operation of pipelines, primarily in compressors, and a small quantity used as vehicle fuel.

SOURCE: *Annual Energy Review 1999*, Energy Information Administration, Washington, D.C., 2000

ifornia in 1896, where about 400 wells were drilled. In the search for oil and gas in offshore areas, the industry has continually improved drilling technology. Today, deep-water petroleum exploration occurs from platforms and drill ships, and from gravel islands and mobile units in shallow water.

Even though natural gas is transported mostly by pipelines instead of by tankers, the 1989 Exxon oil spill in Prince William Sound, Alaska, and other oil spills have focused national attention on all types of offshore drilling. Even before the *Exxon Valdez* oil spill, environmentalists were calling for the curtailment of offshore drilling of both oil and gas.

Natural Gas Reserves

Reserves are estimated volumes of gas in known deposits that are believed to be recoverable in the future. Proved reserves are those gas volumes that geological and engineering data show with reasonable certainty to be recoverable. Proved reserves of natural gas amounted to about 142 trillion cubic feet in 1998. (See Table 3.1.) The reserves in Texas, the Gulf of Mexico, Oklahoma, and Louisiana make up approximately half the total. (See Chapter 5 for further discussion of reserves.)

Natural gas reserves in North America are generally more abundant than crude oil reserves, although historically they have been difficult to estimate accurately. At one time the U.S. Department of Energy had estimated that proven supplies of recoverable gas in the United States would last fewer than eight years. However, new discoveries and technological improvements have increased the estimated recoverable supply of natural gas to 50 years or more.

The North Slope fields of Alaska are estimated to contain reserves amounting to 64 trillion cubic feet, but at the moment there is no easy way to transport Alaska's reserves to the lower 48 states. Low natural gas prices have made building a pipeline from Alaska uneconomical. In 1998, 92 percent of Alaska's North Slope gas was "reinjected," a process in which pressurized natural gas is removed and injected back into the ground to help extract petroleum.

Underground Storage

Because of seasonal, daily, and even hourly changes in gas demand, substantial natural gas storage facilities have been created to meet supply needs. Many of these storage centers are depleted gas reservoirs located near

FIGURE 3.2

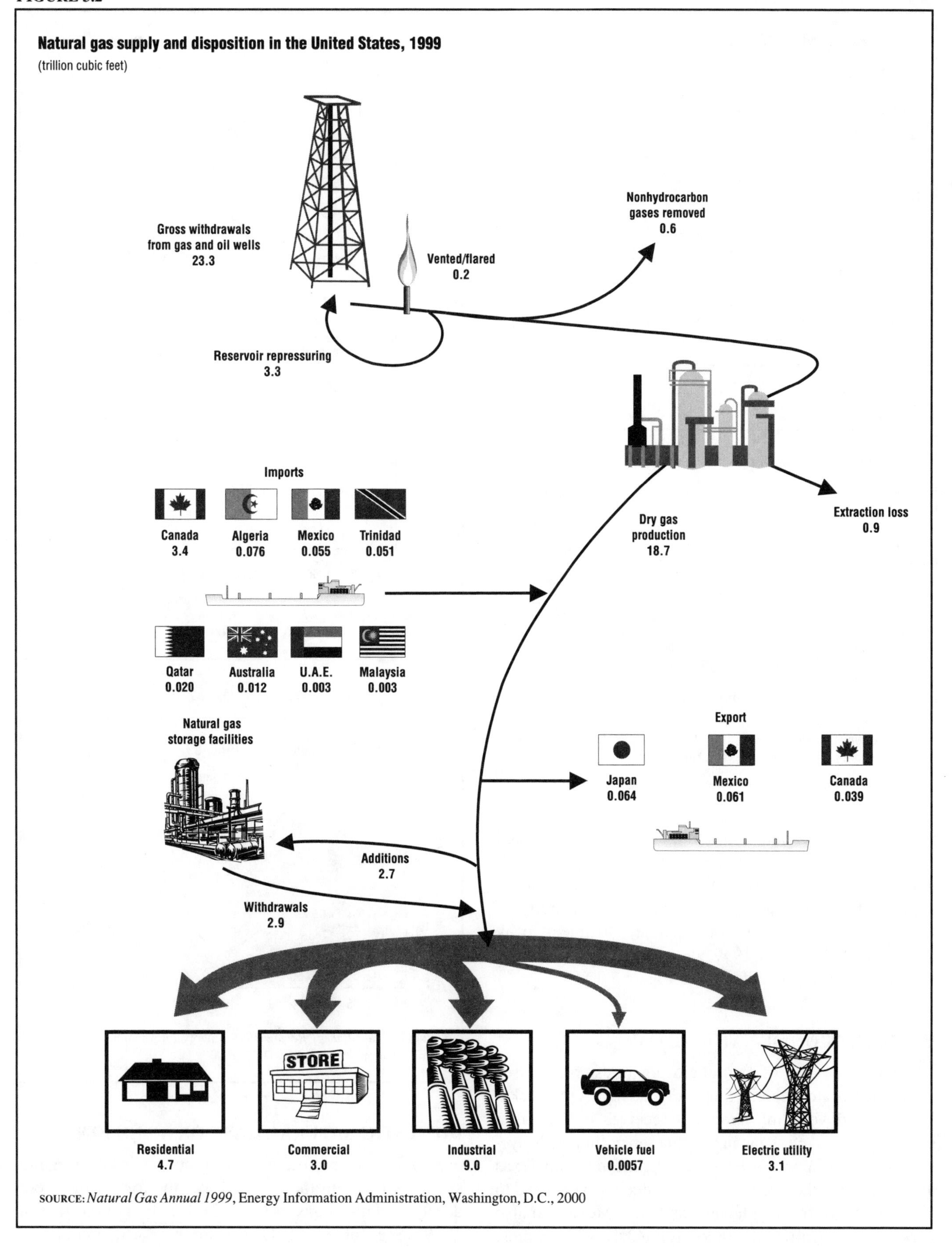

SOURCE: *Natural Gas Annual 1999*, Energy Information Administration, Washington, D.C., 2000

FIGURE 3.3

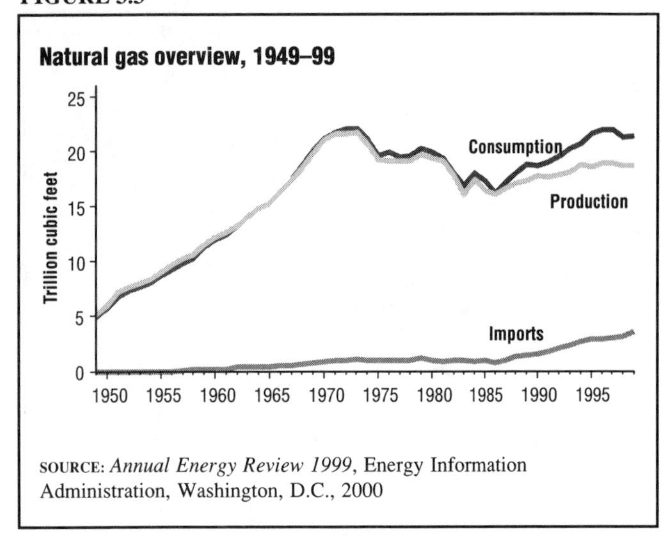

SOURCE: *Annual Energy Review 1999*, Energy Information Administration, Washington, D.C., 2000

FIGURE 3.4

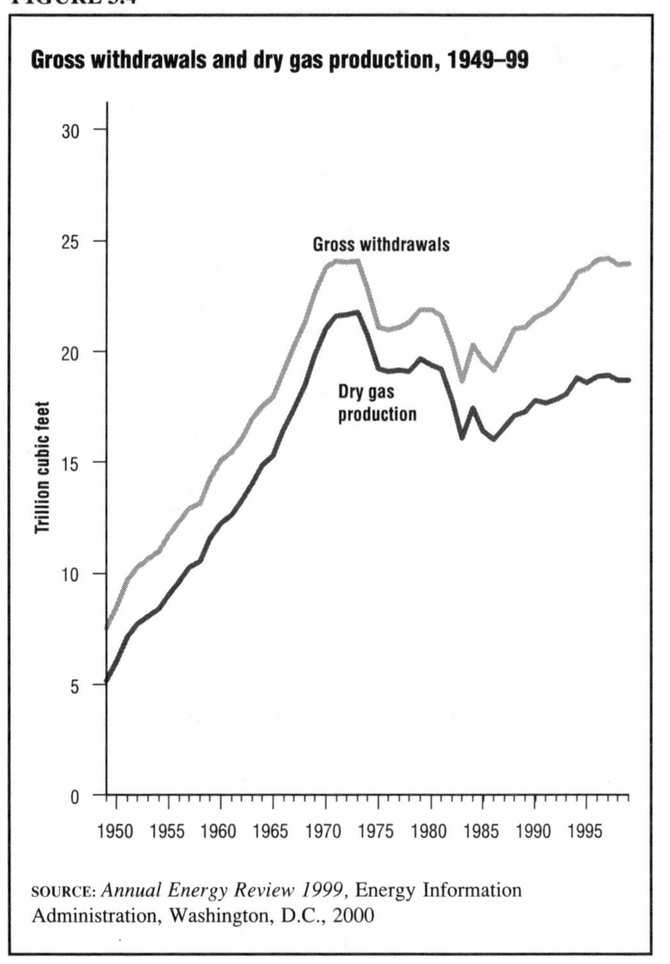

SOURCE: *Annual Energy Review 1999,* Energy Information Administration, Washington, D.C., 2000

FIGURE 3.5

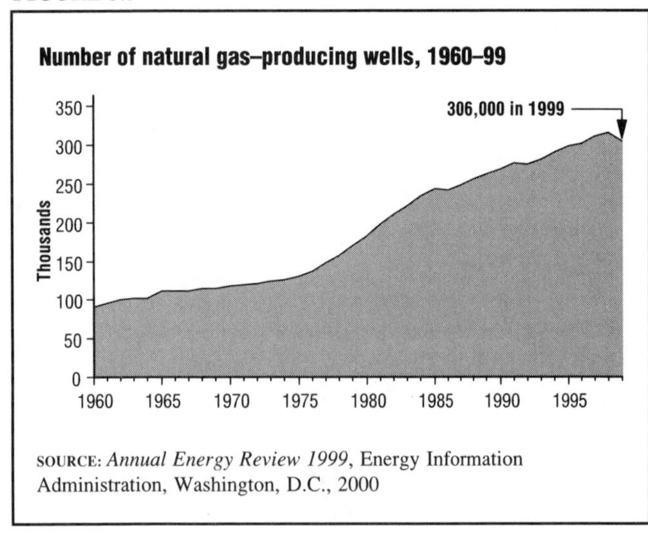

SOURCE: *Annual Energy Review 1999*, Energy Information Administration, Washington, D.C., 2000

FIGURE 3.6

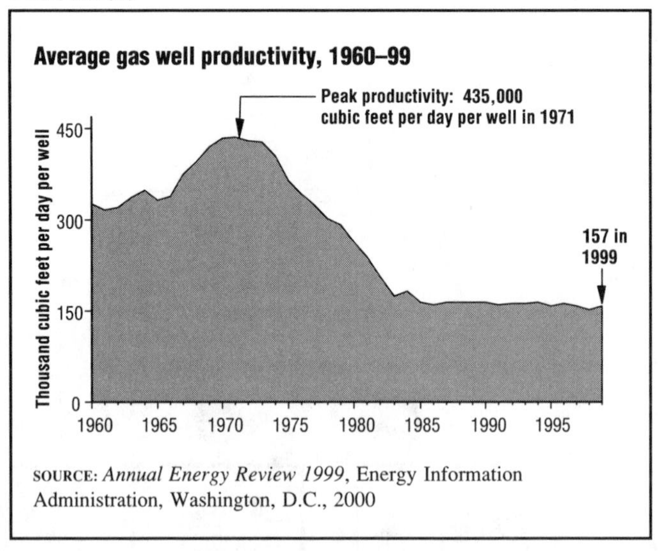

SOURCE: *Annual Energy Review 1999*, Energy Information Administration, Washington, D.C., 2000

transmission lines and marketing areas. Gas is injected into storage when market needs are lower than the available gas flow, and gas is withdrawn from storage when supplies from producing fields and/or the capacity of transmission lines are not adequate to meet peak demands. At the end of 1999, gas in underground storage totaled 6.9 trillion cubic feet. (See Figure 3.9.)

TRANSMISSION OF NATURAL GAS

A vast network of natural gas pipelines crisscrosses the United States, connecting every state except Alaska, Hawaii, and Vermont. (Vermont receives its gas from Canada.) The natural gas in this quarter-million mile system generally flows northeastward, primarily from Texas and Louisiana, the two major gas-producing states, and to a lesser extent, from Oklahoma and New Mexico. It also flows west to California. (See Figure 3.10.) The largest users of natural gas in 1999 were California, Illinois, Michigan, New York, Ohio, and Pennsylvania.

DOMESTIC NATURAL GAS CONSUMPTION

Nationally, natural gas consumption has been rising since 1986, when the demand was the lowest in recent decades. Natural gas use was 21.4 trillion cubic feet in 1999. (See Table 3.2.) Of this amount, 46 percent was used by industry, 22 percent by residences, 15 percent by electric utilities, 14 percent by commercial customers, and 3 percent was used as pipeline fuel in the gas transporting process.

Natural gas fills an important part of the country's energy needs. It is an attractive fuel, not only because its price is relatively low, but also because it is clean and efficient and can help the country meet both its environmental goals and its energy needs.

The residential sector used 4.7 trillion cubic feet in 1999. (See Table 3.2.) Energy consumption by residences depends heavily on weather-related home heating demands. The colder it gets, the more gas is used. According to the U.S. Department of Commerce, approximately one-half of all residential energy consumers in the United States use gas to heat their homes. Conservation practices and efficiency of gas appliances such as water heaters and stoves also affect residential consumption patterns.

The use of natural gas in the commercial sector was 3.1 trillion cubic feet in 1999. Like residential consumption, use in the commercial sector depends heavily on seasonal requirements, as well as the number of users and conservation measures taken by commercial establishments. Commercial customers are particularly sensitive to changes in gas prices and to changes in the economy.

The industrial sector has historically been the largest consumer of natural gas. Consumption in this sector in 1999 was 9.9 trillion cubic feet, approaching the highs reached in the early 1970s. Industrial natural gas consumption peaked at 10.2 trillion cubic feet in 1973. Substitution of natural gas for petroleum for some industrial purposes has increased consumption in the industrial sector.

FIGURE 3.7

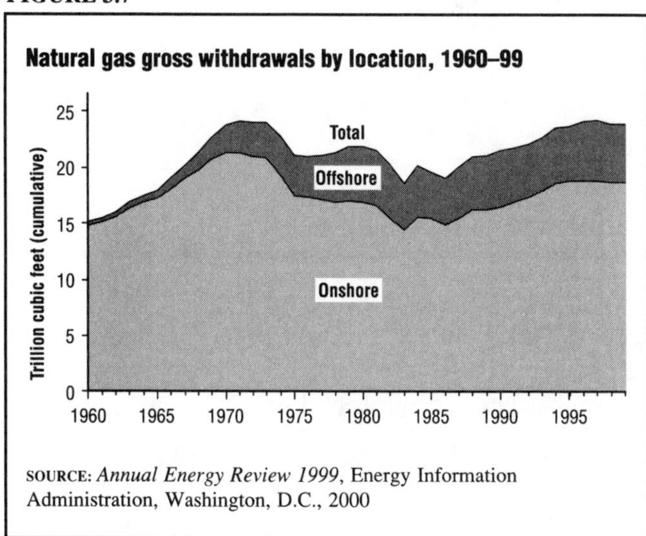

SOURCE: *Annual Energy Review 1999*, Energy Information Administration, Washington, D.C., 2000

FIGURE 3.8

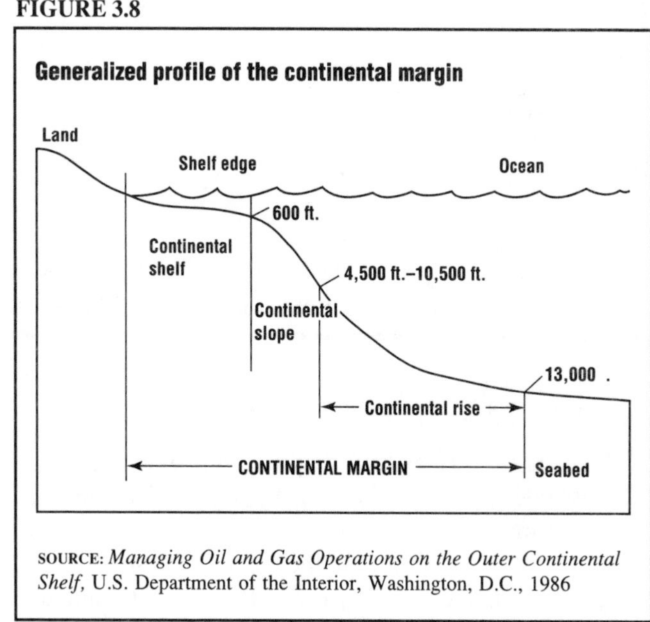

SOURCE: *Managing Oil and Gas Operations on the Outer Continental Shelf,* U.S. Department of the Interior, Washington, D.C., 1986

FIGURE 3.9

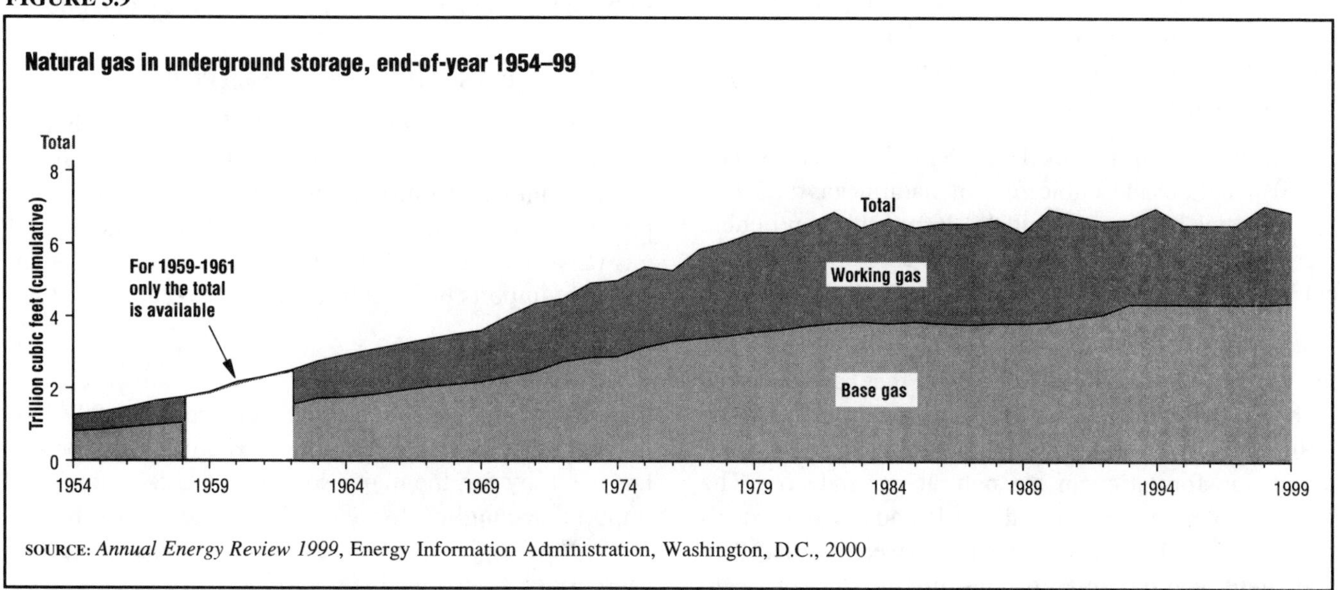

SOURCE: *Annual Energy Review 1999*, Energy Information Administration, Washington, D.C., 2000

TABLE 3.1

Crude oil and natural gas field counts, cumulative production, proved reserves, and ultimate recovery, end-of-year 1977–98

Year	Cumulative number of fields with crude oil and/or natural gas	Cumulative number of fields with crude oil	Crude oil and lease condensate (billion barrels) Cumulative production	Proved reserves	Proved ultimate recovery	Cumulative number of fields with natural gas	Natural gas[1] (trillion cubic feet) Cumulative production	Proved reserves	Proved ultimate recovery
1977	31,360	27,835	121.4	33.6	155.0	23,883	558.3	209.5	767.8
1978	32,430	28,683	124.6	33.1	157.6	24,786	578.4	210.1	788.5
1979	33,644	29,671	127.7	31.2	158.9	25,823	599.1	208.3	807.4
1980	34,999	30,766	130.8	31.3	162.2	26,919	619.4	206.3	825.6
1981	36,621	32,111	133.9	31.0	165.0	28,213	639.4	209.4	848.9
1982	38,123	33,375	137.1	29.5	166.6	29,375	658.1	209.3	867.4
1983	39,489	34,495	140.3	29.3	169.6	30,419	675.1	209.0	884.1
1984	41,038	35,784	143.5	30.0	173.5	31,595	693.5	206.0	899.5
1985	42,317	36,849	146.8	29.9	176.7	32,595	710.9	202.2	913.1
1986	43,076	37,464	150.0	28.3	178.3	33,151	727.8	201.1	928.9
1987	43,742	37,982	153.0	28.7	181.7	33,657	745.4	196.4	941.8
1988	44,414	38,506	156.0	28.2	184.2	34,196	763.4	177.0	940.4
1989	44,883	38,858	158.8	27.9	186.7	34,579	781.7	175.4	957.1
1990	45,385	39,244	161.5	27.6	189.0	34,975	800.4	177.6	978.0
1991	45,776	39,558	164.2	25.9	190.1	35,254	819.1	175.3	994.4
1992	46,149	39,843	166.8	25.0	191.8	35,539	838.0	173.3	1,011.3
1993	46,513	40,124	169.3	24.1	193.4	35,798	857.2	170.5	1,027.7
1994	46,922	40,417	171.7	23.6	195.3	36,142	877.1	171.9	1,049.1
1995	47,296	40,694	174.1	23.5	197.7	36,433	896.9	173.5	1,070.4
1996	47,557	40,875	176.5	23.3	199.8	36,612	917.0	175.1	1,092.1
1997	47,854	40,977	178.9	23.9	202.8	36,830	937.1	175.7	1,112.8
1998	[2]47,664	[2]35,143	181.2	22.7	203.9	[2]32,458	957.0	141.8	1,098.8

[1] Wet, after lease separation.
[2] There is a discontinuity in this time series between 1997 and 1998 due to the absence of updates for a subset of the data used in the past.

SOURCE: *Annual Energy Review 1999,* Energy Information Administration, Washington, D.C., 2000

NATURAL GAS PRICES

Natural gas prices can vary due to differing federal and state rate structures as well as by region. For example, prices are lower in major producing areas where transmission costs are lower. Through the early 1970s natural gas prices were relatively stable. Thereafter, deregulation and industry restructuring brought about a period of price fluctuations, with wellhead prices reaching a high in 1983, declining until 1991, and then varying through the rest of the 1990s. (See Figure 3.11.) The average price of all categories of natural gas at the wellhead was $1.98 per 1,000 cubic feet in 1999.

At the retail price level, residential customers paid $6.60 per thousand cubic feet of natural gas in 1999. Commercial consumers paid $5.26 per thousand cubic feet, while industrial consumers paid $3.04 per thousand cubic feet. (See Table 3.3.)

Much of the variation in natural gas prices through the years can be attributed to changes that have occurred in the natural gas industry. The passage of the Natural Gas Policy Act of 1978 (NGPA, PL 95-621) triggered a dramatic transformation in the natural gas industry. The NGPA allowed gas prices at the wellhead to rise gradually. On January 1, 1985, new gas prices were decontrolled, and additional volumes of onshore production were decontrolled on July 1, 1987. In 1988 President Ronald Reagan signed legislation removing all remaining natural gas wellhead price controls by 1993.

The NGPA allowed prices to go up, but it also opened the market to the forces of supply and demand. Now that prices are decontrolled and the industry is no longer constrained by federal regulations, the natural gas industry has become more sensitive to market signals and able to respond more quickly to changes in economic conditions.

NATURAL GAS IMPORTS AND EXPORTS

U.S. natural gas trade was limited to the neighboring countries of Mexico and Canada until shipping natural gas in liquefied form became a feasible alternative to pipelines. In 1969 the first shipments of liquefied natural gas (LNG) were sent from the United States to Japan, and U.S. imports of LNG from Algeria began the following year.

In 1999 U.S. net imports of natural gas by all routes totaled a record 3.6 trillion cubic feet, approximately 15.8 percent of domestic consumption. Historically, Canada has been, by far, the major supplier of U.S. natural gas imports, accounting for about 3.3 trillion cubic feet in 1999. Natural gas imports have been increasing significantly since 1986. (See Figure 3.12.)

FIGURE 3.10

Principal interstate natural gas flow summary, 1999

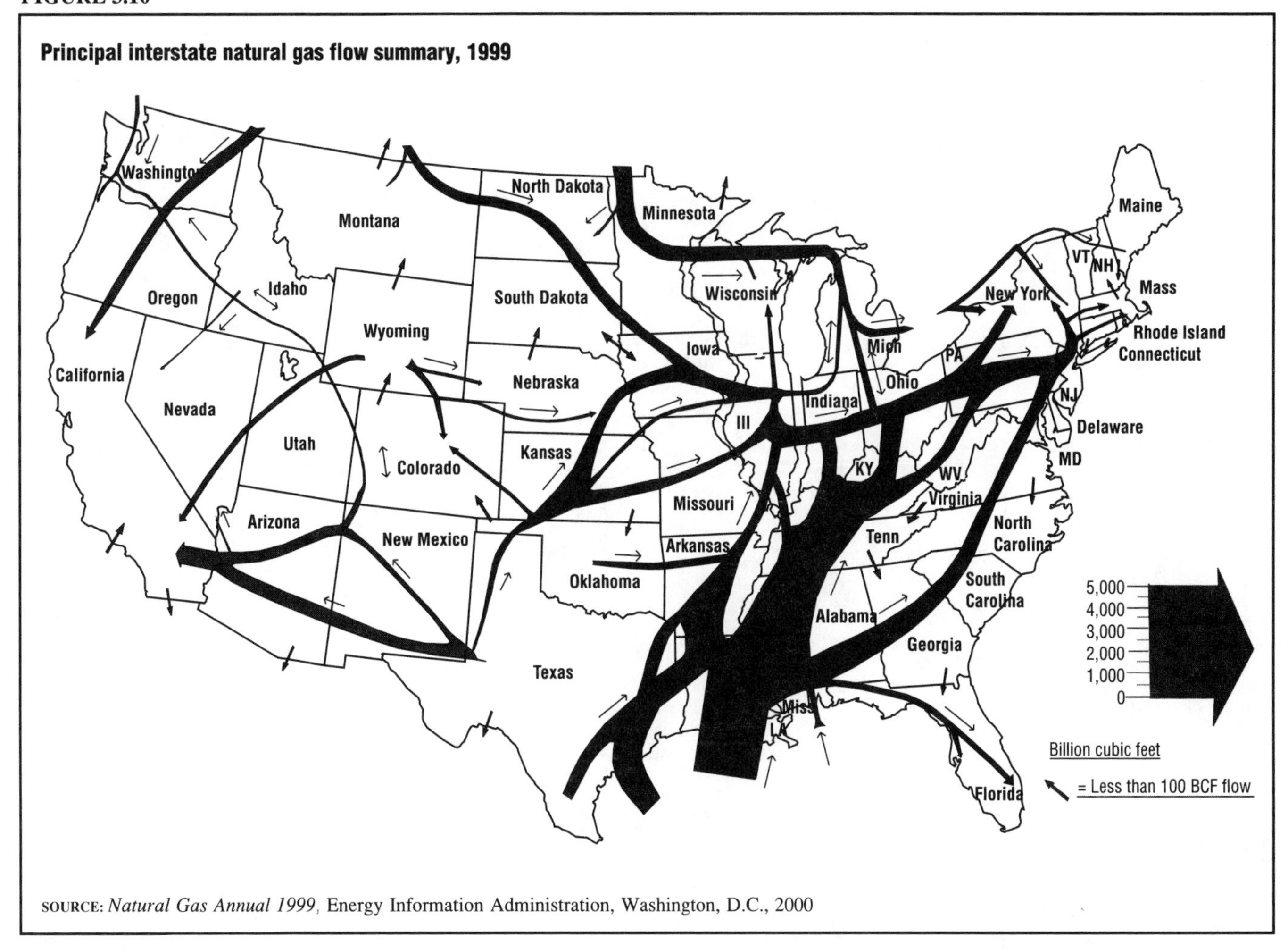

SOURCE: *Natural Gas Annual 1999*, Energy Information Administration, Washington, D.C., 2000

According to the Energy Information Administration, the United States exported 159 billion cubic feet of natural gas in 1999. Of these exports, Japan and Mexico bought the largest amount (64 billion cubic feet), while Canada purchased 32 billion cubic feet.

FIGURE 3.11

Natural gas wellhead, 1949–99

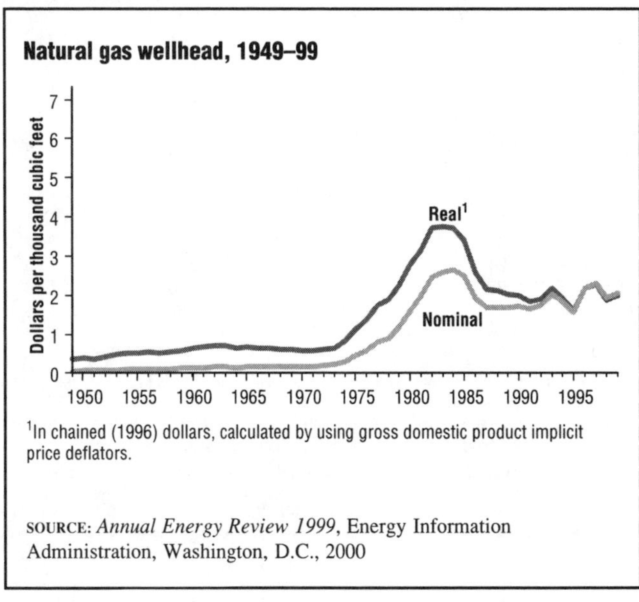

[1]In chained (1996) dollars, calculated by using gross domestic product implicit price deflators.

SOURCE: *Annual Energy Review 1999*, Energy Information Administration, Washington, D.C., 2000

INTERNATIONAL NATURAL GAS USAGE

World Production

The world's gross production of dry natural gas totaled an all-time high of 82.8 trillion cubic feet in 1998, approximately 22 percent of world energy production. Russia and America were by far the largest producers, with Russia accounting for 20.9 trillion cubic feet and the United States producing 18.7 trillion cubic feet. (See Table 3.4.)

World Consumption

The world's consumption of natural gas has continued to increase since 1980, from 53 trillion cubic feet to 82 trillion cubic feet in 1998. The United States consumed the largest amount of natural gas, followed by Russia. Combined, they accounted for 43 percent of world consumption. (See Figure 3.13.)

FUTURE TRENDS IN THE GAS INDUSTRY

The Energy Information Administration (EIA) of the U.S. Department of Energy, in its *Annual Energy Outlook 2000* (1999), projects energy supply, demand, and prices

TABLE 3.2

Natural gas consumption by sector, 1949–99
(trillion cubic feet)

	Residential	Commercial	Industrial [1]			Transportation			Electric utilties	
Year	Delivered to residences	Delivered to commercial facilities [2]	Delivered to industrial facilities	Lease and plant fuel	Total	Pipeline fuel [3]	Delivered for vehicle fuel use	Total	Delivered to electric utilities	Total
1949	0.99	0.35	2.25	0.84	3.08	NA	NA	NA	0.55	4.97
1950	1.20	0.39	2.50	0.93	3.43	0.13	NA	0.13	0.63	5.77
1951	1.47	0.46	2.77	1.15	3.91	0.19	NA	0.19	0.76	6.81
1952	1.62	0.52	2.87	1.16	4.04	0.21	NA	0.21	0.91	7.29
1953	1.69	0.53	3.03	1.13	4.16	0.23	NA	0.23	1.03	7.64
1954	1.89	0.58	3.07	1.10	4.17	0.23	NA	0.23	1.17	8.05
1955	2.12	0.63	3.41	1.13	4.54	0.25	NA	0.25	1.15	8.69
1956	2.33	0.72	3.71	1.00	4.71	0.30	NA	0.30	1.24	9.29
1957	2.50	0.78	3.89	1.05	4.93	0.30	NA	0.30	1.34	9.85
1958	2.71	0.87	3.89	1.15	5.03	0.31	NA	0.31	1.37	10.30
1959	2.91	0.98	4.22	1.24	5.46	0.35	NA	0.35	1.63	11.32
1960	3.10	1.02	4.53	1.24	5.77	0.35	NA	0.35	1.72	11.97
1961	3.25	1.08	4.67	1.29	5.96	0.38	NA	0.38	1.83	12.49
1962	3.48	1.21	4.86	1.37	6.23	0.38	NA	0.38	1.97	13.27
1963	3.59	1.27	5.13	1.41	6.55	0.42	NA	0.42	2.14	13.97
1964	3.79	1.37	5.52	1.37	6.89	0.44	NA	0.44	2.32	14.81
1965	3.90	1.44	5.96	1.16	7.11	0.50	NA	0.50	2.32	15.28
1966	4.14	1.62	6.51	1.03	7.55	0.54	NA	0.54	2.61	16.45
1967	4.31	1.96	6.65	1.14	7.79	0.58	NA	0.58	2.75	17.39
1968	4.45	2.08	7.13	1.24	8.37	0.59	NA	0.59	3.15	18.63
1969	4.73	2.25	7.61	1.35	8.96	0.63	NA	0.63	3.49	20.06
1970	4.84	2.40	7.85	1.40	9.25	0.72	NA	0.72	3.93	21.14
1971	4.97	2.51	8.18	1.41	9.59	0.74	NA	0.74	3.98	21.79
1972	5.13	2.61	8.17	1.46	9.62	0.77	NA	0.77	3.98	22.10
1973	4.88	2.60	8.69	1.50	10.18	0.73	NA	0.73	3.66	22.05
1974	4.79	2.56	8.29	1.48	9.77	0.67	NA	0.67	3.44	21.22
1975	4.92	2.51	6.97	1.40	8.36	0.58	NA	0.58	3.16	19.54
1976	5.05	2.67	6.96	1.63	8.60	0.55	NA	0.55	3.08	19.95
1977	4.82	2.50	6.82	1.66	8.47	0.53	NA	0.53	3.19	19.52
1978	4.90	2.60	6.76	1.65	8.40	0.53	NA	0.53	3.19	19.63
1979	4.97	2.79	6.90	1.50	8.40	0.60	NA	0.60	3.49	20.24
1980	4.75	2.61	7.17	1.03	8.20	0.63	NA	0.63	3.68	19.88
1981	4.55	2.52	7.13	0.93	8.06	0.64	NA	0.64	3.64	19.40
1982	4.63	2.61	5.83	1.11	6.94	0.60	NA	0.60	3.23	18.00
1983	4.38	2.43	5.64	0.98	6.62	0.49	NA	0.49	2.91	16.83
1984	4.56	2.52	6.15	1.08	7.23	0.53	NA	0.53	3.11	17.95
1985	4.43	2.43	5.90	0.97	6.87	0.50	NA	0.50	3.04	17.28
1986	4.31	2.32	5.58	0.92	6.50	0.49	NA	0.49	2.60	16.22
1987	4.31	2.43	5.95	1.15	7.10	0.52	NA	0.52	2.84	17.21
1988	4.63	2.67	6.38	1.10	7.48	0.61	NA	0.61	2.64	18.03
1989	4.78	2.72	6.82	1.07	7.89	0.63	NA	0.63	2.79	18.80
1990	4.39	2.62	7.02	1.24	8.25	0.66	(s)	0.66	2.79	18.72
1991	4.56	2.73	7.23	1.13	8.36	0.60	(s)	0.60	2.79	19.04
1992	4.69	2.80	7.53	1.17	8.70	0.59	(s)	0.59	2.77	19.54
1993	4.96	2.86	7.98	1.17	9.15	0.62	(s)	0.63	2.68	20.28
1994	4.85	2.90	8.17	1.12	9.29	0.69	(s)	0.69	2.99	20.71
1995	4.85	3.03	8.58	1.22	9.80	0.70	(s)	0.70	3.20	21.58
1996	5.24	3.16	8.87	1.25	10.12	0.71	(s)	0.71	2.73	21.97
1997	4.98	R3.21	R8.83	1.20	R10.04	0.75	(s)	0.76	2.97	R21.96
1998	R4.52	R3.00	R8.69	R1.16	R9.84	R0.64	0.01	R0.64	3.26	R21.26
1999P	4.67	3.09	8.67	1.23	9.90	0.64	NA	0.64	3.13	21.43

[1] Most deliveries to nonutility power producers are included in the industrial sector. In instances where the nonutility is primarily a commercial establishment, deliveries are included in the commercial sector.
[2] Includes deliveries to municipalities and public authorities for institutional heating and other purposes.
[3] Natural gas consumed in the operation of pipelines, primarily in compressors.
R=Revised. P=Preliminary. NA=Not available. (s)=Less than 0.005 trillion cubic feet.

SOURCE: *Annual Energy Review 1999,* Energy Information Administration, Washington, D.C., 2000

through 2020. The EIA predicts that natural gas demand will increase steadily, as will consumption, pipeline expansion, and imports. (See Figure 3.14.)

Annual consumption of natural gas is estimated to grow at a rate of 1.8 percent per year to between 29.5 and 32.7 trillion cubic feet in 2020, depending on economic conditions. Demand for natural gas by electricity manufacturers is expected to more than double, from 3.7 to 9.3 trillion cubic feet. Natural gas will be called upon to replace the nation's aging nuclear electricity plants and to take advantage of its efficiency and low emissions. To

TABLE 3.3

Natural gas prices by sector, 1967–99

(price: dollars per thousand cubic feet; share of total volume delivered: percentage)

Year	Residential		Commercial [1]			Industrial [2]			Vehicle fuel [3]			Electric utilities	
	Prices [4]		Prices		Share of total volume delivered	Prices		Share of total volume delivered	Prices		Share of total volume delivered	Prices [5]	
	Nominal	Real [6]	Nominal	Real [6]		Nominal	Real [6]		Nominal	Real [6]		Nominal	Real [6]
1967	1.04	[R]4.13	0.74	[R]2.94	NA	0.34	[R]1.35	NA	NA	NA	NA	0.28	[R]1.11
1968	1.04	[R]3.95	0.73	[R]2.78	NA	0.34	[R]1.29	NA	NA	NA	NA	0.22	[R]0.84
1969	1.05	[R]3.81	0.74	[R]2.68	NA	0.35	[R]1.27	NA	NA	NA	NA	0.27	[R]0.98
1970	1.09	[R]3.75	0.77	[R]2.65	NA	0.37	[R]1.27	NA	NA	NA	NA	0.29	[R]1.00
1971	1.15	[R]3.77	0.82	[R]2.69	NA	0.41	[R]1.34	NA	NA	NA	NA	0.32	[R]1.05
1972	1.21	[R]3.80	0.88	[R]2.77	NA	0.45	[R]1.41	NA	NA	NA	NA	0.34	[R]1.07
1973	1.29	[R]3.84	0.94	[R]2.80	NA	0.50	[R]1.49	NA	NA	NA	NA	0.38	[R]1.13
1974	1.43	[R]3.90	1.07	[R]2.92	NA	0.67	[R]1.83	NA	NA	NA	NA	0.51	[R]1.39
1975	1.71	[R]4.27	1.35	[R]3.37	NA	0.96	[R]2.40	NA	NA	NA	NA	0.77	[R]1.92
1976	1.98	[R]4.68	1.64	[R]3.88	NA	1.24	[R]2.93	NA	NA	NA	NA	1.06	[R]2.51
1977	2.35	[R]5.22	2.04	[R]4.53	NA	1.50	[R]3.33	NA	NA	NA	NA	1.32	[R]2.93
1978	2.56	[R]5.31	2.23	[R]4.62	NA	1.70	[R]3.52	NA	NA	NA	NA	1.48	[R]3.07
1979	2.98	[R]5.70	2.73	[R]5.22	NA	1.99	[R]3.81	NA	NA	NA	NA	1.81	[R]3.46
1980	3.68	[R]6.45	3.39	[R]5.94	NA	2.56	[R]4.49	NA	NA	NA	NA	2.27	[R]3.98
1981	4.29	[R]6.88	4.00	[R]6.41	NA	3.14	[R]5.03	NA	NA	NA	NA	2.89	[R]4.63
1982	5.17	[R]7.80	4.82	[R]7.28	NA	3.87	[R]5.84	85.1	NA	NA	NA	3.48	[R]5.25
1983	6.06	[R]8.80	5.59	[R]8.12	NA	4.18	[R]6.07	80.7	NA	NA	NA	3.58	[R]5.20
1984	6.12	[R]8.57	5.55	[R]7.77	NA	4.22	[R]5.91	74.7	NA	NA	NA	3.70	[R]5.18
1985	6.12	[R]8.31	5.50	[R]7.46	NA	3.95	[R]5.36	68.8	NA	NA	NA	3.55	[R]4.82
1986	5.83	[R]7.74	5.08	[R]6.75	NA	3.23	[R]4.29	59.8	NA	NA	NA	2.43	[R]3.23
1987	5.54	[R]7.14	4.77	[R]6.15	93.1	2.94	[R]3.79	47.4	NA	NA	NA	2.32	[R]2.99
1988	5.47	[R]6.82	4.63	[R]5.77	90.7	2.95	[R]3.68	42.6	NA	NA	NA	2.33	[R]2.90
1989	5.64	[R]6.77	4.74	[R]5.69	89.1	2.96	[R]3.55	36.9	NA	NA	NA	2.43	[R]2.92
1990	5.80	[R]6.70	4.83	[R]5.58	86.6	2.93	[R]3.39	35.2	3.39	[R]3.92	NA	2.38	[R]2.75
1991	5.82	[R]6.49	4.81	[R]5.36	85.1	2.69	[R]3.00	32.7	3.96	[R]4.42	NA	2.18	[R]2.43
1992	5.89	[R]6.41	4.88	[R]5.31	83.2	2.84	[R]3.09	30.3	4.05	[R]4.41	NA	2.36	[R]2.57
1993	6.16	[R]6.55	5.22	[R]5.55	83.9	3.07	[R]3.26	29.7	4.27	[R]4.54	87.8	2.61	[R]2.78
1994	6.41	[R]6.68	5.44	[R]5.67	79.3	3.05	[R]3.18	25.5	4.11	[R]4.28	86.9	2.28	[R]2.37
1995	6.06	[R]6.18	5.05	[R]5.15	76.7	2.71	[R]2.76	24.5	3.98	[R]4.06	86.6	2.02	[R]2.06
1996	6.34	[R]6.34	5.40	[R]5.40	77.6	3.42	[R]3.42	19.4	4.34	[R]4.34	94.0	2.69	[R]2.69
1997	6.94	[R]6.81	[R]5.80	[R]5.69	70.8	3.59	[R]3.52	[R]18.1	4.44	[R]4.36	89.7	[R]2.78	[R]2.73
1998	6.82	[R]6.61	[R]5.48	[R]5.31	[R]67.0	[R]3.14	[R]3.05	[R]16.1	[R]4.59	[R]4.45	[R]85.4	[R]2.40	[R]2.33
1999[P]	6.60	6.31	5.26	5.03	65.1	3.04	2.91	16.9	NA	NA	NA	2.56	2.45

[1] Includes deliveries to municipalities and public authorities for institutional heating and other purposes.
[2] Most volumes and associated revenues for deliveries to nonutility power producers are included in the industrial sector. In instances where the nonutility is primarily a commercial establishment, volumes and associated revenues are included in the calculation of commercial prices.
[3] Much of the natural gas delivered for vehicle fuel represents deliveries to fueling stations that are used primarily or exclusively by respondents' fleet vehicles. Thus, the prices are often those associated with the operation of fleet vehicles.
[4] Based on 100 percent of volume delivered.
[5] Based on all steam-electric utility plants with a combined capacity of 50 megawatts or greater.
[6] In chained (1996) dollars, calculated by using gross domestic product implicit price deflators.

R=Revised. P=Preliminary. NA=Not available.

SOURCE: *Annual Energy Review 1999*, Energy Information Administration, Washington, D.C., 2000

FIGURE 3.12

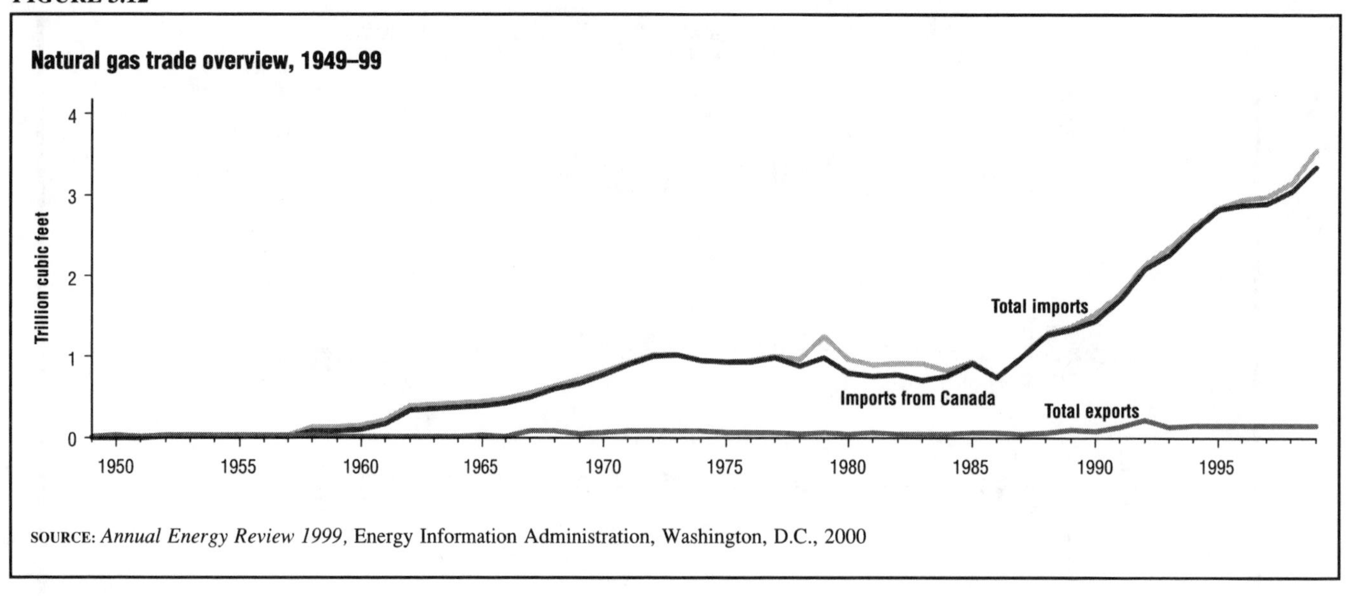

SOURCE: *Annual Energy Review 1999,* Energy Information Administration, Washington, D.C., 2000

FIGURE 3.13

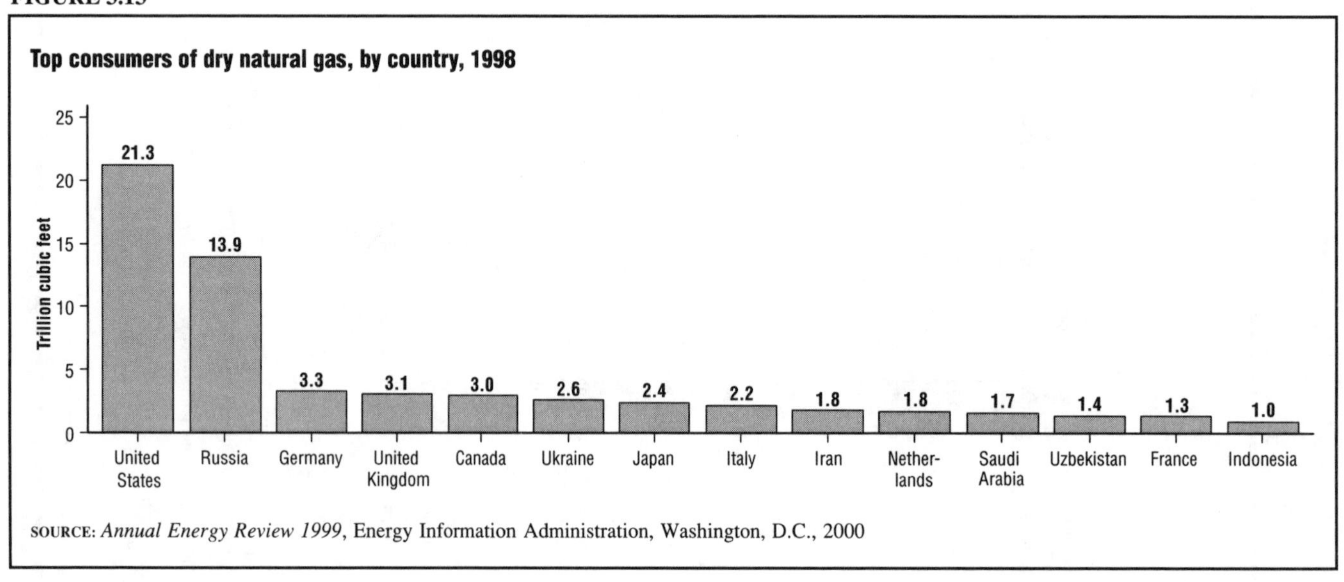

SOURCE: *Annual Energy Review 1999,* Energy Information Administration, Washington, D.C., 2000

meet growing demand, natural gas pipeline capacity will have to be expanded by an estimated 1.5 percent per year, particularly along the corridors that move Canadian and Gulf of Mexico supplies to eastern and midwestern states.

Domestic natural gas production is projected to increase from 1998 levels of 18.9 trillion cubic feet to 26.4 trillion cubic feet in 2020. Domestic production will be boosted by improvements in exploration and production technology, as well as by increased Gulf of Mexico offshore finds. Net imports of natural gas will increase to meet demand, from 3.1 to 5.1 trillion cubic feet, coming mostly from Canada. The percentage of natural gas supplied from imports is projected to increase from the 1998 percentage of 14.6 to 16.3 percent in 2020. Natural gas exports to Mexico will continue, with exports increasing due to mandated conversions of power plants from heavy fuel oil to natural gas, in compliance with new environmental regulations.

The EIA predicts that although natural gas prices to the industrial, electricity, and transportation sectors will generally increase, prices to the residential and commercial sectors will remain somewhat steady through 2020. The average wellhead gas price is projected to increase 1.7 percent per year, from $1.96 per thousand cubic feet in 1998 to $2.81 in 2020.

Imports of liquefied natural gas (LNG) are not expected to serve as a major source of gas imports, growing to roughly 0.3 trillio n cubic feet in 2020. Liquefied gas is the only economically feasible non-pipeline method of transporting natural gas in large amounts.

TABLE 3.4

World dry natural gas production, 1989–98

(trillion cubic feet)

Region and country	1989	1990	1991	1992	1993	1994	1995	1996	1997	1998 [P]
North, Central, and South America	**[R]24.03**	**[R]24.58**	**[R]24.81**	**[R]25.38**	**[R]26.26**	**[R]27.50**	**[R]27.74**	**[R]28.46**	**[R]28.83**	**29.10**
Argentina	0.72	0.63	0.70	0.71	0.76	0.79	0.88	0.94	0.97	1.05
Canada	[R]3.73	3.85	4.06	4.52	4.91	[R]5.27	[R]5.60	5.78	5.85	6.04
Mexico	[R]0.87	[R]0.90	[R]0.90	[R]0.88	[R]0.95	[R]0.97	[R]0.96	[R]1.06	[R]1.16	1.27
United States	17.31	17.81	17.70	17.84	18.10	18.82	18.60	[R]18.85	18.90	18.71
Venezuela	0.77	0.76	0.79	0.76	0.82	0.88	0.89	0.96	[R]0.99	0.99
Other	0.64	0.62	0.65	0.66	0.73	0.78	0.81	0.86	0.96	1.06
Western Europe	**[R]7.31**	**7.24**	**7.83**	**[R]7.92**	**[R]8.33**	**[R]8.44**	**[R]8.80**	**[R]10.09**	**[R]9.72**	**9.66**
Germany [1]	0.86	0.72	0.67	0.68	0.68	[R]0.70	[R]0.74	[R]0.80	[R]0.79	0.77
Italy	0.60	0.61	0.61	0.64	0.69	0.73	0.72	0.71	0.68	0.67
Netherlands	2.67	2.69	3.04	3.06	3.11	2.95	[R]2.98	3.37	2.99	2.84
Norway	1.09	0.98	0.97	1.04	0.97	1.04	1.08	1.45	[R]1.62	1.63
United Kingdom	1.58	1.75	2.01	[R]1.96	2.31	2.47	2.67	3.18	[R]3.03	3.17
Other	0.51	[R]0.49	0.53	0.54	[R]0.57	[R]0.55	0.61	0.59	[R]0.60	0.58
Eastern Europe and former U.S.S.R.	**[R]29.71**	**30.13**	**29.85**	**28.58**	**[R]27.98**	**26.47**	**25.93**	**26.28**	**[R]24.85**	**25.16**
Romania	1.13	1.00	0.88	0.78	0.75	0.69	0.68	0.63	[R]0.61	0.52
Former U.S.S.R.	28.11	28.78	28.62	—	—	—	—	—	—	—
Russia	—	—	—	22.62	21.81	21.45	21.01	21.23	20.17	20.87
Turkmenistan	—	—	—	2.02	2.29	1.26	1.14	1.31	0.90	0.47
Ukraine	—	—	—	0.74	0.68	0.64	0.62	0.64	[R]0.64	0.64
Uzbekistan	—	—	—	1.51	1.59	1.67	1.70	1.70	1.74	1.94
Other	[R]0.47	0.35	0.35	0.91	[R]0.86	0.76	0.78	0.76	0.79	0.74
Middle East and Africa	**6.08**	**6.17**	**6.52**	**6.91**	**7.24**	**7.41**	**7.99**	**8.76**	**[R]9.74**	**10.31**
Algeria	1.71	1.79	1.93	1.97	1.90	1.81	2.05	2.19	[R]2.43	2.60
Egypt	0.27	0.29	0.32	0.35	0.40	0.42	0.44	0.47	0.48	0.49
Iran	0.78	0.84	0.92	0.88	0.96	1.12	1.25	1.42	[R]1.66	1.77
Qatar	0.22	0.28	0.33	0.40	0.48	0.48	0.48	0.48	[R]0.61	0.69
Saudi Arabia	1.05	1.08	1.13	1.20	1.27	1.33	1.34	1.46	[R]1.60	1.65
United Arab Emirates	0.81	0.78	0.92	1.02	0.94	0.91	1.11	1.19	[R]1.28	1.31
Other	1.24	1.13	0.98	1.08	1.30	1.34	1.33	1.53	[R]1.67	1.80
Far East and Oceania	**4.98**	**5.44**	**5.76**	**[R]6.06**	**6.55**	**[R]7.11**	**[R]7.50**	**8.11**	**[R]8.48**	**8.58**
Australia	0.57	0.72	0.75	[R]0.80	0.86	[R]0.93	1.03	1.05	[R]1.07	1.10
China	0.51	0.51	0.53	0.53	0.56	0.59	0.60	0.67	0.75	0.78
India	0.32	0.40	0.45	0.48	0.53	0.59	0.63	0.70	[R]0.72	0.76
Indonesia	1.42	1.53	1.72	1.79	1.97	2.21	2.24	2.35	2.37	2.24
Malaysia	0.61	0.65	0.75	0.80	0.88	0.92	1.02	1.23	1.36	1.44
Pakistan	0.47	0.48	0.53	0.55	0.58	0.63	0.65	0.70	0.70	0.71
Other	1.09	1.15	1.03	1.10	1.16	1.23	1.33	1.42	1.52	1.55
World	**[R]72.13**	**[R]73.57**	**[R]74.78**	**[R]74.84**	**[R]76.36**	**[R]76.93**	**[R]77.96**	**[R]81.70**	**[R]81.61**	**82.81**

[1] Through 1990, this is East and West Germany. Beginning in 1991, this is unified Germany.
R=Revised. P=Preliminary. — = Not applicable.

SOURCE: *Annual Energy Review 1999,* Energy Information Administration, Washington, D.C., 2000

FIGURE 3.14

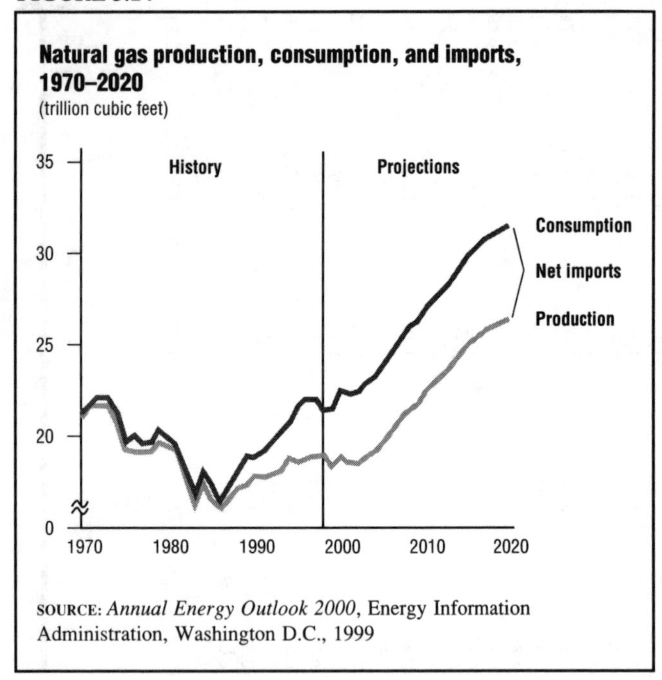

SOURCE: *Annual Energy Outlook 2000*, Energy Information Administration, Washington D.C., 1999

CHAPTER 4
COAL

A HISTORICAL PERSPECTIVE

The first large-scale use of coal occurred during the Industrial Revolution in England. At that time, the sky was filled with billowing columns of black smoke, soot covered the towns and cities, and workers breathed the thick coal dust swirling around them. Most people then were not concerned with environmental issues, since the Industrial Revolution meant jobs to the workers, and factory owners had little desire to control the pollution their factories were creating. Environmental and public health considerations were not understood as they are today.

In America early colonists used wood to heat their homes because it was so plentiful. Coal was less common. Prior to the Civil War (1861–65), some industries used coal as a source of energy, but its major use began with the building of the railroads. After the Civil War ended, America began to expand its railway system westward and increase its manufacturing capacity. Coal became such a fundamental part of American industrialization that some have called this era "The Coal Age." As in England, Americans considered the development of industry a source of national pride. Photographs and postcards of the era proudly featured railroad trains and steel mills with smokestacks belching dark smoke into gray skies.

By the 1900s coal had become the nation's major fuel source, accounting for nearly 90 percent of the nation's energy requirements. By the end of World War II, however, coal accounted for only 38 percent of the energy supply, as oil began to heat homes and offices and the growing numbers of cars used gasoline. Coal fell further out of favor as an energy source in the 1950s and 1960s as oil became more attractive. This decline continued, with coal producing as little as 18 percent of the yearly energy used during the early 1970s, because of concerns about environmental pollution and because of the emergence of nuclear power as an energy source.

By 1973 Americans recognized they could no longer rely on imported oil for their energy. The Arab oil embargo clearly demonstrated the nation's heavy reliance on foreign sources of energy, and its potentially crippling effect on the American economy. Consequently, the nation revived its interest in domestic coal as a plentiful and economical energy source.

After the 1973 embargo, coal and nuclear fuel received more attention, especially in the electric utility sector. In 1977 President Jimmy Carter called for a two-thirds annual increase in national coal production by 1985. He also asked utility companies and other large industries to convert their operations to coal, and proposed a 10-year, $10 billion program to encourage domestic coal production. In 1999 coal contributed more to America's energy production than any other source, generating 23 quadrillion Btu and 32 percent of all energy. (See Figure 4.1.)

WHAT IS COAL?

Coal is a black, combustible, mineral solid that develops over a period of millions of years from the partial decomposition of plant matter in an airless space, under increased temperature and pressure. Coal beds, sometimes called seams, are found in the earth between beds of sandstone, shale, and limestone, and range in thickness from less than an inch to more than a hundred feet. Approximately five to ten feet of ancient, composted plant material have been compressed to create each foot of coal.

Coal is used as a fuel and in the production of coke (the solid substance left after coal gas and coal tar have been extracted), coal gas, water gas, and many coal-tar compounds. When coal is burned, its fossil energy—sunlight converted and stored by plants over millions of years—is released. One ton of coal produces an average of 22 million Btu, about the same heating value as 22,000 cubic feet of natural gas, 159 gallons of distillate fuel oil, or one cord of seasoned firewood.

FIGURE 4.1

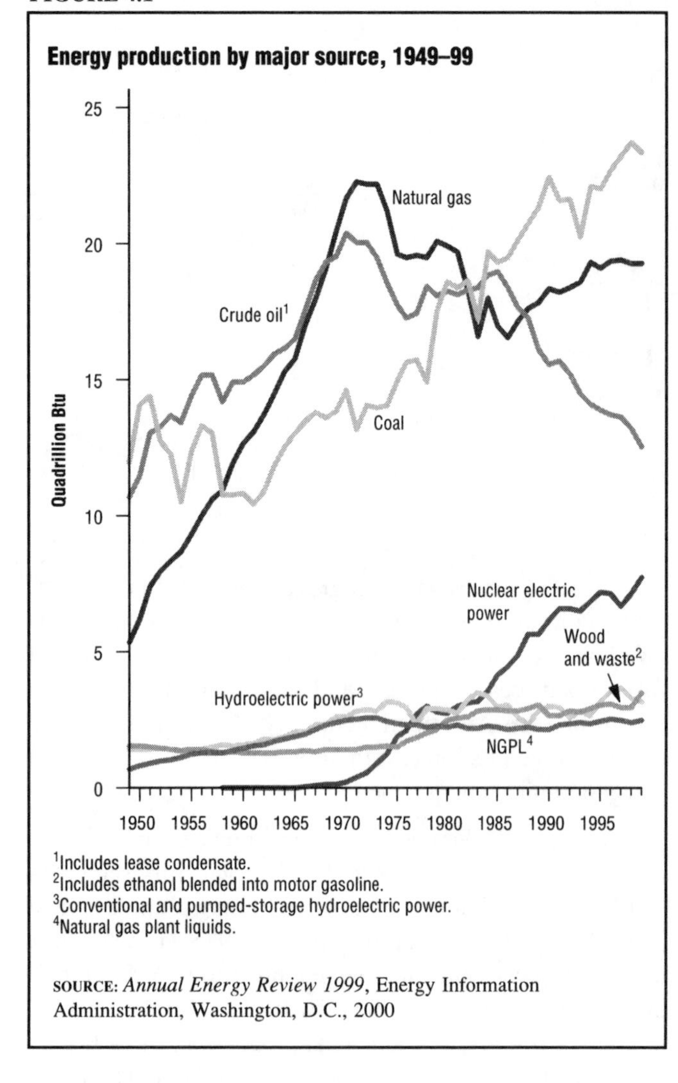

[1]Includes lease condensate.
[2]Includes ethanol blended into motor gasoline.
[3]Conventional and pumped-storage hydroelectric power.
[4]Natural gas plant liquids.

SOURCE: *Annual Energy Review 1999*, Energy Information Administration, Washington, D.C., 2000

CLASSIFICATIONS OF COAL

There are four basic types of coal. Classifications, or "coal ranks," are based on how much carbon, volatile matter, and heating value are contained in the coal.

- Anthracite, or hard coal, is the highest ranked coal. It is hard and jet black, with a moisture content of less than 15 percent. Anthracite is used mainly for generating electricity and for space heating. It contains approximately 22–28 million Btu per ton, with an ignition temperature of approximately 925–970 degrees Fahrenheit. Anthracite is mined mainly in northeastern Pennsylvania. (See Figure 4.2.)

- Bituminous, or soft coal, is the most common coal. It is dense and black, with a moisture content of less than 20 percent. Soft coal has an ignition range of 700–900 degrees Fahrenheit. Bituminous coal is used to generate electricity, for space heating, and to produce coke. Bituminous coal contains a heating value range of 19–30 million Btu per ton. It is mined chiefly in the Appalachian and Interior regions of the United States.

- Subbituminous coal, or black lignite, is dull black in color and generally contains 20–30 percent moisture. Black lignite is used for generating electricity and for space heating. It contains 16–24 million Btu per ton. Black lignite is primarily mined in the western United States.

- Lignite, the lowest rank of coal, is brownish-black in color and has a high moisture content. It tends to disintegrate when exposed to weather. Lignite is used mainly to generate electricity and contains about 9–17 million Btu per ton. Lignite has an ignition temperature of approximately 600 degrees Fahrenheit and is mined in North Dakota, Montana, Texas, California, and Louisiana.

Bituminous coal accounts for, by far, the largest share of all coal production. (See Table 4.1.) In 1999 production of all types of coal totaled 1.1 billion short tons (a short ton of coal is a unit of weight equal to 2,000 pounds). Of that, over 1 billion short tons were bituminous and subbituminous coal. Lignite and anthracite accounted for the remainder.

LOCATIONS OF COAL DEPOSITS

Coal is found in about 13 percent, or 458,600 square miles, of the U.S. total land area. Geologists have geographically divided U.S. coal fields into three zones: the Appalachian, Interior, and Western regions. (See Figure 4.2.) The Appalachian region is subdivided into three areas: Northern (Ohio, Pennsylvania, Maryland, and northern West Virginia); Central (Virginia, southern West Virginia, eastern Kentucky, and Tennessee); and Southern Appalachia (Alabama). Coal production in the Interior region occurs in Illinois, Indiana, western Kentucky, Iowa, Missouri, Kansas, Arkansas, Oklahoma, Louisiana, and Texas. The Western region includes the Northern Great Plains (Montana, Wyoming, northern Colorado, and North and South Dakota), the Rocky Mountains, the Southwest (southern Colorado, Utah, Arizona, and New Mexico), and the Northwest (Washington and Alaska).

More coal is mined east of the Mississippi than in the West, but the West's proportion of total production has increased almost every year since 1965. (See Table 4.1.) In 1965 the production of coal in the West was 27 million short tons, only 5 percent of total production. By 1999 western production had increased almost 20-fold, to 538 million short tons, or 49 percent of the total. The amount of coal mined east of the Mississippi was 561 million short tons.

The growth in coal production in the Western region has been partly due to environmental concerns that have led to an increased demand for low-sulfur coal, which is concentrated in the West. In addition, surface mining, which is cheaper and more efficient, is more prevalent in the West. Finally, improved rail service has made it easier

FIGURE 4.2

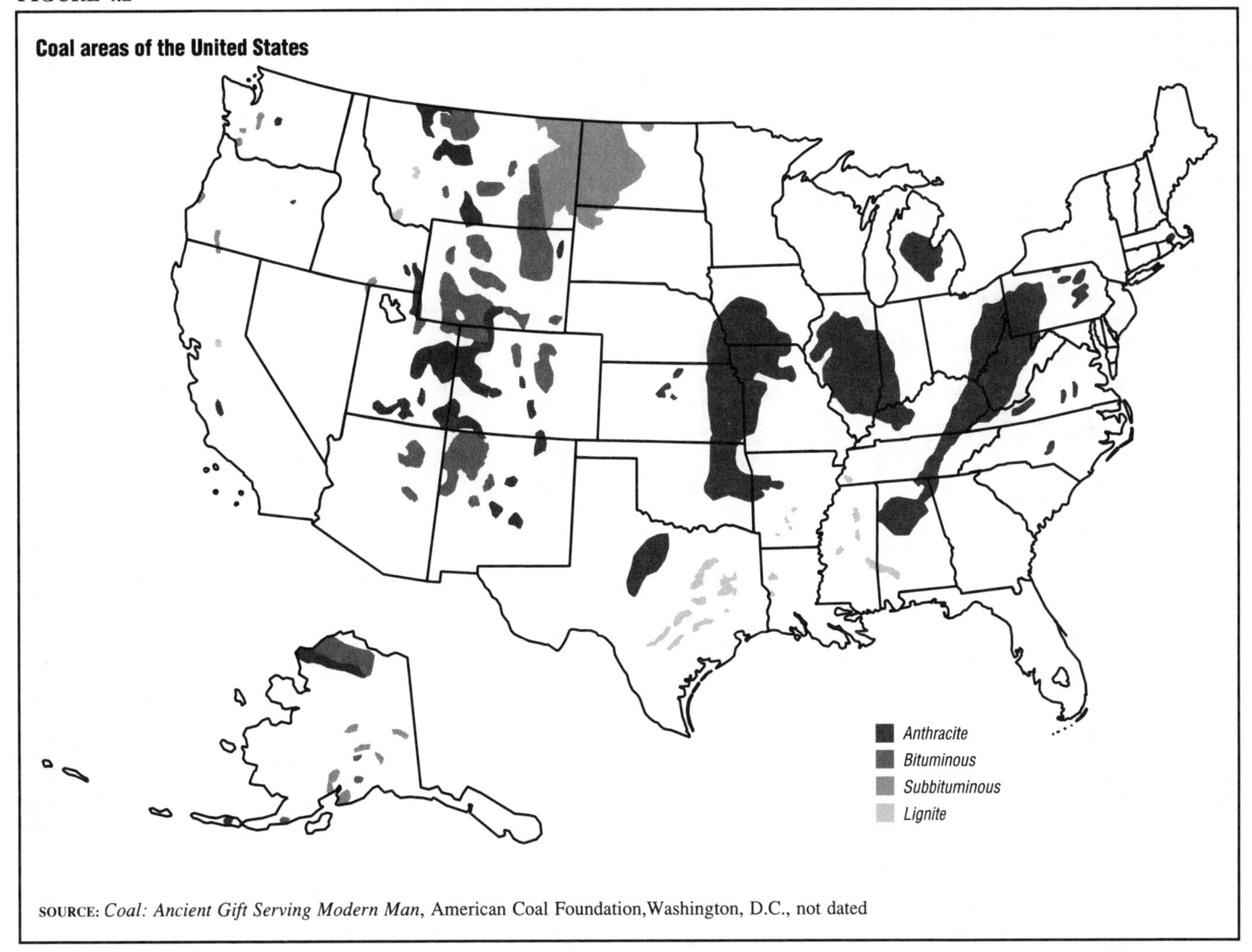

SOURCE: *Coal: Ancient Gift Serving Modern Man*, American Coal Foundation,Washington, D.C., not dated

to deliver this low-sulfur coal to utility plants located east of the Mississippi River.

COAL MINING METHODS

Historically, most coal has been taken from underground mines. Since the early 1970s, however, coal production has shifted from underground mines to surface mines. (See Figure 4.3 and Table 4.1.)

FIGURE 4.3

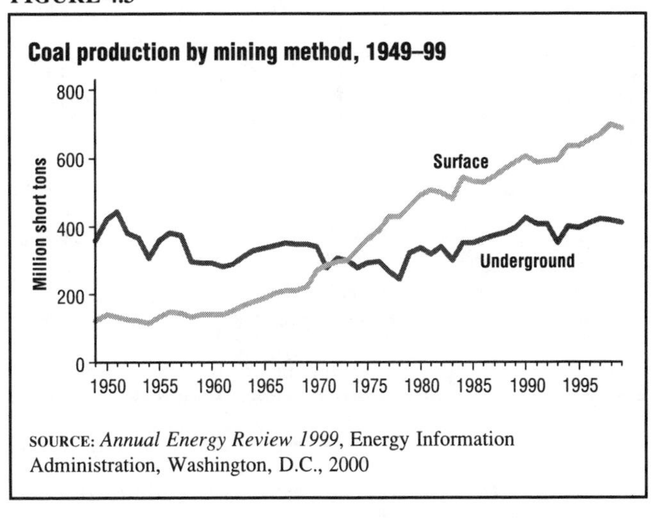

SOURCE: *Annual Energy Review 1999*, Energy Information Administration, Washington, D.C., 2000

The method used to mine coal depends on the terrain and the depth of the coal. (See Figure 4.4.) Underground mining is required when the coal lies deeper than 200 feet below ground level. The depth of most underground mines is less than 1,000 feet, but a few go down as far as 1,500–2,000 feet. In underground mines, some coal must be left untouched in order to form pillars that prevent the mine from caving in. In both underground mines and surface mines, natural features such as folded, faulted, and interlaid rock strata reduce the amount of coal that can actually be recovered.

Surface mines are usually less than 200 feet deep and can be developed in flat or hilly terrain. Area surface mining is practiced in large areas on relatively flat ground, while contour surface mining follows coal beds along hillsides. Open pit mining is used to mine thick, steeply inclined coal beds and utilizes a combination of contour and area mining methods.

TABLE 4.1

Coal production, 1949–99
(million short tons)

	Rank				Mining method		Location		
Year	Bituminous coal	Subbituminous coal	Lignite	Anthracite	Underground	Surface	West of the Mississippi	East of the Mississippi	Total
1949	437.9	([1])	([1])	42.7	358.9	121.7	36.4	444.2	480.6
1950	516.3	([1])	([1])	44.1	421.0	139.4	36.0	524.4	560.4
1951	533.7	([1])	([1])	42.7	442.2	134.2	34.6	541.7	576.3
1952	466.8	([1])	([1])	40.6	381.2	126.3	32.7	474.8	507.4
1953	457.3	([1])	([1])	30.9	367.4	120.8	30.6	457.7	488.2
1954	391.7	([1])	([1])	29.1	306.0	114.8	25.4	395.4	420.8
1955	464.6	([1])	([1])	26.2	358.0	132.9	26.6	464.2	490.8
1956	500.9	([1])	([1])	28.9	380.8	148.9	25.8	504.0	529.8
1957	492.7	([1])	([1])	25.3	373.6	144.5	24.7	493.4	518.0
1958	410.4	([1])	([1])	21.2	297.6	134.0	20.3	411.3	431.6
1959	412.0	([1])	([1])	20.6	292.8	139.8	20.3	412.4	432.7
1960	415.5	([1])	([1])	18.8	292.6	141.7	21.3	413.0	434.3
1961	403.0	([1])	([1])	17.4	279.6	140.9	21.8	398.6	420.4
1962	422.1	([1])	([1])	16.9	287.9	151.1	21.4	417.6	439.0
1963	458.9	([1])	([1])	18.3	309.0	168.2	23.7	453.5	477.2
1964	487.0	([1])	([1])	17.2	327.7	176.5	25.7	478.5	504.2
1965	512.1	([1])	([1])	14.9	338.0	189.0	27.4	499.5	527.0
1966	533.9	([1])	([1])	12.9	342.6	204.2	28.0	518.8	546.8
1967	552.6	([1])	([1])	12.3	352.4	212.5	28.9	536.0	564.9
1968	545.2	([1])	([1])	11.5	346.6	210.1	29.7	527.0	556.7
1969	547.2	8.3	5.0	10.5	349.2	221.7	33.3	537.7	571.0
1970	578.5	16.4	8.0	9.7	340.5	272.1	44.9	567.8	612.7
1971	521.3	22.2	8.7	8.7	277.2	283.7	51.0	509.9	560.9
1972	556.8	27.5	11.0	7.1	305.0	297.4	64.3	538.2	602.5
1973	543.5	33.9	14.3	6.8	300.1	298.5	76.4	522.1	598.6
1974	545.7	42.2	15.5	6.6	278.0	332.1	91.9	518.1	610.0
1975	577.5	51.1	19.8	6.2	293.5	361.2	110.9	543.7	654.6
1976	588.4	64.8	25.5	6.2	295.5	389.4	136.1	548.8	684.9
1977	581.0	82.1	28.2	5.9	266.6	430.6	163.9	533.3	697.2
1978	534.0	96.8	34.4	5.0	242.8	427.4	183.0	487.2	670.2
1979	612.3	121.5	42.5	4.8	320.9	460.2	221.4	559.7	781.1
1980	628.8	147.7	47.2	6.1	337.5	492.2	251.0	578.7	829.7
1981	608.0	159.7	50.7	5.4	316.5	507.3	269.9	553.9	823.8
1982	620.2	160.9	52.4	4.6	339.2	499.0	273.9	564.3	838.1
1983	568.6	151.0	58.3	4.1	300.4	481.7	274.7	507.4	782.1
1984	649.5	179.2	63.1	4.2	352.1	543.9	308.3	587.6	895.9
1985	613.9	192.7	72.4	4.7	350.8	532.8	324.9	558.7	883.6
1986	620.1	189.6	76.4	4.3	360.4	529.9	325.9	564.4	890.3
1987	636.6	200.2	78.4	3.6	372.9	545.9	336.8	581.9	918.8
1988	638.1	223.5	85.1	3.6	382.2	568.1	370.7	579.6	950.3
1989	659.8	231.2	86.4	3.3	393.8	586.9	381.7	599.0	980.7
1990	693.2	244.3	88.1	3.5	424.5	604.5	398.9	630.2	1,029.1
1991	650.7	255.3	86.5	3.4	407.2	588.8	404.7	591.3	996.0
1992	651.8	252.2	90.1	3.5	407.2	590.3	409.0	588.6	997.5
1993	576.7	274.9	89.5	4.3	351.1	594.4	429.2	516.2	945.4
1994	640.3	300.5	88.1	4.6	399.1	634.4	467.2	566.3	1,033.5
1995	613.8	328.0	86.5	4.7	396.2	636.7	488.7	544.2	1,033.0
1996	630.7	340.3	88.1	4.8	409.8	654.0	500.2	563.7	1,063.9
1997	653.8	345.1	86.3	4.7	420.7	669.3	510.6	579.4	1,089.9
1998	[R]631.7	[R]394.8	[R]85.8	[R]5.3	[R]417.7	[R]699.8	[R]547.0	570.6	[R]1,117.5
1999	[E]621.3	[E]388.3	[E]84.4	[E]5.2	[E]410.8	[E]688.3	[E]537.9	[E]561.2	[P]1,099.1

[1] Included in bituminous coal.
R=Revised. P=Preliminary. E=Estimated.

SOURCE: *Annual Energy Review 1999,* Energy Information Administration, Washington, D.C., 2000

The growing prevalence of surface coal mining and the closing of nonproductive mines led to increases in coal mining productivity in the 1980s and 1990s. (See Figure 4.5.) In 1999 average productivity reached an all-time high of 6.22 short tons per miner hour. Because surface mines are easier to work, they average up to three times the productivity of underground mines. In 1999 the productivity for surface mines was 9.85 short tons of coal per miner hour, while underground mines produced 3.84 short tons per miner hour.

COAL IN THE DOMESTIC MARKET

Overall Production and Consumption

Before 1951, coal was the leading source of energy produced in the United States. In the years from 1952 to

FIGURE 4.4

Coal mining methods

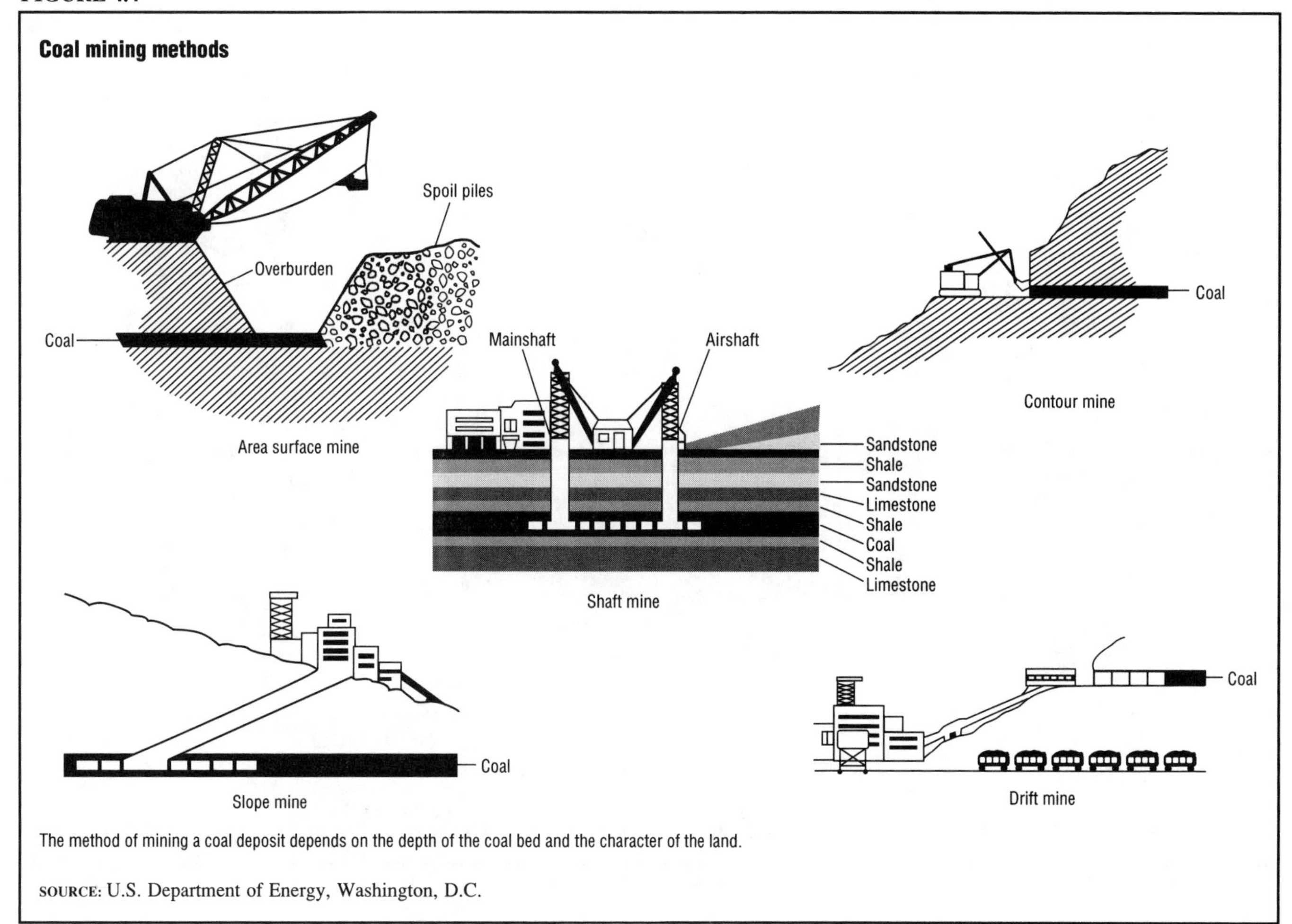

The method of mining a coal deposit depends on the depth of the coal bed and the character of the land.

SOURCE: U.S. Department of Energy, Washington, D.C.

1984, crude oil and natural gas often exchanged first place in energy production. (See Figure 4.1.) Sparked by the 1973 oil embargo and the resulting energy price increases, however, U.S. coal consumption began increasing steadily. In 1984 coal regained the top position and has remained the leader since. The share of coal in total national energy consumption grew from 18 percent in 1975 to 32 percent in 1999, when coal supplied 23 quadrillion Btu of energy.

According to the Energy Information Administration, the nation consumed 558 million short tons of coal in 1974. Twenty-five years later, in 1999, this number had grown to 1,045 million short tons. (Figure 4.6 shows the flow of coal in 1999.) The increases in coal consumption were greatest in the electric utility sector, as many existing electric power plants switched to coal from more expensive oil and gas, and many new, coal-fired power plants were constructed during the 1970s.

FIGURE 4.5

Coal mining productivity, 1949–98

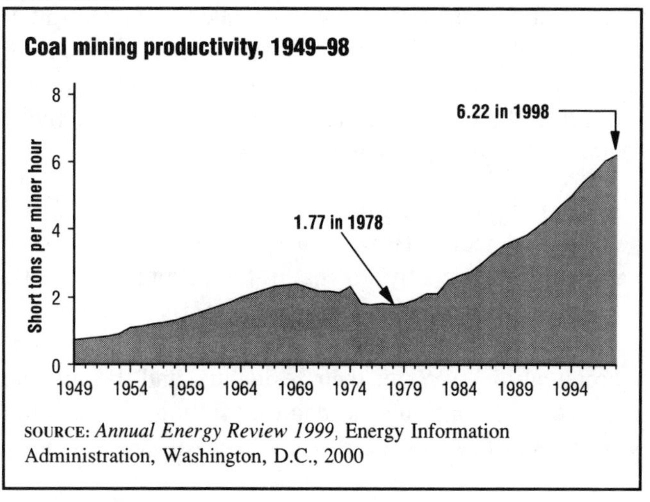

SOURCE: *Annual Energy Review 1999*, Energy Information Administration, Washington, D.C., 2000

Coal Consumption by Sector

To make electricity, coal is pulverized and burned to produce steam, which then drives electric generators. Each ton of coal used by an electric generator produces about 2,000 kilowatt hours of electricity. In household terms, each pound of coal produces enough electricity to light ten 100-watt light bulbs for one hour.

Electric utility companies are by far the largest consumers of coal. (See Figures 4.6 and 4.7.) Electricity producers accounted for 90 percent of domestic coal consumption, or 944 million short tons, in 1999. Coal-fired plants produced 1.89 trillion kilowatt-hours of

FIGURE 4.6

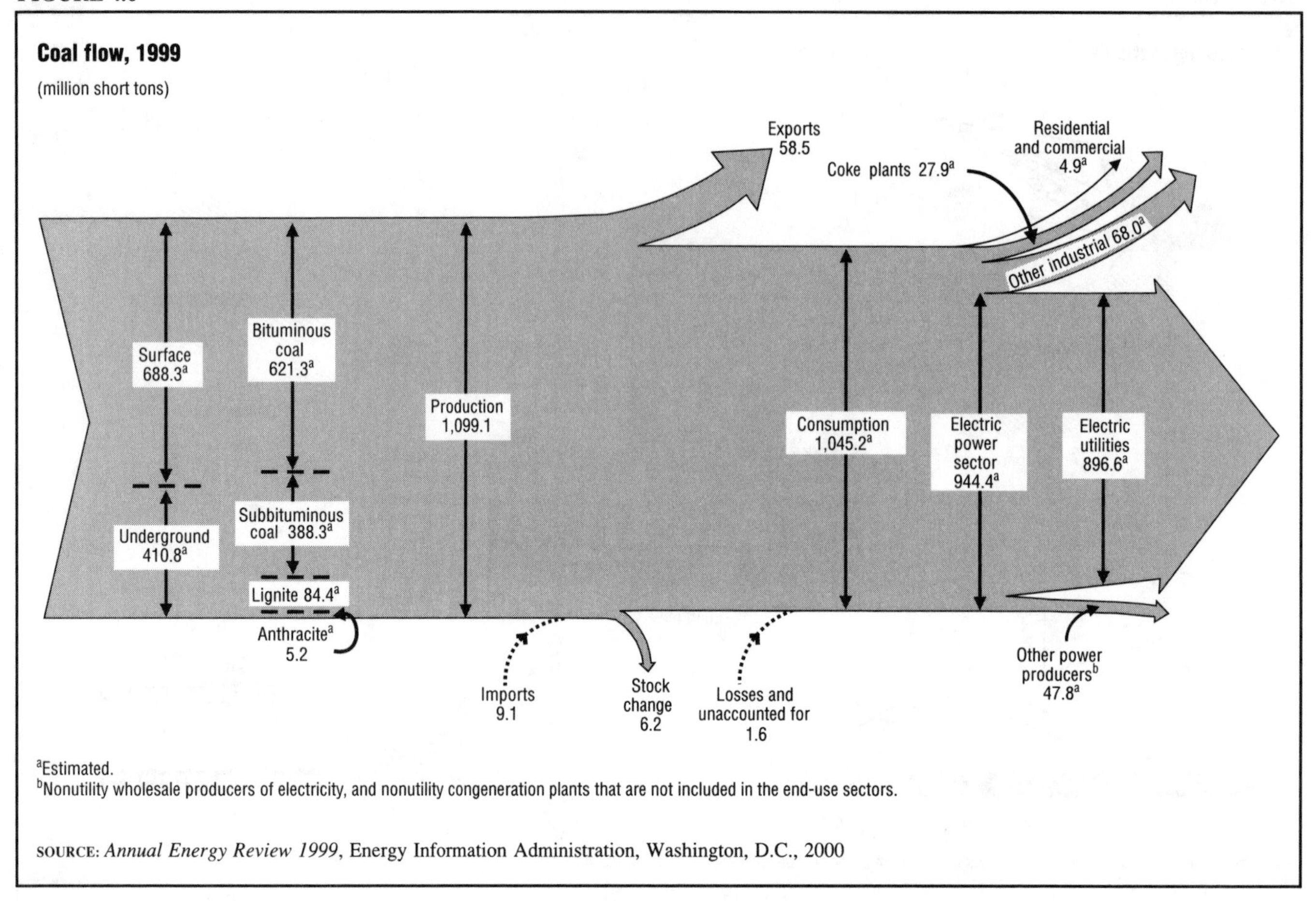

SOURCE: *Annual Energy Review 1999*, Energy Information Administration, Washington, D.C., 2000

electricity, or 51 percent of U.S. electricity net generation, in 1999.

The industrial sector was the second-largest consumer of coal in 1999, accounting for 9 percent of coal use, or 96 million short tons. Coal is used in many applications, including the chemical, cement, paper, synthetic fuels, metals, and food-processing industries.

Coal was once the major fuel in the residential and commercial sector. However, since the late 1940s, coal has been replaced by oil, natural gas, and electricity, which are cleaner and more convenient. By 1970 only 16 million tons of coal were used for residential and commercial buildings. Since then, residential and commercial coal use has continued to decline, falling to 5 million short tons in 1999, or less than 1 percent of total coal use.

The Price of Coal

In 1998 the average price of coal fell to $17.14 per short ton, down for the 17th year in a row and less than half the 1982 price. (See Table 4.2.) On a Btu basis, coal remains the least expensive fossil fuel. In 1999 the average cost for coal was 80 cents per million Btu, compared with $1.86 per million Btu for natural gas and $2.56 per million Btu for crude oil.

ENVIRONMENTAL CONCERNS ABOUT COAL

The negative side of energy use—pollution of the environment—is not a recent problem. King Edward I of England, in 1306, so objected to the noxious smoke from London's coal-burning fires that he banned coal's use by everyone except blacksmiths. The enormous scale of today's energy use has increased concerns.

Coal-fired electric power plants emit gases that are harmful to the environment. Scientists believe that burning huge quantities of fossil fuels causes the "greenhouse effect," in which "greenhouse" gases from fuels trap heat in the earth's atmosphere and cause increased warming, which may threaten the environment. Burning coal also contributes to the formation of acid rain and to public health concerns. Sulfur dioxide, for instance, has been shown to cause respiratory problems.

Carbon dioxide accounts for the largest share of greenhouse gas emissions. In 1998 the combustion of coal in the United States produced more than half-a-billion metric tons of carbon, 36 percent of total carbon dioxide emissions from all sources. (See Figure 4.8.)

Acid Rain

Acid rain is any form of precipitation that contains a greater-than-normal amount of acid. Chemicals such as

FIGURE 4.7

Coal consumption, shares by sector, 1949 and 1999

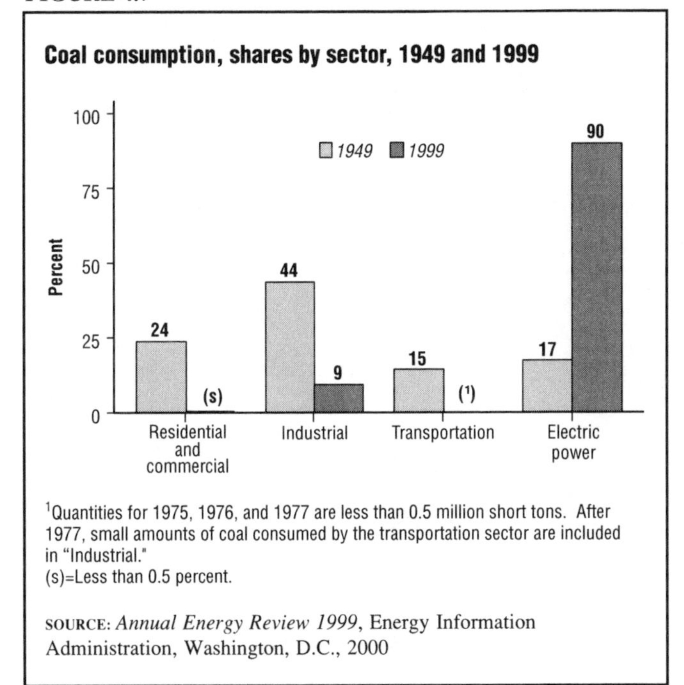

[1]Quantities for 1975, 1976, and 1977 are less than 0.5 million short tons. After 1977, small amounts of coal consumed by the transportation sector are included in "Industrial."
(s)=Less than 0.5 percent.

SOURCE: *Annual Energy Review 1999*, Energy Information Administration, Washington, D.C., 2000

FIGURE 4.8

Carbon dioxide emissions from energy consumption, by fuel, 1998

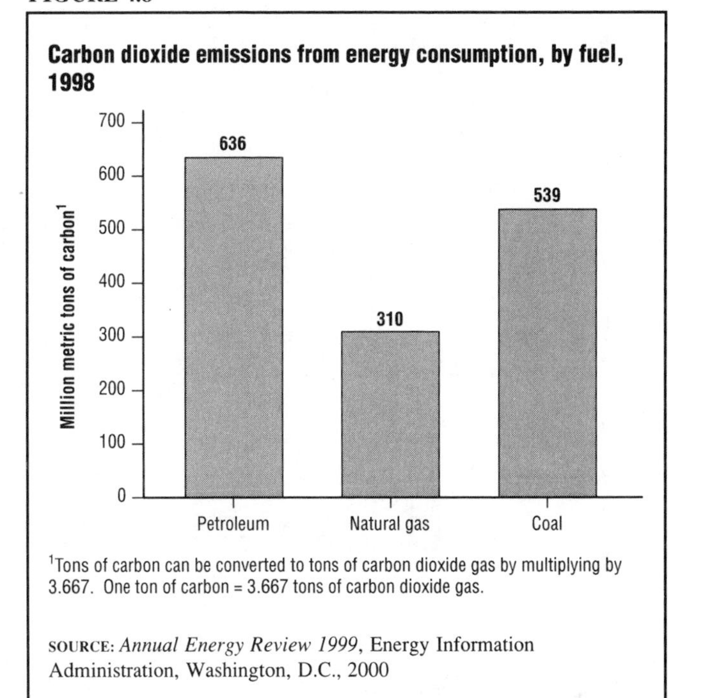

[1]Tons of carbon can be converted to tons of carbon dioxide gas by multiplying by 3.667. One ton of carbon = 3.667 tons of carbon dioxide gas.

SOURCE: *Annual Energy Review 1999*, Energy Information Administration, Washington, D.C., 2000

oxides of sulfur and nitrogen are emitted from coal-fired power plants. These emissions chemically change as they travel through the atmosphere and eventually fall to the earth as acidic rain, snow, fog, or dust, creating air pollution. In nature the combination of rain and oxides is part of a natural balance that nourishes plants and aquatic life. When oxides in the atmosphere increase because of automobile exhaust, industrial and power plant pollution, or other fossil fuel combustion processes, the balance of nature is upset. In many parts of the world acid rain has caused significant damage to forests, lakes, and other ecosystems.

The Clean Coal Technology Law

In 1984 Congress established the Department of Energy's (DOE) Clean Coal Technology (CCT) program (PL 98-473). Congress directed the DOE to administer cost-shared projects (financed by both industry and government) to demonstrate clean coal technologies. The demonstration projects had the goal of using coal in more environmentally and economically efficient ways.

Clean Coal Technology and the Clean Air Act

The stated goal of both Congress and the DOE has been to develop cost-effective ways to burn coal more cleanly, both to control acid rain and to improve the nation's energy security by reducing dependence on imported fuels. One strategy is a slow, phased-in approach in which utility companies and states reduce their emissions in stages.

Under the Clean Air Act of 1990 (PL 101-549), restrictions on sulfur dioxide and nitrogen oxide emissions, which contribute heavily to air pollution, took effect in 1995 and tightened in 2000. Each round of regulation requires coal-burning utilities to find lower-sulfur coal or to install cleaner technology such as "scrubbers," which reduce smokestack emissions. When the first Clean Air Act was passed in 1970, it was aimed at changing the air quality standards at new generating stations, and older coal-using plants were exempt. Under the new act, older plants are also covered by the regulations.

An Environmental Protection Agency (EPA) analysis, prepared by ICF Resources, Inc., predicted that the 1990 Clean Air Act would reduce coal production from the high-sulfur areas of the Midwest and northern Appalachia by 2005. Low-sulfur replacement coal will come mostly from central Appalachia and the West. The only major drawback for western coal is the cost of transportation, but recent improvements in rail service have reduced this disadvantage.

Cleaner Coal Use

The coal-burning process can be cleaned by physical or chemical methods. Scrubbers, which are a commonly used physical method to reduce sulfur dioxide emissions, filter coal emissions by spraying lime or a calcium compound and water across the emission stream before it leaves the smokestack. The sulfur dioxide bonds to the spray and settles as a mud-like substance that can be pumped out for disposal. Scrubbers are expensive to operate.

Particulate collectors are the most common emissions cleaners for coal. They are cheaper to operate than scrubbers, but are less effective. Cooling towers reduce heat

TABLE 4.2

Coal prices, 1950–98
(dollars per short ton)

Year	Bituminous coal		Subbituminous coal		Lignite		Subtotal [1]		Anthracite		Total	
	Nominal	Real [2]	Nominal	Real [2]	Nominal	Real [2]	Nominal	Real [2]	Nominal	Real [2]	Nominal	Real [2]
1950	[3]4.86	[3]27.85	([4])	([4])	2.41	[R]13.81	4.84	[R]27.74	9.34	[R]53.52	5.19	[R]29.74
1960	[3]4.71	[3]21.23	([4])	([4])	2.29	[R]10.32	4.69	[R]21.14	8.01	[R]36.10	4.83	[R]21.77
1970	[3]6.30	[3]21.68	([4])	([4])	1.86	[R]6.40	6.26	[R]21.54	11.03	[R]37.96	6.34	[R]21.82
1975	[3]19.79	[3]49.44	([4])	([4])	3.17	[R]7.92	19.23	[R]48.04	32.26	[R]80.59	19.35	[R]48.34
1980	29.17	[R]51.14	11.08	[R]19.42	W	W	24.52	[R]42.99	42.51	[R]74.53	24.65	[R]43.22
1981	31.51	[R]50.52	12.18	[R]19.53	W	W	26.29	[R]42.15	44.28	[R]71.00	26.40	[R]42.33
1982	32.15	[R]48.53	13.37	[R]20.18	W	W	27.14	[R]40.97	49.85	[R]75.25	27.25	[R]41.13
1983	31.11	[R]45.17	13.03	[R]18.92	W	W	25.85	[R]37.53	52.29	[R]75.91	25.98	[R]37.72
1984	30.63	[R]42.88	12.41	[R]17.37	10.45	[R]14.63	25.51	[R]35.71	48.22	[R]67.50	25.61	[R]35.85
1985	30.78	[R]41.77	12.57	[R]17.06	10.68	[R]14.49	25.10	[R]34.06	45.80	[R]62.15	25.20	[R]34.20
1986	28.84	[R]38.30	12.26	[R]16.28	10.64	[R]14.13	23.70	[R]31.47	44.12	[R]58.58	23.79	[R]31.59
1987	28.19	[R]36.34	11.32	[R]14.59	10.85	[R]13.99	23.00	[R]29.65	43.65	[R]56.26	23.07	[R]29.74
1988	27.66	[R]34.48	10.45	[R]13.03	10.06	[R]12.54	22.00	[R]27.43	44.16	[R]55.06	22.07	[R]27.52
1989	27.40	[R]32.91	10.16	[R]12.20	9.91	[R]11.90	21.76	[R]26.13	42.93	[R]51.56	21.82	[R]26.20
1990	27.43	[R]31.71	9.70	[R]11.21	10.13	[R]11.71	21.71	[R]25.10	39.40	[R]45.54	21.76	[R]25.15
1991	27.49	[R]30.66	9.68	[R]10.80	10.89	[R]12.15	21.45	[R]23.92	36.34	[R]40.53	21.49	[R]23.97
1992	26.78	[R]29.16	9.68	[R]10.54	10.81	[R]11.77	20.99	[R]22.85	34.24	[R]37.28	21.03	[R]22.90
1993	26.15	[R]27.80	9.33	[R]9.92	11.11	[R]11.81	19.79	[R]21.04	32.94	[R]35.02	19.85	[R]21.11
1994	25.68	[R]26.75	8.37	[R]8.72	10.77	[R]11.22	19.34	[R]20.14	36.07	[R]37.57	19.41	[R]20.22
1995	25.56	[R]26.06	8.10	[R]8.26	10.83	[R]11.04	18.74	[R]19.10	39.78	[R]40.55	18.83	[R]19.19
1996	25.17	[R]25.17	7.87	[R]7.87	10.92	[R]10.92	18.42	[R]18.42	36.78	[R]36.78	18.50	[R]18.50
1997	24.64	[R]24.18	7.42	[R]7.28	10.91	[R]10.71	[R]18.07	[R]17.73	35.12	[R]34.46	18.14	[R]17.80
1998	[R]24.87	[R]24.12	[R]6.96	[R]6.75	[R]11.08	[R]10.75	[R]17.55	[R]17.02	[R]42.91	[R]41.62	[R]17.67	[R]17.14

[1] Subtotal of bituminous coal, subbituminous coal, and lignite.
[2] In chained (1996) dollars, calculated by using gross domestic product implicit price deflators.
[3] Includes subbituminous coal.
[4] Included in bituminous coal.

R=Revised. W=Withheld to avoid disclosure of individual company data.

Note: Prices are free-on-board (f.o.b.) mine prices.

SOURCE: *Annual Energy Review 1999*, Energy Information Administration, Washington, D.C., 2000

TABLE 4.3

Coal exports by country of destination, 1960–99

(million short tons)

Year	Canada	Brazil	Europe: Belgium and Luxembourg	Denmark	France	Germany [1]	Italy	Netherlands	Spain	United Kingdom	Other	Total	Japan	Other	Total
1960	12.8	1.1	1.1	0.1	0.8	4.6	4.9	2.8	0.3	0.0	2.4	17.1	5.6	1.3	38.0
1961	12.1	1.0	1.0	0.1	0.7	4.3	4.8	2.6	0.2	0.0	2.0	15.7	6.6	1.0	36.4
1962	12.3	1.3	1.3	(s)	0.9	5.1	6.0	3.3	0.8	(s)	1.8	19.1	6.5	1.0	40.2
1963	14.6	1.2	2.7	(s)	2.7	5.6	7.9	5.0	1.5	0.0	2.4	27.7	6.1	0.9	50.4
1964	14.8	1.1	2.3	(s)	2.2	5.2	8.1	4.2	1.4	0.0	2.6	26.0	6.5	1.1	49.5
1965	16.3	1.2	2.2	(s)	2.1	4.7	9.0	3.4	1.4	(s)	2.3	25.1	7.5	0.9	51.0
1966	16.5	1.7	1.8	(s)	1.6	4.9	7.8	3.2	1.2	(s)	2.5	23.1	7.8	1.0	50.1
1967	15.8	1.7	1.4	0.0	2.1	4.7	5.9	2.2	1.0	0.0	2.1	19.4	12.2	1.0	50.1
1968	17.1	1.8	1.1	0.0	1.5	3.8	4.3	1.5	1.5	0.0	1.9	15.5	15.8	0.9	51.2
1969	17.3	1.8	0.9	0.0	2.3	3.5	3.7	1.6	1.8	0.0	1.3	15.2	21.4	1.2	56.9
1970	19.1	2.0	1.9	0.0	3.6	5.0	4.3	2.1	3.2	(s)	1.8	21.8	27.6	1.2	71.7
1971	18.0	1.9	0.8	0.0	3.2	2.9	2.7	1.6	2.6	1.7	1.1	16.6	19.7	1.1	57.3
1972	18.7	1.9	1.1	0.0	1.7	2.4	3.7	2.3	2.1	2.4	1.1	16.9	18.0	1.2	56.7
1973	16.7	1.6	1.2	0.0	2.0	1.6	3.3	1.8	2.2	0.9	1.3	14.4	19.2	1.6	53.6
1974	14.2	1.3	1.1	0.0	2.7	1.5	3.9	2.6	2.0	1.4	0.9	16.1	27.3	1.8	60.7
1975	17.3	2.0	0.6	0.0	3.6	2.0	4.5	2.1	2.7	1.9	1.6	19.0	25.4	2.6	66.3
1976	16.9	2.2	2.2	(s)	3.5	1.0	4.2	3.5	2.5	0.8	2.1	19.9	18.8	2.1	60.0
1977	17.7	2.3	1.5	0.1	2.1	0.9	4.1	2.0	1.6	0.6	2.1	15.0	15.9	3.5	54.3
1978	15.7	1.5	1.1	0.0	1.7	0.6	3.2	1.1	0.8	0.4	2.2	11.0	10.1	2.5	40.7
1979	19.5	2.8	3.2	0.2	3.9	2.6	5.0	2.0	1.4	1.4	4.4	23.9	15.7	4.1	66.0
1980	17.5	3.3	4.6	1.7	7.8	2.5	7.1	4.7	3.4	4.1	6.0	41.9	23.1	6.0	91.7
1981	18.2	2.7	4.3	3.9	9.7	4.3	10.5	6.8	6.4	2.3	8.8	57.0	25.9	8.7	112.5
1982	18.6	3.1	4.8	2.8	9.0	2.3	11.3	5.9	5.6	2.0	7.6	51.3	25.8	7.5	106.3
1983	17.2	3.6	2.5	1.7	4.2	1.5	8.1	4.2	3.3	1.2	6.4	33.1	17.9	6.1	77.8
1984	20.4	4.7	3.9	0.6	3.8	0.9	7.6	5.5	2.3	2.9	5.3	32.8	16.3	7.2	81.5
1985	16.4	5.9	4.4	2.2	4.5	1.1	10.3	6.3	3.5	2.7	10.3	45.1	15.4	9.9	92.7
1986	14.5	5.7	4.4	2.1	5.4	0.8	10.4	5.6	2.6	2.9	8.4	42.6	11.4	11.4	85.5
1987	16.2	5.8	4.6	0.9	2.9	0.5	9.5	4.1	2.5	2.6	6.6	34.2	11.1	12.3	79.6
1988	19.2	5.3	6.5	2.8	4.3	0.7	11.1	5.1	2.5	3.7	8.5	45.1	14.1	11.3	95.0
1989	16.8	5.7	7.1	3.2	6.5	0.7	11.2	6.1	3.3	4.5	8.9	51.6	13.8	12.9	100.8
1990	15.5	5.8	8.5	3.2	6.9	1.1	11.9	8.4	3.8	5.2	9.5	58.4	13.3	12.7	105.8
1991	11.2	7.1	7.5	4.7	9.5	1.7	11.3	9.6	4.7	6.2	10.4	65.5	12.3	13.0	109.0
1992	15.1	6.4	7.2	3.8	8.1	1.0	9.3	9.1	4.5	5.6	8.5	57.3	12.3	11.4	102.5
1993	8.9	5.2	5.2	0.3	4.0	0.5	6.9	5.6	4.1	4.1	6.9	37.6	11.9	11.0	74.5
1994	9.2	5.5	4.9	0.5	2.9	0.3	7.5	4.9	4.1	3.4	7.3	35.8	10.2	10.7	71.4
1995	9.4	6.4	4.5	2.1	3.7	2.0	9.1	7.3	4.7	4.7	10.7	48.6	11.8	12.4	88.5
1996	12.0	6.5	4.6	1.3	3.9	1.1	9.2	7.1	4.1	6.2	9.8	47.2	10.5	14.2	90.5
1997	15.0	7.5	4.3	0.4	3.4	0.9	7.0	4.8	4.1	7.2	9.2	41.3	8.0	11.8	83.5
1998	R20.7	6.5	3.2	0.3	3.2	1.2	5.3	4.5	3.2	5.9	6.9	33.8	7.7	9.4	R78.0
1999	19.8	4.4	2.1	0.0	2.5	0.6	4.0	3.4	2.5	3.2	4.3	22.5	5.0	6.7	58.5

[1] Through 1990, the data for Germany are for the former West Germany only. Beginning with 1991, the data for Germany are for the unified Germany, i.e., the former East Germany and West Germany.
R=Revised. (s)=Less than 0.05 million short tons.

SOURCE: *Annual Energy Review 1999*, Energy Information Administration, Washington, D.C., 2000

FIGURE 4.9

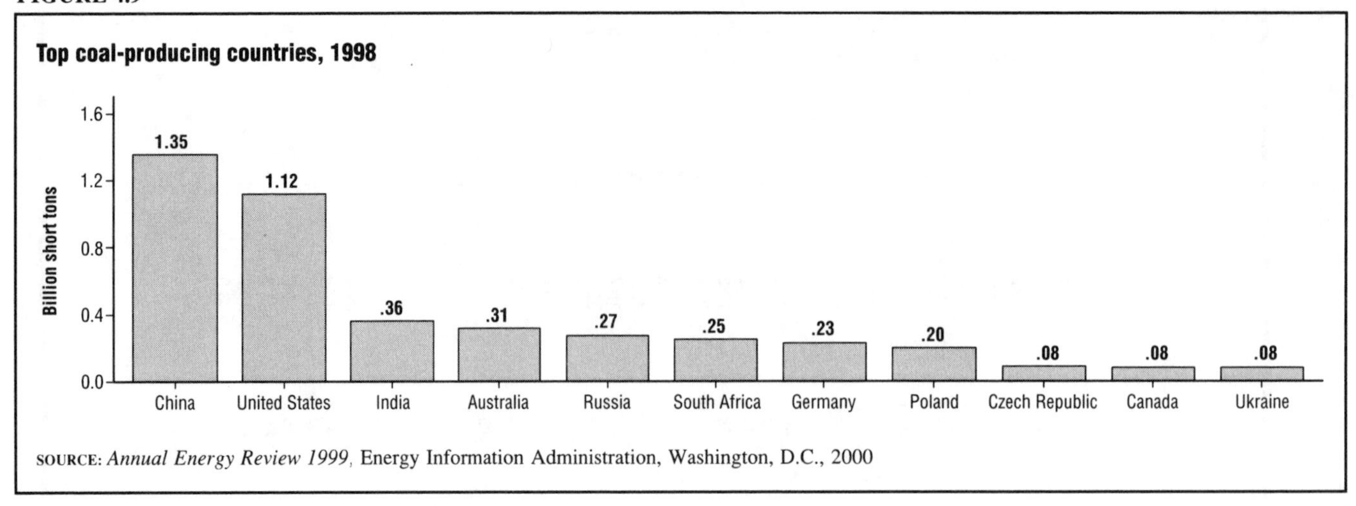

SOURCE: *Annual Energy Review 1999*, Energy Information Administration, Washington, D.C., 2000

FIGURE 4.10

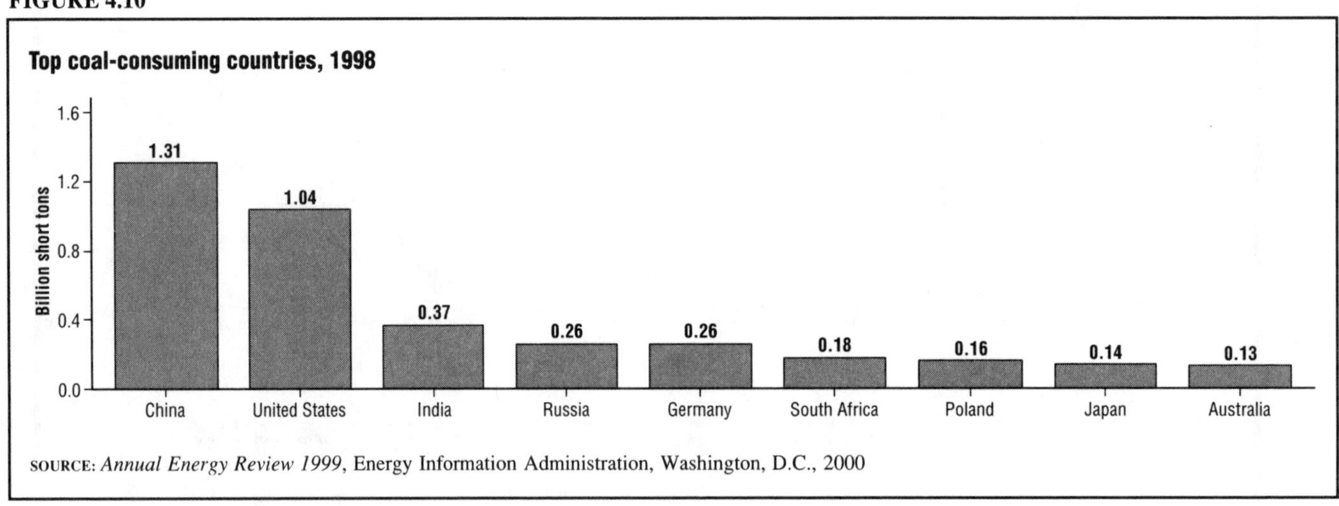

SOURCE: *Annual Energy Review 1999*, Energy Information Administration, Washington, D.C., 2000

released into the atmosphere and reduce some pollutants. Chemical cleaning, a relatively new technology, involves the use of biological or chemical agents to clean emissions, but is not in widespread use.

Under new environmental regulations, plants with coal-generated boilers must be built to reduce sulfur emissions by 70–90 percent. New, high-sulfur coal electricity plants, designed to meet emission standards, use 30 percent of their construction costs on pollution control equipment, as well as up to 5 percent of their power output to operate this equipment. Research to lower these costs is important because of the quantity of electricity produced in the United States with coal.

In 1998, according to the Energy Information Administration (EIA) of the DOE, coal-fired electricity plants that had environmental equipment installed had a production capacity of 321 gigawatts. Of this capacity, nearly all was treated with particulate collectors, 43 percent with cooling towers, and 28 percent with scrubbers. The use of scrubbers is projected to increase as new regulations take effect.

COAL EXPORTS

Since 1950 the United States has produced more coal than it has consumed. The excess production has allowed the United States to become a significant exporter of coal to other nations. Exports amounted to 58.5 million short tons in 1999, down from 78.0 million short tons in 1998. (See Table 4.3.)

In 1999 coal provided 43 percent of all U.S. energy exports. Europe received 38 percent of U.S. coal exports. The individual countries that bought the most U.S. coal were Canada, Japan, Brazil, Italy, and the United Kingdom. While the amount of coal leaving the nation is considerable, it still represents only about 7 percent of the Btu content of the petroleum coming into the United States each year.

INTERNATIONAL COAL USAGE

The EIA disclosed that world coal production exceeded 5.0 billion short tons in 1998 and accounted for 23 percent of world energy production. China led the world in

TABLE 4.4

Comparison of coal forecasts

(million short tons, except where noted)

Projection	AEO2000 Reference	AEO2000 Low economic growth	AEO2000 High economic growth	Other forecasts WEFA	Other forecasts GRI/Hill	Other forecasts DRI
			2015			
Production	**1,269**	**1,229**	**1,325**	**1,082**	**965**	**1,224**
Consumption by sector						
Electricity generation[a]	1,129	1,094	1,182	887	855	1,033
Coking plants	21	22	21	24	19	24
Industrial/other	81	78	85	61	60	87
Total	**1,232**	**1,193**	**1,288**	**972**	**934**	**1,144**
Net coal exports	**38**	**38**	**38**	**109**	**31**	**80**
Minemouth price						
(1998 dollars per short ton)	13.34	13.09	13.52	13.30	NA	NA
(1998 dollars per million Btu)	0.64	0.62	0.64	0.61	NA	NA
Average delivered price, electricity						
(1998 dollars per short ton)	21.19	20.74	21.60	20.29[b]	21.88	21.11
(1998 dollars per million Btu)	1.03	1.01	1.05	0.99	NA	1.03
			2020			
Production	**1,316**	**1,256**	**1,429**	**1,129**	**786**	**1,210**
Consumption by sector						
Electricity generation[a]	1,177	1,123	1,286	919	678	1,018
Coking plants	20	20	19	23	16	23
Industrial/other	82	77	87	63	57	88
Total	**1,279**	**1,219**	**1,393**	**1,005**	**751**	**1,128**
Net coal exports	**38**	**38**	**38**	**125**	**35**	**83**
Minemouth price						
(1998 dollars per short ton)	12.54	12.40	12.58	12.84	NA	NA
(1998 dollars per million Btu)	0.60	0.60	0.61	0.59	NA	NA
Average delivered price, electricity						
(1998 dollars per short ton)	20.01	19.61	20.32	19.47[b]	21.03	19.84
(1998 dollars per million Btu)	0.98	0.96	1.00	0.95	NA	0.97

[a] The DRI and *AEO2000* forecasts for electricity generation include nonutility generators. Consumption by industrial cogenerators is included in industrial consumption. The WEFA values for electricity consumption have been adjusted by including consumption by nonutility generators.

[b] Computed using a conversion factor of 20.495 million Btu per short ton.

NA = Not available.

Btu = British thermal unit.

SOURCE: *Annual Energy Outlook 2000,* Energy Information Administration, Washington, D.C., 1999

coal production, mining 1.35 billion short tons, followed by the United States at 1.12 billion short tons. Other major producers were India, Australia, Russia, South Africa, Germany, and Poland. (See Figure 4.9.)

World consumption of coal in 1998 totaled 5 billion short tons. China was also the largest consumer of coal in 1998, using 1.31 billion short tons, followed by the United States, which consumed 1.04 billion short tons. Other major consumers included India, Russia, and Germany. (See Figure 4.10.)

FUTURE TRENDS IN THE COAL INDUSTRY

The EIA, in its *Annual Energy Outlook 2000* (1999), forecasts that domestic coal production will increase to 1.27 billion short tons by 2015 and 1.32 billion short tons by 2020. (See Table 4.4.) Domestic consumption is projected to increase by 0.9 percent each year, reaching 1.28 billion short tons in 2020. Electricity generation will still use the majority of coal in 2020 (1.18 billion short tons), but coal's share in total electricity generation will decrease slightly because of the increased use of cleaner and more efficient natural gas.

Coal prices are projected to decline from $17.67 per short ton in 1998 to $13.34 per short ton in 2015 and $12.54 per short ton in 2020 (all in 1998 dollars). These price decreases will occur because of improvements in productivity and from the lower costs of mining in the West. Eastern mines will likely lose market share to less expensive, low-sulfur coal from western mines, and coal transportation and labor costs are expected to decline.

Environmental concerns about acid rain and global warming may continue to grow. The outlook for the U.S. coal industry could be affected by acid rain legislation, the development of clean coal technologies, and, over the longer term, the problem of global warming. Environmental issues will increasingly become worldwide issues. China, with nearly five times the population of the United States and a growing economy, may surpass the United States in carbon emissions by 2020.

The EIA predicts an expansion of world coal trade in the future. Increased growth in coal demand is projected in both Western Europe and Asia, particularly China. U.S. coal exports are expected to remain relatively steady, although the U.S. share in world coal trade is projected to decline from 14 percent in 1998 to 8 percent in 2020, as other countries reduce costs and gain advantages against the strong U.S. dollar.

CHAPTER 5
ENERGY RESERVES—OIL, GAS, COAL, AND URANIUM

Fossil fuels are nonrenewable resources. Nonrenewable resources are defined as concentrations of solid, liquid, or gaseous hydrocarbons that occur naturally in or near the earth's surface. These resources must be currently or potentially recoverable for economic use. Once used, these substances cannot be replaced. It is important to know the recoverable quantities of crude oil, natural gas, coal, and uranium resources in the United States and on Earth. Such estimates are essential to the development, implementation, and evaluation of national energy policies and legislation. In the U.S., Congress requires the Department of Energy to prepare estimates of energy reserves.

Proved reserves are reserves from known locations that geological and engineering data demonstrate, with reasonable certainty, to be recoverable with current technological means and economic conditions. Undiscovered recoverable resources are quantities of fuel that are thought to exist in favorable geologic settings but are not yet discovered. These resources would be feasible to retrieve with existing technological means, although they may not be feasible to recover under current economic conditions.

FIGURE 5.1

U.S. crude oil proved reserves, 1988–98

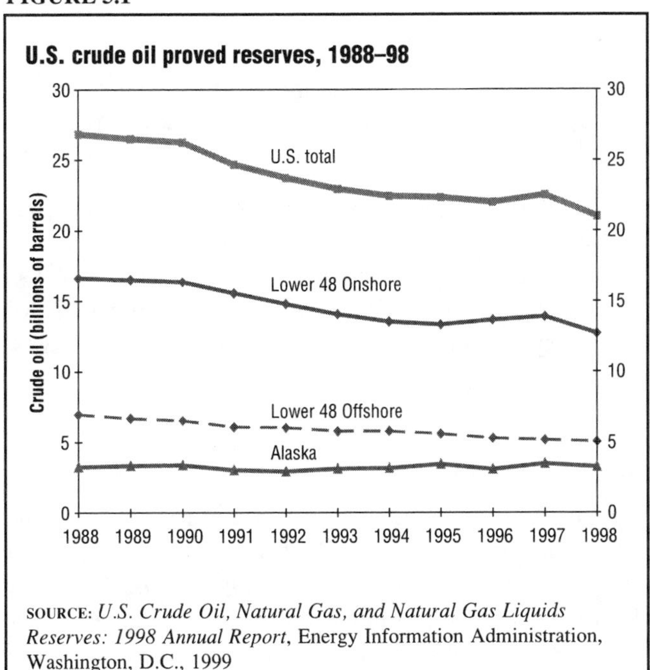

SOURCE: *U.S. Crude Oil, Natural Gas, and Natural Gas Liquids Reserves: 1998 Annual Report*, Energy Information Administration, Washington, D.C., 1999

CRUDE OIL

Over the past decade, U.S. crude oil proved reserves have been declining. (See Figure 5.1.) The U.S. had 21 billion barrels of crude oil proved reserves as of December 31, 1998. This is about 7 percent less than in 1997, the largest percentage drop in 53 years, and the 11th consecutive year that crude oil proved reserves have declined, by an average of 2 percent each year. Alaska, Texas, California, and the Gulf of Mexico offshore areas account for 79 percent of U.S. proved reserves. (See Table 5.1.) Of these four regions, only California reported increased proved reserves in 1998. In California, increased reserves were added from the San Joachin Basin Onshore.

TABLE 5.1

U.S. proved reserves of crude oil as of December 31, 1998, by selected states

Area	Percent of U.S. oil reserves
Alaska	24
Texas	23
California	18
Gulf of Mexico Federal Offshore	13
Area Total	**79**

SOURCE: *U.S. Crude Oil, Natural Gas, and Natural Gas Liquids Reserves: 1998 Annual Report*, Energy Information Administration, Washington, D.C., 1999

FIGURE 5.2

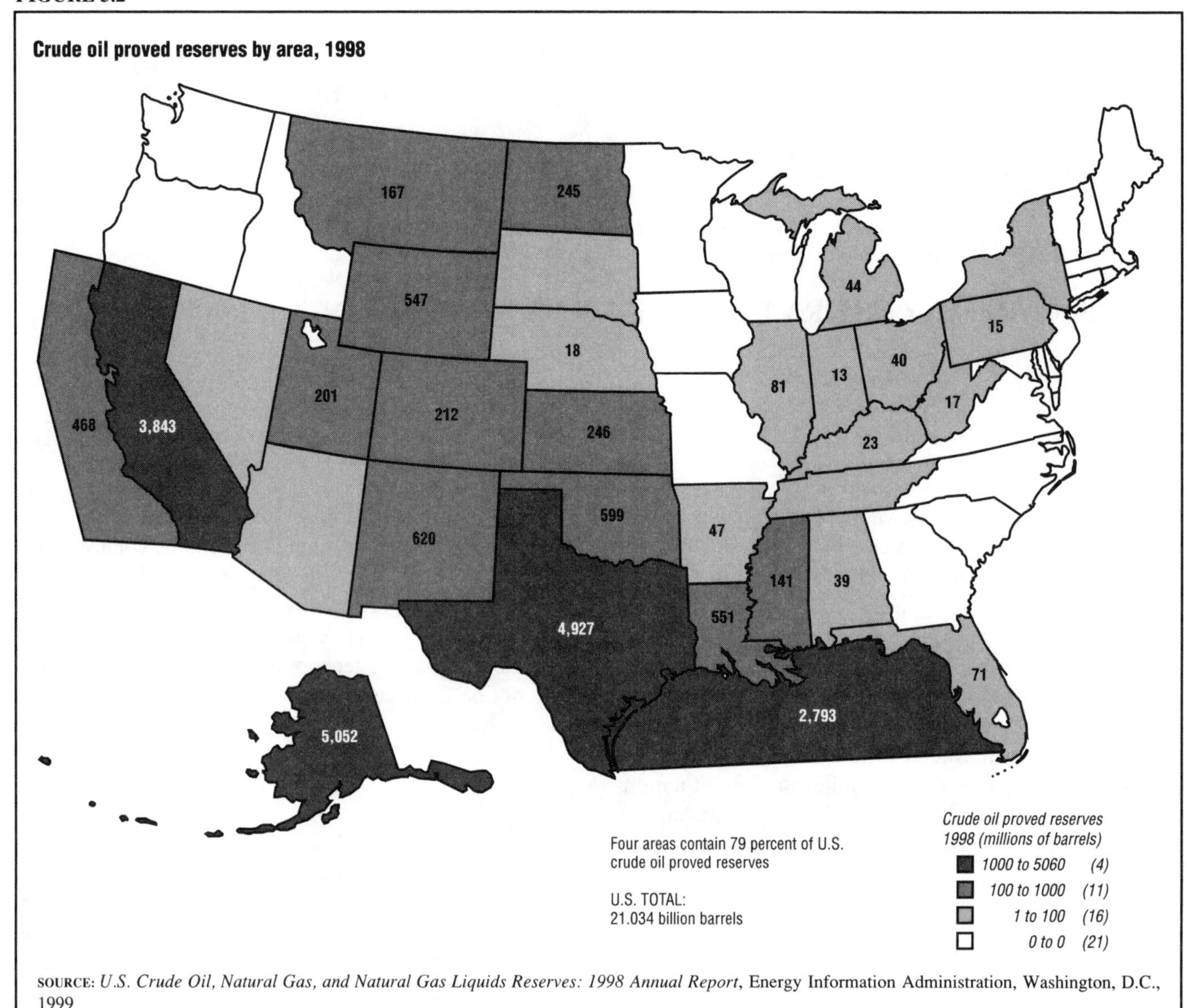

SOURCE: *U.S. Crude Oil, Natural Gas, and Natural Gas Liquids Reserves: 1998 Annual Report*, Energy Information Administration, Washington, D.C., 1999

Proved reserves of crude oil and natural gas rose in 1970 with the inclusion of Alaska's North Slope oil fields. Since then, Alaskan reserves have steadily declined. In 1987 Alaska was estimated to have 13.2 billion barrels of crude oil; by 1998 it had only 5 billion barrels. (See Figure 5.2.) The discovery of a new field in 1996, the Arco-owned Alpine field on Alaska's North Slope, was expected to raise recovery by 70,000 barrels per day by 2001. In 1998 most of the new discoveries of oil in Alaska were in "satellite" (nearby) fields of developed areas in the North Slope.

The Gulf of Mexico Federal Offshore areas, which are in U.S. territorial waters, had about 2.8 billion barrels of crude oil proved reserves in 1998, down slightly from 3.0 billion barrels in 1997. (See Figure 5.2.) The improvement in deep-water drilling systems—floating platforms and subsea wells—has allowed the industry to continually expand into deeper Gulf waters in search of crude oil. Despite the drop in 1998 reserves, the Gulf still holds much promise for future reserves discoveries.

In 1996 scientists turned their attention to the Permian Basin of west Texas and southeastern New Mexico, where plenty of dry-land potential for crude oil was found, making it one of the most active onshore areas for recent exploration. In 1998 Texas experienced the largest decline in reserves of any state.

NATURAL GAS

In 1998 the United States had about 164 trillion cubic feet of dry natural gas and 7.5 billion barrels of natural gas liquid proved reserves. (See Figures 5.3 and 5.4.) This represented a slight drop in natural gas reserves from 1997, before which there were three consecutive yearly increases in natural gas resources.

FIGURE 5.3

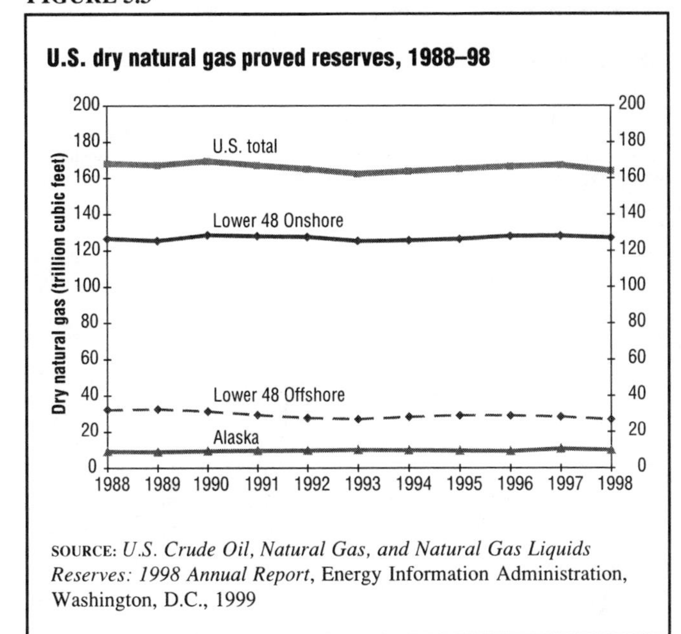

SOURCE: *U.S. Crude Oil, Natural Gas, and Natural Gas Liquids Reserves: 1998 Annual Report*, Energy Information Administration, Washington, D.C., 1999

FIGURE 5.4

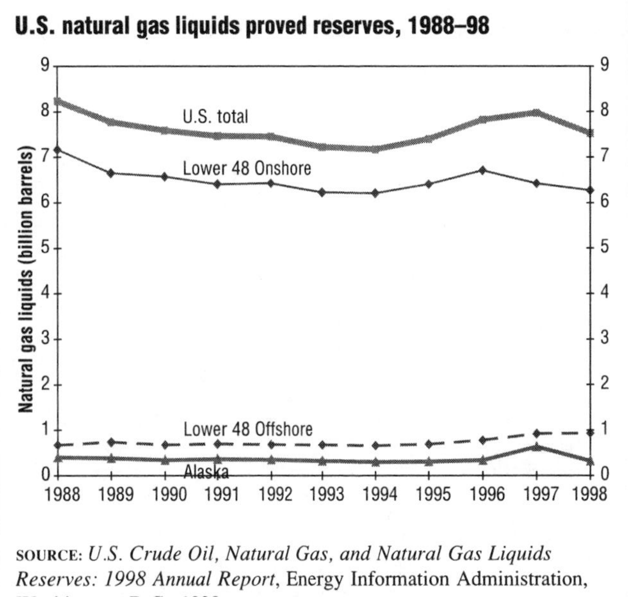

SOURCE: *U.S. Crude Oil, Natural Gas, and Natural Gas Liquids Reserves: 1998 Annual Report*, Energy Information Administration, Washington, D.C., 1999

UNDISCOVERED RESOURCES

In addition to those proved resources, other resources are believed to exist based on past geological experience, although they are not yet proved. For 1998 the Energy Information Administration (EIA) reported there were an estimated 77.9 billion barrels of crude oil, 884.7 quadrillion cubic feet of natural gas, and up to 11.1 billion barrels of natural gas liquids of undiscovered resources in the United States. (See Table 5.2.)

Looking for Oil and Gas

Finding oil and gas is usually a two-step process. First, geological and geophysical (primarily seismic, which measures the movement of the earth) exploration identifies the areas where oil and gas may most likely be found. Then exploratory wells are drilled to determine if oil or gas is present.

Market conditions and technological developments shape exploration for oil and gas. The economic problems of the oil industry can be seen in the drop in the number of exploratory oil and gas wells completed. Drilling activity for exploratory wells has declined dramatically since the early 1980s. (See Table 5.3.) In 1981 a peak of 91,550 exploratory wells were drilled, with nearly 70 percent of those successful. In 1999 only 19,080 were attempted, with 80 percent successful. In 1981, 3,970 rotary rigs were in operation; by 1999 only 624 were operating. Of this number, 496 rigs drilled for gas, while 128 drilled for oil. (See Figure 5.5.) There were 519 onshore rigs and 105 offshore rigs. The average depth of wells has steadily increased, from 3,635 feet in 1949 to 6,105 feet in 1999. Gas wells (6,640 feet in 1999) are typically deeper than oil wells (5,004 feet).

FIGURE 5.5

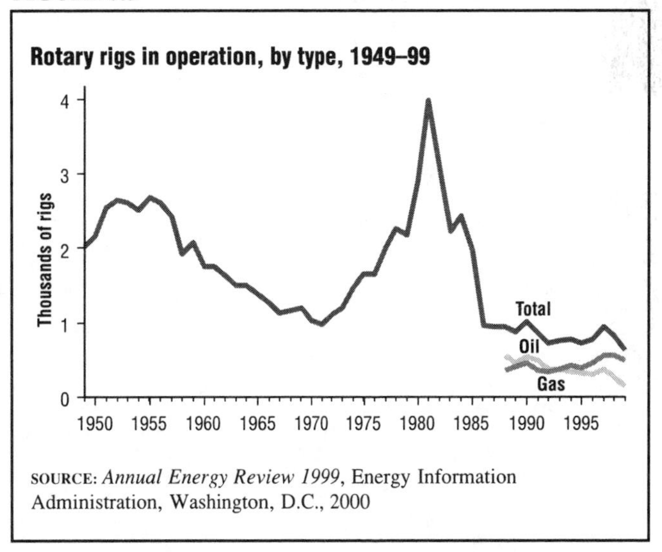

SOURCE: *Annual Energy Review 1999*, Energy Information Administration, Washington, D.C., 2000

The Cost to Drill

In 1998 an average well cost nearly $769,000 to drill, or $129 per foot. (See Table 5.4.) A gas well ($816,000) costs more than an oil well ($566,000) to drill because it is deeper. The average cost per foot of a gas well ($128) is also more than an oil well ($109). Although drilling costs have fluctuated in recent years, it costs considerably more to drill a well today than it did in the 1960s and early 1970s, not only because of inflation but also because wells must now be drilled deeper. Even between 1996 and 1998, the cost per well increased significantly, at nearly $273,000 more per well.

Spending on Exploration and Development

The estimated expenditures on exploration for, and development of, oil and gas by major U.S. energy-produc-

TABLE 5.2

Estimated oil and gas reserves and mean estimates of technically recoverable oil and gas resources

Categories	Crude oil[a] (million barrels)	Natural gas (dry) (billion cubic feet)	Natural gas liquids (millionbarrels)
Lower 48 states			
Discovered			
Proved Reserves (EIA, 1998)	15,982	[b]154,114	7,204
Reserve Growth - conventional, onshore[c] (USGS, 1991)	[d]47,000	290,000	12,900
Reserve Growth - conventional, Federal Offshore (MMS, 1995)	[e]2,238	[e]32,719	NE
Unproved Reserves, Federal Offshore (MMS, 1996)	1,643	4,436	NE
Undiscovered, Technically Recoverable			
Conventional, onshore[c] (USGS, 1993)	21,810	190,280	6,080
Continuous-type - sandstone, shale, chalk; onshore[c] (USGS, 1993)	2,066	308,080	2,119
Continuous-type - coalbeds, onshore[c] (USGS, 1993)	NA	49,910	NA
Federal Offshore - conventional (MMS, 1994)	21,300	142,100	[f]<1,800
Subtotal	**112,039**	**1,171,639**	**NA**
Alaska			
Discovered			
Proved Reserves (EIA, 1998)	5,052	9,927	320
Reserve Growth - conventional, onshore[c] (USGS, 1991)	[g]13,000	32,000	500
Reserve Growth conventional, Federal Offshore (MMS, 1994)	0	0	NE
Unproved Reserves, Federal Offshore (MMS, 1994)	400	700	NE
Undiscovered, Technically Recoverable			
Conventional onshore[c] (USGS, 1993)	8,440	68,410	1,120
Continuous-type - sandstone, shale, chalk; onshore[c] (USGS, 1993)	NE	NE	NE
Continuous-type - coalbeds, onshore[c] (USGS, 1993)	NA	NE	NA
Federal Offshore - conventional (MMS, 1994)	24,300	125,900	[f]<1,800
Subtotal	**51,192**	**236,937**	**NA**
Total lower 48 states and Alaska	**163,231**	**1,408,576**	**32,043**
Deductions for production and proved reserves changes, 1991–1998	**–10,778**	**–107,646**	**–5,251**
U.S. Total, 1998	**152,453**	**1,330,930**	**26,792**

[a] Condensate is included with crude oil for MMS estimates in Federal Offshore regions.
[b] Includes 12,179 billion cubic feet of coalbed methane (EIA, 1998).
[c] Includes USGS estimates for all onshore plus State Offshore (near-shore and shallow-water areas under state jurisdiction).
[d] Using USGS definition, 1,924 million barrels of indicated additional oil reserves in the lower 48 states were included (EIA, 1996).
[e] Reserve growth in the Pacific Federal Offshore is not included and was not estimated by the MMS. This volume is not dry gas, but wet, after lease separation.
[f] Total undiscovered natural gas liquids for Federal Offshore are 1,800 million barrels; MMS source did not separate lower 48 and Alaska estimates of undiscovered natural gas liquids (1986).
[g] Using USGS definition, 952 million barrels of indicated additional oil reserves in Alaska were included (EIA, 1996).

NE = not estimated.

NA = not applicable.

Notes: Federal Offshore indicates MMS estimates for Federal Offshore jurisdictions (outer continental shelf and deeper water areas seaward of State Offshore). Energy Information Administration (EIA), onshore and offshore estimated reserves. U.S. Geological Survey (USGS): 1995 National Assessment mean estimates as of the end of 1993 (onshore and State Offshore). Minerals Management Service (MMS): 1996 National Assessment mean estimates as of the end of 1994. The MMS also has end-1994 estimates for economically recoverable resources. Probable and possible reserves are considered by USGS definition to be part of USGS Reserve Growth, but are separately considered by the MMS as its Unproved Reserves term. The USGS did not set a time limit for the duration of Reserve Growth; the MMS set the year 2020 as the time limit in its estimates of Reserve Growth in existing fields of the Gulf of Mexico. Excluded from the estimates are undiscovered oil resources in tar deposits and oil shales, and undiscovered gas resources in geopressured brines and gas hydrates.

SOURCE: *U.S. Crude Oil, Natural Gas, and Natural Gas Liquids Reserves: 1998 Annual Report*, Energy Information Administration, Washington, D.C., 1999.

ing companies peaked at $65.3 billion in 1984. (See Table 5.5.) U.S. energy companies spent $50.8 billion in 1998 on oil and gas exploration around the world, with about half of that, $24.4 billion, spent for exploration in the U.S.

The chance of finding a major oil or gas field has become increasingly small. Oil explorers are spending more of their time looking for less oil. This does not necessarily imply that the nation will soon be facing a shortage of gas or oil: reserves are still considerable, especially of gas. It does show, however, that the oil and gas industries are mature, or highly developed. Finding new reserves will become more expensive, and until the prices of oil and gas rise sharply, fewer companies will go looking for it.

Drilling in the Arctic National Wildlife Refuge

Much controversy developed over opening the Arctic National Wildlife Refuge (ANWR) in Alaska to oil drilling. Prudhoe Bay, directly to the west of the refuge, supplies about 60 percent of Alaska's oil and 20 percent of the country's domestic oil, although production is dropping steadily as the oil is used up. In 1980 Congress passed the Alaska National Interest Lands Conservation Act (PL 96-487), which set aside more than 104 million acres for parks, refuges, and wilderness areas, including the Arctic Refuge (18 million acres), but the conservation act did not include the coastal plain.

TABLE 5.3

Oil and gas exploratory and development wells, 1950–99

Year	Wells drilled (thousands)				Successful wells (percent)	Footage drilled (million feet)				Average depth (feet per well)			
	Oil	Gas	Dry holes	Total		Oil	Gas	Dry holes	Total	Oil	Gas	Dry holes	Total
1950	23.81	3.44	14.80	42.05	64.8	92.7	13.7	51.0	157.4	3,893	3,979	3,445	3,742
1960	22.26	5.15	18.21	45.62	60.1	86.6	28.2	77.4	192.2	3,889	5,486	4,248	4,213
1970	12.97	[R]4.01	11.03	[R]28.01	60.6	56.9	23.6	58.1	138.6	4,385	5,860	5,265	4,943
1975	16.95	8.13	13.65	38.72	64.8	66.8	44.4	69.3	180.5	3,944	5,462	5,076	4,661
1980	32.64	17.33	20.64	70.61	70.8	124.3	[R]91.5	98.9	314.7	[R]3,807	[R]5,278	[R]4,792	4,456
1981	43.60	20.17	27.79	91.55	69.6	171.1	107.8	134.2	413.1	3,925	5,346	4,828	4,512
1982	39.20	18.98	26.22	84.40	68.9	148.8	106.7	122.8	378.3	3,795	[R]5,621	4,685	4,482
1983	37.12	14.56	24.15	75.84	68.2	136.1	77.6	104.3	318.0	[R]3,667	[R]5,325	4,320	4,193
1984	42.61	17.13	25.68	85.41	69.9	[R]161.8	90.6	[R]119.0	371.4	[R]3,797	5,289	4,636	4,348
1985	35.12	14.17	21.06	70.34	70.1	137.3	[R]75.8	[R]99.9	313.0	3,911	[R]5,353	[R]4,743	4,450
1986	19.10	[R]8.52	[R]12.68	[R]40.29	[R]68.5	76.6	44.7	60.5	181.9	[R]4,013	[R]5,255	[R]4,770	[R]4,514
1987	16.16	8.06	[R]11.11	[R]35.33	[R]68.5	66.3	42.5	53.4	162.2	4,104	[R]5,273	[R]4,803	[R]4,590
1988	13.64	8.56	10.04	32.23	68.8	58.7	[R]45.3	52.3	156.4	[R]4,305	[R]5,298	[R]5,211	4,851
1989	10.20	9.54	8.19	27.93	70.7	43.3	49.2	41.9	134.4	4,243	[R]5,157	[R]5,123	4,813
1990	12.20	11.04	8.31	[R]31.56	73.7	54.4	56.2	43.1	153.7	[R]4,459	[R]5,091	[R]5,183	[R]4,871
1991	11.77	9.53	7.60	[R]28.89	73.7	54.1	[R]50.0	[R]38.9	143.0	4,597	[R]5,251	[R]5,121	4,950
1992	8.76	8.21	6.12	23.08	73.5	43.8	46.1	31.2	121.1	[R]4,999	[R]5,619	[R]5,103	5,247
1993	[R]8.41	10.02	[R]6.33	[R]24.75	74.4	42.4	60.1	[R]32.6	135.1	[R]5,046	[R]6,000	[R]5,150	[R]5,459
1994	[R]6.72	9.54	[R]5.31	[R]21.57	[R]75.4	[R]36.0	59.6	[R]29.2	[R]124.8	[R]5,355	[R]6,251	[R]5,502	[R]5,787
1995[E]	7.63	[R]8.35	5.08	[R]21.06	75.9	[R]38.2	[R]51.6	[R]28.1	[R]117.8	[R]5,007	[R]6,171	[R]5,535	[R]5,596
1996[E]	[R]8.31	[R]9.30	[R]5.28	[R]22.90	76.9	[R]40.6	[R]58.5	[R]30.0	[R]129.0	[R]4,885	[R]6,286	[R]5,672	[R]5,636
1997[E]	[R]10.44	[R]11.33	[R]5.70	[R]27.47	79.2	[R]52.3	[R]71.1	[R]33.3	[R]156.7	[R]5,009	[R]6,279	[R]5,833	[R]5,704
1998[E]	[R]7.06	[R]12.11	[R]4.91	[R]24.08	[R]79.6	[R]37.1	[R]81.0	[R]31.5	[R]149.6	[R]5,256	[R]6,693	[R]6,406	[R]6,213
1999[E]	4.80	10.51	3.76	19.08	80.3	24.0	69.8	22.6	116.4	5,004	6,640	6,013	6,105

R=Revised. E=Estimated.

Notes: Service wells, stratigraphic tests, and core tests are excluded. For 1950, data represents wells completed in a given year. For 1960, data are for well completion reports received by the American Petroleum Institute during the reporting year. For 1970 forward, the data represent wells completed in a given year.

SOURCE: *Annual Energy Review 1999*, Energy Information Administration, Washington, D.C., 2000

TABLE 5.4

Costs of oil and gas wells drilled, 1960–98

Year	Costs per well (thousand dollars)					Costs per foot (dollars)				
	Oil	Gas	Dry holes	All		Oil	Gas	Dry holes	All	
	(nominal)	(nominal)	(nominal)	(nominal)	(real) [1]	(nominal)	(nominal)	(nominal)	(nominal)	(real) [1]
1960	52.2	102.7	44.0	54.9	[R]247.6	13.22	18.57	10.56	13.01	[R]58.63
1961	51.3	94.7	45.2	54.5	[R]243.0	13.11	17.65	10.56	12.85	[R]57.26
1962	54.2	97.1	50.8	58.6	[R]257.9	13.41	18.10	11.20	13.31	[R]58.53
1963	51.8	92.4	48.2	55.0	[R]239.2	13.20	17.19	10.58	12.69	[R]55.17
1964	50.6	104.8	48.5	55.8	[R]239.2	13.12	18.57	10.64	12.86	[R]55.10
1965	56.6	101.9	53.1	60.6	[R]255.0	13.94	18.35	11.21	13.44	[R]56.52
1966	62.2	133.8	56.9	68.4	[R]279.6	15.04	21.75	12.34	14.95	[R]61.12
1967	66.6	141.0	61.5	72.9	[R]289.2	16.61	23.05	12.87	15.97	[R]63.35
1968	79.1	148.5	66.2	81.5	[R]309.7	18.63	24.05	12.88	16.83	[R]63.99
1969	86.5	154.3	70.2	88.6	[R]321.0	19.28	25.58	13.23	17.56	[R]63.65
1970	86.7	160.7	80.9	94.9	[R]326.5	19.29	26.75	15.21	18.84	[R]64.83
1971	78.4	166.6	86.8	94.7	[R]310.3	18.41	27.70	16.02	19.03	[R]62.35
1972	93.5	157.8	94.9	106.4	[R]334.5	20.77	27.78	17.28	20.76	[R]65.24
1973	103.8	155.3	105.8	117.2	[R]348.7	22.54	27.46	19.22	22.50	[R]66.96
1974	110.2	189.2	141.7	138.7	[R]378.8	27.82	34.11	26.76	28.93	[R]79.00
1975	138.6	262.0	177.2	177.8	[R]444.1	34.17	46.23	33.86	36.99	[R]92.41
1976	151.1	270.4	190.3	191.6	[R]453.0	37.35	49.78	36.94	40.46	[R]95.65
1977	170.0	313.5	230.2	227.2	[R]504.6	41.16	57.57	43.49	46.81	[R]103.98
1978	208.0	374.2	281.7	280.0	[R]580.4	49.72	68.37	52.55	56.63	[R]117.42
1979	243.1	443.1	339.6	331.4	[R]634.2	58.29	80.66	64.60	67.70	[R]129.57
1980	272.1	536.4	376.5	367.7	[R]644.6	66.36	95.16	73.70	77.02	[R]135.03
1981	336.3	698.6	464.0	453.7	[R]727.4	80.40	122.17	90.03	94.30	[R]151.19
1982	347.4	864.3	515.4	514.4	[R]776.4	86.34	146.20	104.09	108.73	[R]164.12
1983	283.8	608.1	366.5	371.7	[R]539.7	72.65	108.37	79.10	83.34	[R]120.99
1984	262.1	489.8	329.2	326.5	[R]457.0	66.32	88.80	67.18	71.90	[R]100.64
1985	270.4	508.7	372.3	349.4	[R]474.1	66.78	93.09	73.69	75.35	[R]102.25
1986	284.9	522.9	389.2	364.6	[R]484.1	68.35	93.02	76.53	76.88	[R]102.08
1987	246.0	380.4	259.1	279.6	[R]360.4	58.35	69.55	51.05	58.71	[R]75.68
1988	279.4	460.3	366.4	354.7	[R]442.2	62.28	84.65	66.96	70.23	[R]87.56
1989	282.3	457.8	355.4	362.2	[R]435.0	64.92	86.86	67.61	73.55	[R]88.33
1990	321.8	471.3	367.5	383.6	[R]443.4	69.17	90.73	67.49	76.07	[R]87.93
1991	346.9	506.6	441.2	421.5	[R]470.1	73.75	93.10	83.05	82.64	[R]92.17
1992	362.3	426.1	357.6	382.6	[R]416.6	69.50	72.83	67.82	70.27	[R]76.51
1993	356.6	521.2	387.7	426.8	[R]453.8	67.52	83.15	72.56	75.30	[R]80.06
1994	409.5	535.1	491.5	483.2	[R]503.3	70.57	81.90	86.60	79.49	[R]82.79
1995	415.8	629.7	481.2	513.4	[R]523.4	78.09	95.97	84.60	87.22	[R]88.91
1996	341.0	616.0	541.0	496.1	[R]496.1	70.60	98.67	95.74	88.92	[R]88.92
1997	445.6	728.6	655.6	603.9	[R]592.6	90.48	117.55	115.09	107.83	[R]105.81
1998	566.0	815.6	973.2	769.1	745.9	108.88	127.94	157.79	128.97	125.08

[1] In chained (1996) dollars, calculated by using gross domestic product implicit price deflators.

R=Revised.

Notes: The information reported for 1965 and prior years is not strictly comparable to that in the more recent surveys. Average cost is the arithmetic mean and includes all costs for drilling and equipping wells and for surface-producing facilities. Wells drilled include exploratory and development wells; excludes service wells, stratigraphic tests, and core tests.

SOURCE: *Annual Energy Review 1999*, Energy Information Administration, Washington, D.C., 2000

TABLE 5.5

Major U.S. energy companies' expenditures for oil and gas exploration and development, by region, 1974–98

(billion dollars)

	United States			Foreign								
Year	Onshore	Offshore	Total	Canada	OECD [2] Europe	Eastern Europe and former U.S.S.R.	Africa	Middle East	Other Eastern Hemisphere [3]	Other Western Hemisphere [4]	Total	Total
1974	NA	NA	8.7	NA	NA	—	NA	NA	NA	NA	3.8	12.5
1975	NA	NA	7.8	NA	NA	—	NA	NA	NA	NA	5.3	13.1
1976	NA	NA	9.5	NA	NA	—	NA	NA	NA	NA	5.2	14.7
1977	6.7	4.0	10.7	1.5	2.5	—	0.7	0.2	0.3	0.4	5.6	16.3
1978	7.5	4.3	11.8	1.6	2.6	—	0.8	0.3	0.4	0.6	6.4	18.2
1979	13.0	8.3	21.3	2.3	3.0	—	0.8	0.2	0.5	0.8	7.8	29.1
1980	16.8	9.4	26.2	3.1	4.3	—	1.4	0.2	0.8	1.0	11.0	37.2
1981	19.9	13.0	33.0	1.8	5.0	—	2.1	0.3	1.9	1.3	12.4	45.4
1982	27.2	11.9	39.1	1.9	6.3	—	2.1	0.4	2.4	1.1	14.2	53.3
1983	16.0	11.1	27.1	1.6	4.3	—	1.7	0.5	2.0	0.6	10.7	37.7
1984	32.1	16.0	48.1	5.4	5.5	—	3.4	0.5	2.0	0.5	17.3	65.3
1985	20.0	8.5	28.5	1.9	3.7	—	1.6	0.9	1.3	0.7	10.1	38.6
1986	12.5	4.9	17.4	1.1	3.2	—	1.1	0.3	1.2	0.6	7.5	24.9
1987	9.7	4.5	14.3	1.9	3.0	—	0.8	0.4	2.8	0.5	9.2	23.5
1988	12.9	8.1	21.0	5.4	4.3	—	0.8	0.4	1.4	0.7	13.0	34.1
1989	9.0	6.0	15.0	6.3	3.5	—	1.0	0.4	2.3	0.6	14.1	29.1
1990	10.2	4.9	15.1	1.8	6.6	—	1.4	0.6	2.4	0.7	13.6	28.7
1991	9.6	4.6	14.2	1.7	6.8	—	1.5	0.5	2.4	0.7	13.7	27.9
1992	7.3	3.0	10.3	1.1	6.8	—	1.4	0.6	2.4	0.6	12.9	23.2
1993	7.2	3.7	10.9	1.6	5.5	0.3	1.5	0.7	2.5	0.6	12.5	23.5
1994	7.8	4.8	12.6	1.8	4.4	0.3	1.4	0.4	2.8	0.7	11.9	24.5
1995	7.7	4.7	12.4	1.9	5.2	0.4	2.0	0.4	2.4	0.9	13.2	25.6
1996	7.9	6.7	14.6	1.6	5.6	0.5	2.8	0.5	4.1	1.6	16.6	31.3
1997	R13.0	8.8	R21.8	2.0	7.1	0.6	3.0	0.6	3.0	1.6	17.9	R39.8
1998	13.5	11.0	24.4	4.8	8.6	1.3	3.1	0.9	3.9	3.7	26.4	50.8

[1] Nominal dollars.
[2] Organization for Economic Cooperation and Development.
[3] This region includes areas that are eastward of the Greenwich prime meridian to 180° longitude and that are not included in other domestic or foreign classifications.
[4] This region includes areas that are westward of the Greenwich prime meridian to 180° longitude and that are not included in other domestic or foreign classifications.

R=Revised. — = Not applicable. NA=Not available.

Notes: "Major U.S. energy companies" are the top publicly-owned, U.S.-based crude oil and natural gas producers and petroleum refiners that form the Financial Reporting System (FRS).

SOURCE: *Annual Energy Review 1999,* Energy Information Administration, Washington, D.C., 2000

TABLE 5.6

Coal demonstrated reserve base, January 1, 1999
(billion short tons)

		Bituminous coal		Subbituminous coal		Lignite	Total		
Region and state	**Anthracite**	**Underground**	**Surface**	**Underground**	**Surface**	**Surface** [1]	**Underground**	**Surface**	**Total**
Appalachian	**7.3**	**74.0**	**24.0**	**0.0**	**0.0**	**1.1**	**78.0**	**28.5**	**106.5**
Alabama	0.0	1.2	2.2	0.0	0.0	1.1	1.2	3.2	4.5
Kentucky, Eastern	0.0	2.0	9.7	0.0	0.0	0.0	2.0	9.7	11.7
Ohio	0.0	17.7	5.8	0.0	0.0	0.0	17.7	5.8	23.6
Pennsylvania	7.2	20.2	1.0	0.0	0.0	0.0	24.0	4.4	28.4
Virginia	0.1	1.3	0.7	0.0	0.0	0.0	1.4	0.7	2.1
West Virginia	0.0	30.5	4.3	0.0	0.0	0.0	30.5	4.3	34.8
Other [2]	0.0	1.2	0.4	0.0	0.0	0.0	1.2	0.4	1.5
Interior	**0.1**	**118.1**	**27.6**	**0.0**	**0.0**	**13.3**	**118.2**	**40.9**	**159.1**
Illinois	0.0	88.3	16.6	0.0	0.0	0.0	88.3	16.6	104.9
Indiana	0.0	8.8	1.0	0.0	0.0	0.0	8.8	1.0	9.8
Iowa	0.0	1.7	0.5	0.0	0.0	0.0	1.7	0.5	2.2
Kentucky, Western	0.0	16.2	3.7	0.0	0.0	0.0	16.2	3.7	19.8
Missouri	0.0	1.5	4.5	0.0	0.0	0.0	1.5	4.5	6.0
Oklahoma	0.0	1.2	0.3	0.0	0.0	0.0	1.2	0.3	1.6
Texas	0.0	0.0	0.0	0.0	0.0	12.8	0.0	12.8	12.8
Other [3]	0.1	0.3	1.1	0.0	0.0	0.5	0.4	1.6	2.0
Western	**(s)**	**22.5**	**2.4**	**121.4**	**62.8**	**29.7**	**143.9**	**94.8**	**238.7**
Alaska	0.0	0.6	0.1	4.8	0.6	(s)	5.4	0.7	6.1
Colorado	(s)	8.0	0.6	3.8	0.0	4.2	11.9	4.8	16.7
Montana	0.0	1.4	0.0	69.6	32.9	15.8	71.0	48.6	119.6
New Mexico	(s)	2.7	0.9	3.5	5.2	0.0	6.2	6.2	12.4
North Dakota	0.0	0.0	0.0	0.0	0.0	9.3	0.0	9.3	9.3
Utah	0.0	5.5	0.3	(s)	0.0	0.0	5.5	0.3	5.7
Washington	0.0	0.3	0.0	1.0	(s)	(s)	1.3	(s)	1.4
Wyoming	0.0	3.8	0.5	38.7	24.1	0.0	42.5	24.6	67.1
Other [4]	0.0	0.1	0.0	(s)	(s)	0.4	0.1	0.4	0.5
U.S. Total	**7.5**	**214.6**	**54.0**	**121.4**	**62.8**	**44.0**	**340.1**	**164.2**	**504.3**
States east of the Mississippi River	7.3	187.5	45.3	0.0	0.0	1.1	191.5	49.7	241.2
States west of the Mississippi River	0.1	27.1	8.7	121.4	62.8	42.9	148.6	114.5	263.1

[1] Lignite resources are not mined underground in the United States.
[2] Georgia, Maryland, North Carolina, and Tennessee.
[3] Arkansas, Kansas, Louisiana, and Michigan.
[4] Arizona, Idaho, Oregon, and South Dakota.
(s)=Less than 0.05 billion short tons.

Notes: Data represent known measured and indicated coal resources meeting minimum seam and depth criteria, in the ground as of January 1, 1999. These coal resources are not totally recoverable. Net recoverability ranges from 0 percent to more than 90 percent. Fifty-four percent of the demonstrated reserve base of coal in the United States is estimated to be recoverable.

SOURCE: *Annual Energy Review 1999,* Energy Information Administration, Washington, D.C., 2000

The Department of the Interior, in a 1987 report to Congress, recommended that the 1.5 million acre coastal plain be opened for exploration and extraction. Alaskan corporations supported the proposal in hopes of sharing in the proceeds. Environmentalists strongly opposed the plan because of the destruction that drilling could do to native wildlife, such as caribou, polar and grizzly bears, musk oxen, wolves, Arctic foxes, and millions of nesting birds.

The Department of the Interior's Fish and Wildlife Service estimated that 3.2 billion barrels of recoverable oil exists in the contested area. They also estimated that there was a 46 percent chance of recovering the oil, a high figure by industry standards. Considering the generally declining American production, this could account for a significant portion of American petroleum production in the next century. Vast quantities of natural gas are also likely to be found in the area.

The report further stated, however, that there could be a major effect on the migratory caribou herds, which number about 180,000 animals. While environmentalists estimated that 20–40 percent of the animals would be threatened, Interior Department officials claimed that the caribou would change their migratory habits. If major oil reserves are found, oil companies might operate on the coastal plain for several decades. Possible development of offshore oil fields with onshore support in the ANWR could mean significant human activity in the area for the next century. As of February 2001, the ANWR remains off-limits for oil exploration.

COAL

In 1999 the Energy Information Administration estimated U.S. coal reserves at 504 billion short tons. (See Table 5.6.) Most of this, 215 billion short tons, is underground bituminous coal. Montana, Illinois, and Wyoming have the largest coal reserves.

FIGURE 5.6

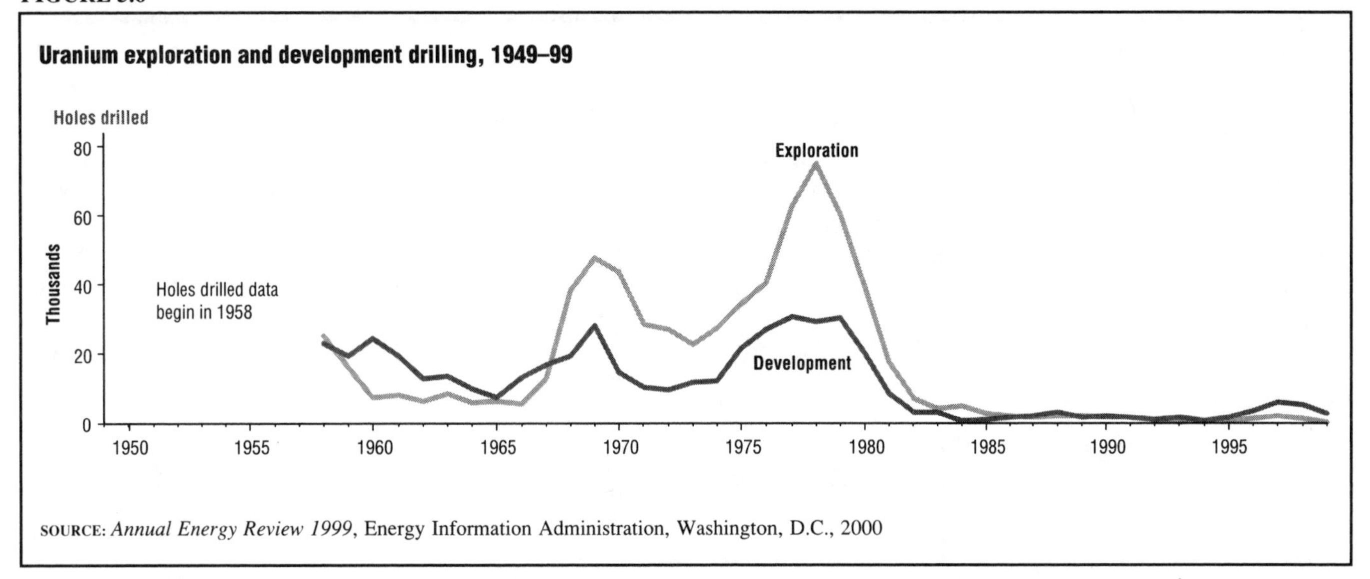

SOURCE: *Annual Energy Review 1999*, Energy Information Administration, Washington, D.C., 2000

In addition to untapped coal reserves, large stockpiles of coal are maintained by coal producers, distributors, and major consumers such as electric utility companies and industrial plants, to compensate for possible interruptions in supply. Although there is little seasonal change in demand for coal, supply can vary owing to factors such as coal miners' strikes and bad weather. According to the Energy Information Administration, coal stockpiles totaled 171 million short tons in 1999. Electric utilities held over 75 percent of this coal, and coal producers and distributors stocked another 20 percent.

URANIUM

The United States possesses enough uranium to fuel existing nuclear reactors for the next 40 years. In 1999 uranium reserves totaled 1.4 billion pounds of uranium oxide, mostly in Wyoming and New Mexico. (See Table 5.7.) Exploration for uranium has reflected that energy markets are moving away from nuclear energy. The number of exploratory and developmental holes drilled peaked at 104,000 in 1978 and declined to just 3,180 in 1999. (See Figure 5.6.)

TABLE 5.7

Uranium reserves and resources, 1999
(million pounds U_3O_8)

Resource category and state	Forward cost category (dollars per pound)[1] $30 or less	$50 or less	$100 or less
Reserves[2]	**274**	**908**	**1,432**
New Mexico	84	341	567
Wyoming	113	377	596
Texas	7	24	40
Arizona, Colorado, Utah	41	115	160
Others[3]	28	51	69
Potential resources[4]			
Estimated additional resources	2,180	3,310	4,850
Speculative resources	1,310	2,230	3,480

[1] Forward costs are all operating and capital costs (in current dollars) yet to be incurred in the production of uranium from estimated resources. Excluded are previous expenditures (such as exploration and land acquisitions), taxes, profit, and the cost of money. Generally, forward costs are lower than market prices. Resource values in forward-cost categories are cumulative; that is, the quantity at each level of forward-cost includes all reserves/resources at the lower cost in that category.
[2] The Energy Information Administration category of uranium reserves is equivalent to the internationally reported category of Reasonably Assured Resources (RAR).
[3] California, Idaho, Nebraska, Nevada, North Dakota, Oregon, South Dakota, and Washington.
[4] Shown are the mean values for the distribution of estimates for each forward-cost category, rounded to the nearest million pounds U_3O_8.
Note: Data are at end of year.

SOURCE: *Annual Energy Review 1999*, Energy Information Administration, Washington, D.C., 2000

INTERNATIONAL RESERVES

Crude Oil

Most of the estimated world crude oil reserves of approximately 1 trillion barrels (as of 1999) are located in the Middle East. (See Table 5.8.) Saudi Arabia, Iraq, United Arab Emirates, Kuwait, and Iran have the largest reserves. With an estimated 21 billion barrels of reserves, or only about 2 percent of the world's oil reserves total, the United States can no longer depend on its own oil reserves unless it drastically lowers consumption. If current consumption rates continue, the world's oil reserves could be used up in 50–90 years.

Oil industry experts say that even though the world, at the moment, is sufficiently supplied with oil, demand will rise in the future as economies such as China and India expand. This knowledge drives oil companies to seek new sources for oil. Companies from many nations have turned their attention to areas in the former Soviet Union, including the Caspian Sea, Azerbaijan, and Kazakhstan. Geologists believe that many billions of barrels of oil may be found there.

TABLE 5.8

World crude oil and natural gas reserves, January 1, 1999

Region and country	Crude oil (billion barrels)		Natural gas (trillion cubic feet)	
	Oil & Gas Journal	*World Oil*	*Oil & Gas Journal*	*World Oil*
North America	**73.8**	**55.0**	**291.4**	**257.9**
Canada	4.9	5.6	63.9	63.6
Mexico	47.8	28.4	63.5	30.3
United States	21.0	21.0	164.0	164.0
Central and South America	**89.5**	**63.4**	**219.1**	**226.1**
Argentina	2.6	2.8	24.1	24.2
Bolivia	0.1	0.2	4.3	5.3
Brazil	7.1	7.5	8.0	8.7
Colombia	2.6	2.6	6.9	8.0
Ecuador	2.1	2.6	3.7	3.7
Peru	0.8	0.8	7.0	7.1
Trinidad and Tobago	0.5	0.6	18.3	19.8
Venezuela	72.6	45.5	142.5	146.6
Other	1.0	0.9	4.2	2.7
Western Europe	**18.9**	**19.8**	**161.5**	**159.8**
Denmark	0.9	0.9	3.9	3.2
Germany	0.4	0.4	12.3	12.0
Italy	0.6	0.6	8.1	7.8
Netherlands	0.1	0.1	63.1	62.5
Norway	10.9	11.9	41.4	43.6
United Kingdom	5.2	5.2	27.0	26.7
Other	0.7	0.8	5.7	4.0
Eastern Europe and former U.S.S.R.	**58.9**	**67.9**	**1,999.4**	**1,916.2**
Hungary	0.1	(s)	3.1	1.4
Kazakhstan	5.4	7.0	65.0	70.6
Romania	1.4	0.9	13.2	4.1
Russia	48.6	55.1	1,700.0	1,705.0
Other [1]	3.3	4.9	218.1	135.1

Region and country	Crude oil (billion barrels)		Natural gas (trillion cubic feet)	
	Oil & Gas Journal	*World Oil*	*Oil & Gas Journal*	*World Oil*
Middle East	**673.6**	**627.1**	**1,749.5**	**1,853.2**
Bahrain	0.2	NA	4.2	NA
Iran	89.7	92.9	812.3	812.2
Iraq	112.5	99.0	109.8	112.6
Kuwait	96.5	94.7	52.7	56.4
Oman	5.3	5.6	28.4	29.1
Qatar	3.7	5.3	300.0	395.0
Saudi Arabia	261.5	261.4	204.5	208.0
Syria	2.5	2.3	8.5	8.4
United Arab Emirates	97.8	63.9	212.0	209.0
Yemen	4.0	1.9	16.9	17.0
Other	(s)	0.2	0.3	5.5
Africa	**75.4**	**77.2**	**361.1**	**377.9**
Algeria	9.2	13.0	130.3	137.5
Angola	5.4	4.0	1.6	1.7
Cameroon	0.4	0.7	3.9	3.9
Congo	1.5	1.7	3.2	4.3
Egypt	3.5	3.7	31.5	37.2
Libya	29.5	26.9	46.4	46.3
Nigeria	22.5	22.5	124.0	124.0
Tunisia	0.3	0.3	2.8	2.3
Other	3.1	4.4	17.4	20.8
Far East and Oceania	**43.0**	**57.1**	**359.6**	**354.0**
Australia	2.9	1.8	44.6	28.4
Brunei	1.4	1.0	13.8	9.6
China	24.0	33.5	48.3	42.4
India	4.0	3.0	19.0	12.9
Indonesia	5.0	8.6	72.3	77.1
Malaysia	3.9	4.6	81.7	85.8
New Zealand	0.1	0.1	2.5	2.2
Pakistan	0.2	0.2	21.6	21.6
Papua New Guinea	0.3	0.6	5.4	14.0
Thailand	0.3	0.4	12.5	14.8
Other	0.9	3.2	37.9	45.3
World	**1,033.2**	**967.5**	**5,141.6**	**5,145.2**

[1] Albania, Azerbaijan, Belarus, Bulgaria, Croatia, Czech Republic, Georgia, Kyrgyzstan, Lithuania, Poland, Serbia, Slovakia, Tajikistan, Turkmenistan, Ukraine, Uzbekistan.

NA=Not available. (s)=Less than 50 million barrels.

Notes: Data for Kuwait and Saudi Arabia include one-half of the reserves in the Neutral Zone between Kuwait and Saudi Arabia. All reserve figures except those for the former U.S.S.R. and natural gas reserves in Canada are proved reserves recoverable with present technology and prices at the time of estimation. Former U.S.S.R. and Canadian natural gas figures include proved, and some probable reserves.

SOURCE: *Annual Energy Review 1999,* Energy Information Administration, Washington, D.C., 2000

Many energy experts predict increased offshore exploration, including in deep water. In the next couple of decades, offshore production is expected to increase near Algeria, Nigeria, and Venezuela, as well as in the North Sea and offshore areas of West Africa, Brazil, Colombia, Mexico, and Canada. Deep-sea exploration is extremely expensive, but if oil prices rise, as they surely must, the oil industry may increasingly launch deep-water explorations.

Natural Gas

Russia and the Middle East have most of the world's estimated 5,140 trillion cubic feet (as of 1999) of natural gas reserves. (See Table 5.8.) Russia has more than twice as much natural gas reserves as any other country, while Iran possesses the largest natural gas reserves in the Middle East by far. Large reserves are also located in Qatar, United Arab Emirates, Saudi Arabia, the United States, Venezuela, Algeria, Nigeria, and Malaysia.

Coal

In 1999 recoverable reserves of coal were estimated at about 1.1 trillion short tons. (See Table 5.9. Note that the data for the United States are for 1999, while for other countries 1996, the latest available.) The four countries with the most plentiful coal reserves are the United States (274 billion short tons), Russia (173 billion short tons), China (126 billion short tons), and Australia (100 billion short tons).

Uranium

The world's supply of uranium is much larger than the capacity for disposing it, should it be used as fuel. The countries with the largest known uranium reserves are Kazakhstan, Australia, South Africa, Nigeria, Canada, the United States, and Namibia.

TABLE 5.9

World recoverable reserves of coal
(million short tons)

Region and country	Anthracite and bituminous coal	Subbituminous coal and lignite	Total
North America	**R131,807**	**R153,390**	**R285,197**
Canada	4,970	4,535	9,505
Greenland	0	202	202
Mexico	948	387	1,335
United States [1]	R125,889	R148,267	R274,156
Central and South America	**8,641**	**15,140**	**23,781**
Brazil	0	13,173	13,173
Chile	34	1,268	1,302
Colombia	7,020	420	7,439
Peru	1,058	110	1,168
Other	529	170	699
Western Europe	**29,022**	**70,636**	**99,658**
Germany	26,455	47,399	73,855
Greece	0	3,168	3,168
Serbia and Montenegro	71	18,087	18,157
Turkey	495	690	1,185
United Kingdom	1,102	551	1,653
Other	898	741	1,639
Eastern Europe and former U.S.S.R.	**124,354**	**164,032**	**288,386**
Bulgaria	14	2,974	2,988
Czech Republic	2,880	3,929	6,809
Hungary	657	4,260	4,917
Kazakhstan	34,172	3,307	37,479
Poland	13,352	2,421	15,773
Romania	1	3,979	3,980
Russia	54,110	118,964	173,074
Ukraine	18,065	19,806	37,871
Uzbekistan	1,102	3,307	4,409
Other	0	1,085	1,085
Africa	**67,420**	**276**	**67,695**
Botswana	4,754	0	4,754
South Africa	60,994	0	60,994
Zimbabwe	809	0	809
Other	862	276	1,138
Middle East, Far East, and Oceania	**203,534**	**118,934**	**322,468**
Australia	52,139	47,510	99,649
China	68,564	57,651	126,215
India	80,174	2,205	82,379
Indonesia	849	4,905	5,754
Japan	865	0	865
Pakistan	0	3,228	3,228
Thailand	(s)	2,205	2,205
Other	942	1,231	2,174
World	**R564,777**	**R522,408**	**R1,087,185**

[1] U.S. data are more current than other data on this table. They represent recoverable reserves as of January 1, 1999; data for the other countries are as of December 31, 1996, the most recent period for which they are available.

R=Revised. (s)=Less than 0.5 million short tons.

Notes: World Energy Council data represent "proved recoverable reserves," which are the tonnage within the "proved amount in place" that can be recovered (extracted from the earth in raw form) under present and expected local economic conditions with existing, available technology. The EIA does not certify the international reserves data but reproduces the information as a matter of convenience for the reader. U. S. reserves represent estimated recoverable reserves from the Demonstrated Reserve Base which includes both measured and indicated tonnage. The U.S. term "measured" approximates the term "proved," used by the World Energy Council. The U.S. "measured and indicated" data have been combined and cannot be recaptured as "measured alone."

SOURCE: *Annual Energy Review 1999,* Energy Information Administration, Washington, D.C., 2000

CHAPTER 6

NUCLEAR ENERGY

CONQUERING OIL DEPENDENCY?

Much environmental contamination is directly related to the use of fossil fuels. To prevent further damage, many energy experts have turned their attention to other means of energy production. Some observers consider nuclear energy an attractive alternative because it does not pollute the atmosphere as fossil fuels do. On the other hand, producing nuclear energy depends on uranium, which is a nonrenewable resource. In addition, safe disposal of radioactive waste, which is a by-product of nuclear energy, has proved difficult. (See Chapter 7 for further discussion of nuclear waste.)

Many citizens do not want nuclear power plants in their neighborhoods, while others oppose nuclear power for broader, environmental reasons. In the four and a half decades since the first commercial nuclear reactor went into operation, the nuclear power industry has not been able to persuade many Americans of the safety of its enterprise. In the early 1970s most Americans favored the use of nuclear power; three decades later, most Americans—and, indeed, most people around the world—oppose building additional nuclear power plants. The 1979 near-disaster at Three Mile Island in Pennsylvania and the 1986 catastrophe at Chernobyl in the former Soviet Union greatly increased public concerns about the safety of nuclear power, as have reports of design flaws, cracks, and leaks in other reactors.

Supporters of nuclear power believe that it is as safe as any other form of energy production, if it is monitored correctly. They point to the growing concern over global warming and acid rain, caused by fossil fuel use, as well as other environmental problems such as damage caused by mining and transporting fossil fuels. In fact, this growing concern over fossil fuels has led a small number of environmentalists who had previously opposed nuclear power to reconsider their position. Nonetheless, environmental, safety, and economic concerns have restrained growth in the nuclear industry since the mid-1970s. Unwillingness to commission new nuclear plants became evident in 1992, when the number of operating units, and nuclear energy's share of electricity production, began to level off and decline in the U.S. (See Figure 6.1.)

When it was introduced in the 1950s, nuclear power was presented as the energy of the future, a source so cheap it would eliminate the need for electric meters. Those dreams never came true. Construction and operating costs skyrocketed and far exceeded early estimates. The technology was considerably more complicated than originally thought, and it turned out to be increasingly expensive to build and run reactors that would meet Nuclear Regulatory Commission (NRC) standards.

THE SOURCE OF NUCLEAR ENERGY

Nuclear energy is currently used for steam generation at electric utilities, for ship propulsion, and for nuclear

FIGURE 6.1

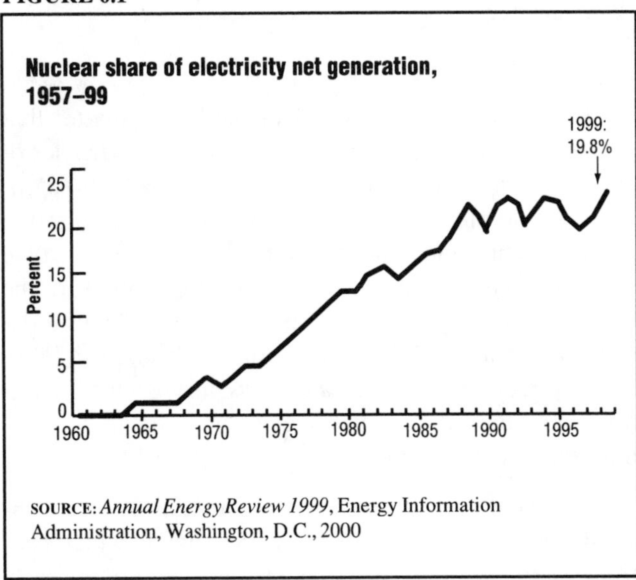

SOURCE: *Annual Energy Review 1999*, Energy Information Administration, Washington, D.C., 2000

FIGURE 6.2

Nuclear chain reaction

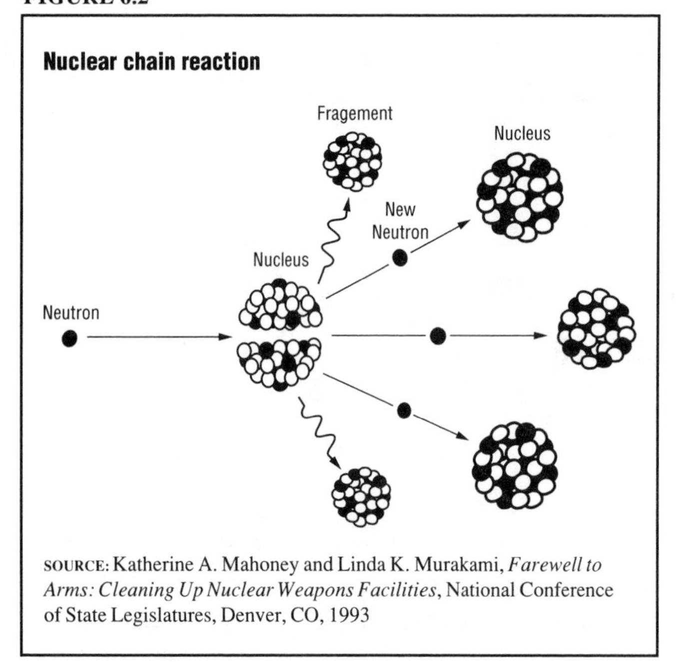

SOURCE: Katherine A. Mahoney and Linda K. Murakami, *Farewell to Arms: Cleaning Up Nuclear Weapons Facilities*, National Conference of State Legislatures, Denver, CO, 1993

FIGURE 6.3

Sources of radiation

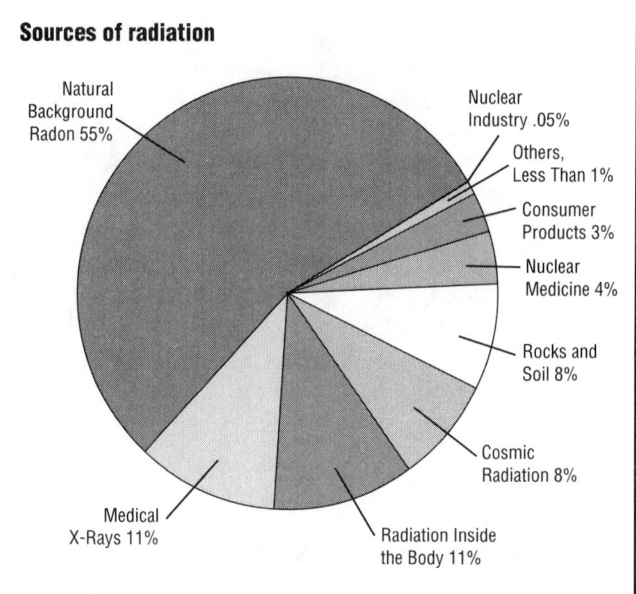

SOURCE: Katherine A. Mahoney and Linda K. Murakami, *Farewell to Arms: Cleaning Up Nuclear Weapons Facilities*, National Conference of State Legislatures, Denver, CO, 1993

weapons. In a nuclear power plant, fuel (uranium in the United States) in the reactor generates a nuclear reaction (fission) that produces heat. The heat from the reaction is carried away by water under high pressure, which heats a second water stream, producing steam. The steam runs through a turbine (similar to a jet engine), making it and the attached electrical generator spin, which produces electricity. The steam is then cooled and recirculated. The large, hourglass-shaped cooling towers associated with nuclear plants are used to cool the steam after it has run through the turbines.

The key problems in operating a nuclear power reactor include finding material (uranium 235) that will sustain a chain reaction, maintaining the reaction at a level that yields heat but does not escalate out of control and explode, and coping with the radiation produced by the chain reaction.

Radioactivity

Radioactivity is a state of instability in matter that occurs when there is an imbalance of protons (particles of matter with a positive electric charge) and neutrons (particles with no electric charge) in the nucleus of an atom. The production and explosion of nuclear weapons involve splitting atoms by bombarding them with neutrons, or the opposite, fusing atoms together. (Figure 6.2 shows a fission chain reaction.) The transformation back to stability is called radioactive decay, a process during which radiation is emitted. All radioactive elements eventually decay into lead, a stable element.

Some radioactivity occurs in nature. This natural radiation makes up the majority of the radiation to which people are exposed. It occurs from elements within the earth and from cosmic rays that filter through Earth's atmosphere from outer space. (See Figure 6.3.) Certain areas experience higher levels of radiation than others. Residents of higher elevations, such as Denver, Colorado, receive roughly twice as much radiation as people who live in Amsterdam, the Netherlands, which is below sea level. Other radiation is created by humans. The largest doses of radiation from non-natural sources come from radiation used for medical diagnosis and treatment. Color television and video games also account for a considerable amount of man-made radiation.

Mining Nuclear Fuel

There are no natural deposits of material that are immediately suitable for use in a reactor. Such nuclear deposits would have immediately started reacting in the ground (exploding) when they were first formed millions or billions of years ago and would have long since been scattered by the resulting explosions. Only a few elements split easily enough to be used as nuclear fuel. All of these fuel sources must first be carefully refined in nuclear fuel processing plants. In the United States, uranium 235, found in the form of ore, is used as nuclear fuel. The majority of uranium in the U.S. is found in two states: Wyoming and New Mexico.

Ore that contains uranium is first located by geological methods such as drilling. Uranium-bearing ores are mined by methods similar to those used for other metal ores. The mining of uranium is dangerous because uranium atoms split by themselves at a slow rate, causing radioactive substances such as radon to slowly accumulate in the deposits.

Nuclear fuel processing plants shape processed uranium into very small cylinders, called pellets. The small pellets are less than half-an-inch in diameter, but each one can make as much energy as 120 gallons of oil. The pellets are stacked in tubes about 12 feet long, called rods. Many rods are bundled together in assemblies, and hundreds of these assemblies make up the core of a nuclear reactor.

DOMESTIC NUCLEAR ENERGY PRODUCTION

The percentage of U.S. electricity supplied by nuclear power grew considerably during the 1970s and 1980s, before leveling off in the 1990s. In 1973 nuclear power supplied only 4.5 percent of the total U.S. electricity generated; by 1999 nuclear power's 19.8 percent share of electricity was down slightly from the 1995 high of 20.1 percent. (See Table 6.1.) In 1999 nuclear energy provided an all-time high of 728 billion kilowatt-hours of electrical energy, which was 10.6 percent of total U.S. energy production.

According to the Energy Information Administration, in 1999, U.S. nuclear units achieved a record overall average capacity factor (the percentage of total output achieved on average) of nearly 86 percent, up from 58 percent in 1979, the year of the Three Mile Island accident. Better training for operators, longer operating cycles between refueling, and control system improvements contributed to improved plant performance.

In 1999, 104 nuclear reactors were operating in 32 states, with a total net capacity of 97.2 million kilowatts (See Figures 6.4 and 6.5.) Most of these reactors are located east of the Mississippi River, where the demand for electricity is high. The number of plants operating are far below the 226 reactors that were in planning, construction, and operation in 1974. Since 1978 no new nuclear power plants have been ordered. (See Table 6.2.)

Several factors have contributed to the slowdown in U.S. nuclear reactor construction. The demand for electricity has grown more slowly than expected. Overall costs have increased as a result of more expensive financing, partly influenced by longer delays for licensing. Expenses have also increased because of regulations that were instituted as a result of the Three Mile Island incident. Operating costs have been higher than expected, as well. Originally, it was projected that plants could run almost 90 percent of the time, with brief pauses for refueling, but for many years this was not the case. For example, Indian Point 3, a nuclear power plant in New York, has been operational only about 50 percent of the time.

OUTLOOK FOR DOMESTIC NUCLEAR ENERGY

In its *Annual Energy Outlook 2000* (1999), the Energy Information Administration of the U.S. Department of Energy predicts that nuclear energy's contribution to U.S. electricity production will decline over the next 20 years, as aging nuclear plants are shut down. Of nuclear energy's 1998 capacity of 97 gigawatts, 40 gigawatts are expected to be retired by 2020 and be replaced by coal, natural gas, and renewable sources.

Nuclear energy's future may vary depending on several factors. Nuclear plants could be granted extensions for their operating licenses if aging problems are not severe and if safety improvements are not too expensive. Nuclear energy could maintain steady output levels if performance improvements are gained, if the waste-disposal problem is alleviated, or if stricter emissions regulations

TABLE 6.1

Nuclear power plant operations, 1957–99

Year	Nuclear electricity net generation	Nuclear share of electricity net generation	Net summer capability of operable units [1]	Capacity factor
	Billion kilowatt–hours	Percent	Million kilowatts	Percent
1957	(s)	(s)	0.1	NA
1958	0.2	(s)	0.1	NA
1959	0.2	(s)	0.1	NA
1960	0.5	0.1	0.4	NA
1961	1.7	0.2	0.4	NA
1962	2.3	0.3	0.7	NA
1963	3.2	0.4	0.8	NA
1964	3.3	0.3	0.8	NA
1965	3.7	0.3	0.8	NA
1966	5.5	0.5	1.7	NA
1967	7.7	0.6	2.7	NA
1968	12.5	0.9	2.7	NA
1969	13.9	1.0	4.4	NA
1970	21.8	1.4	7.0	NA
1971	38.1	2.4	9.0	NA
1972	54.1	3.1	14.5	NA
1973	83.5	4.5	22.7	53.5
1974	114.0	6.1	31.9	47.8
1975	172.5	9.0	37.3	55.9
1976	191.1	9.4	43.8	54.7
1977	250.9	11.8	46.3	63.3
1978	276.4	12.5	50.8	64.5
1979	255.2	11.4	49.7	58.4
1980	251.1	11.0	51.8	56.3
1981	272.7	11.9	56.0	58.2
1982	282.8	12.6	60.0	56.6
1983	293.7	12.7	63.0	54.4
1984	327.6	13.6	69.7	56.3
1985	383.7	15.5	79.4	58.0
1986	414.0	16.6	85.2	56.9
1987	455.3	17.7	93.6	57.4
1988	527.0	19.5	94.7	63.5
1989	[2]529.4	[2]17.8	[2]98.2	[2]62.2
1990	577.0	19.1	99.6	66.0
1991	612.6	19.9	99.6	70.2
1992	618.8	20.1	99.0	70.9
1993	610.4	19.1	99.1	70.5
1994	640.5	19.7	99.1	73.8
1995	673.4	20.1	99.5	77.4
1996	674.7	19.6	100.8	76.2
1997	628.6	18.0	99.7	71.1
1998	673.7	18.6	97.1	78.2
1999[P]	727.9	19.8	97.2	85.5

[1] At end of year.
[2] Beginning in 1989, includes nonutility facilities.
P=Preliminary. NA=Not available. (s)=Less than 0.05 billion kilowatt–hours or less than 0.05 percent.

SOURCE: *Annual Energy Review 1999*, Energy Information Administration, Washington, D.C., 2000

FIGURE 6.4

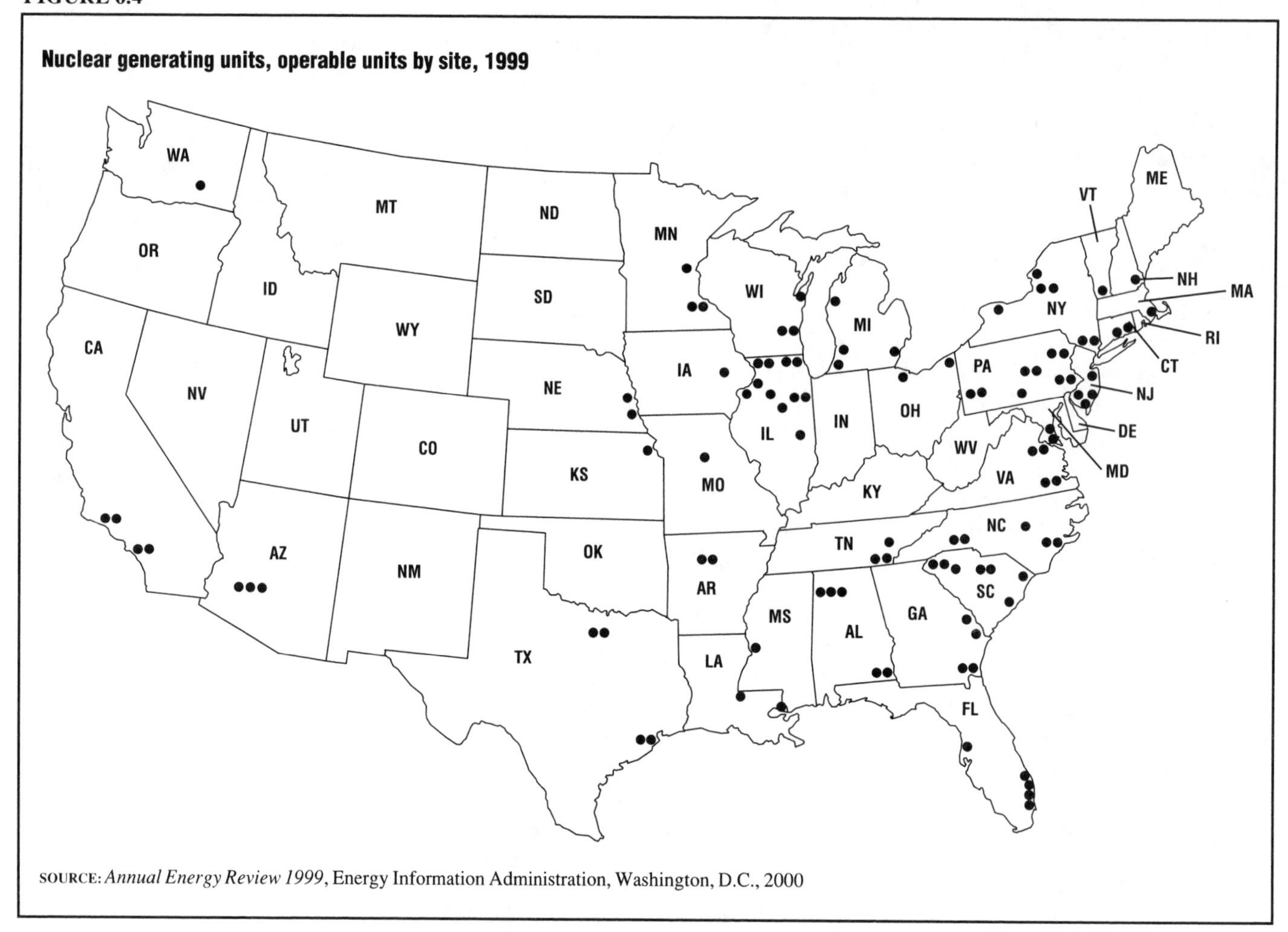

SOURCE: *Annual Energy Review 1999*, Energy Information Administration, Washington, D.C., 2000

FIGURE 6.5

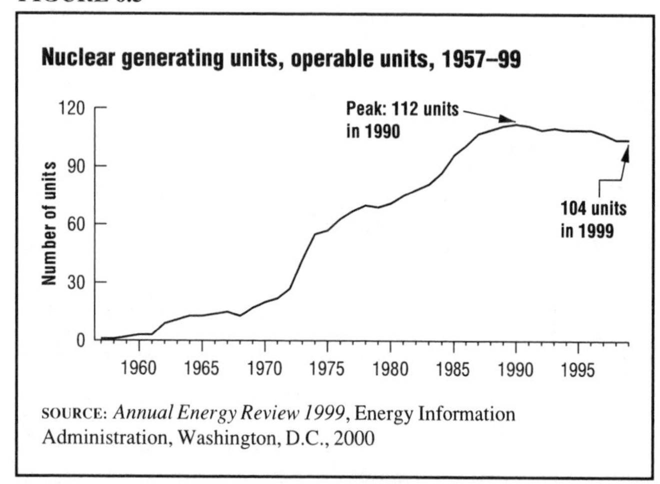

SOURCE: *Annual Energy Review 1999*, Energy Information Administration, Washington, D.C., 2000

for fossil fuels are enacted. If nuclear energy does not solve some of the problems plaguing the industry, nuclear output will decline considerably.

INTERNATIONAL PRODUCTION

Nuclear power provides about 12 percent of the world's electricity and 5 percent of total energy, levels unlikely to increase much in the future. Although the United States is the largest producer of nuclear power, it trails other Western countries in the proportion of national electrical production generated by nuclear power. That is primarily because oil, natural gas, and coal have been more accessible in the United States than in other countries.

In 1998 the United States led the world in nuclear power generation with 674 billion kilowatt-hours, followed by France's 367 billion kilowatt-hours and Japan's 318 billion kilowatt-hours. (See Table 6.3.) These three countries generated 59 percent of the world's nuclear electric power. France had, by far, the highest proportion (approximately 77 percent) of its electrical power produced by nuclear energy, followed by Belgium (56 percent), Sweden (49 percent), and Switzerland (36 percent).

As of the most recent international data available from the Energy Information Administration—from December 31, 1996—442 reactors operated worldwide, while 45 additional plants were under construction, 27 new units were planned, and 6 had been deferred. In Western Europe, France was the only country still building nuclear plants, with three units under construction. Russia and South Korea had seven reactors under construction, of which four began operating by 2000. India had six plants under construction, of which two began operating in 2000 and

TABLE 6.2

Nuclear generating units, 1953–99

Year	Orders [1]	Construction permits [2]	LPOL [3]	New Operable Units [4]	Shutdowns [5]	Total operable units [6]	Cancellations [7]	Cumulative cancellations
1953	1	0	0	0	0	0	0	0
1954	0	0	0	0	0	0	0	0
1955	3	1	0	0	0	0	0	0
1956	1	3	0	0	0	0	0	0
1957	2	1	1	1	0	1	0	0
1958	4	0	0	0	0	1	0	0
1959	4	3	1	1	0	2	0	0
1960	1	7	1	1	0	3	0	0
1961	0	0	0	0	0	3	0	0
1962	2	1	7	6	0	9	0	0
1963	4	1	3	2	0	11	0	0
1964	0	3	2	3	1	13	0	0
1965	7	1	0	0	0	13	0	0
1966	20	5	1	2	1	14	0	0
1967	29	14	3	3	2	15	0	0
1968	16	23	0	0	2	13	0	0
1969	9	7	4	4	0	17	0	0
1970	14	10	4	3	0	20	0	0
1971	21	4	5	2	0	22	0	0
1972	38	8	6	6	1	27	7	7
1973	42	14	12	15	0	42	0	7
1974	28	23	14	15	2	55	9	16
1975	4	9	3	2	0	57	13	29
1976	3	9	7	7	1	63	1	30
1977	4	15	4	4	0	67	10	40
1978	2	13	3	4	1	70	13	53
1979	0	2	0	0	1	69	6	59
1980	0	0	5	2	0	71	15	74
1981	0	0	3	4	0	75	9	83
1982	0	0	6	4	1	78	18	101
1983	0	0	3	3	0	81	6	107
1984	0	0	7	6	0	87	6	113
1985	0	0	7	9	0	96	2	115
1986	0	0	7	5	0	101	2	117
1987	0	0	6	8	2	107	0	117
1988	0	0	1	2	0	109	3	120
1989	0	0	3	4	2	111	0	120
1990	0	0	1	2	1	112	1	121
1991	0	0	0	0	1	111	0	121
1992	0	0	0	0	2	109	0	121
1993	0	0	1	1	0	110	0	121
1994	0	0	0	0	1	109	1	122
1995	0	0	1	0	0	109	2	124
1996	0	0	0	1	1	109	0	124
1997	0	0	0	0	2	107	0	124
1998	0	0	0	0	3	104	0	124
1999	0	0	0	0	0	104	0	124

[1] Placement of an order by a utility or government agency for a nuclear steam supply system.
[2] Issuance by regulatory authority of a permit, or equivalent permission, to begin construction. Numbers reflect permits issued in a given year, not extant permits.
[3] Low-power operating license: Issuance by regulatory authority of license, or equivalent permission, to conduct testing but not to operate at full power.
[4] Issuance by regulatory authority of full-power operating license, or equivalent permission.Units generally did not begin immediate operation.
[5] Ceased operation permanently.
[6] Total of units holding full-power licenses, or equivalent permission to operate, at the end of the year.
[7] Cancellation by utilities of ordered units. Does not include three units (Bellefonte 1 and 2 and Watts Bar 2) where construction has been stopped indefinitely.
R=Revised.
Note: Data are at end of year.

SOURCE: *Annual Energy Review 1999,* Energy Information Administration,Washington, D.C., 2000

two were expected to be operational by the end of 2000. Taiwan is currently constructing two large plants, and Argentina and Brazil are also building nuclear facilities.

Asia, excluding Japan, is the only region where nuclear power is expected to expand significantly, accounting for a projected 70 percent of the world's new nuclear capacity in the next two decades. In Japan, political support for nuclear power declined in 1996 after several accidents. It sank even further when, in 1999, the third-worst nuclear accident in history—behind Chernobyl and Three Mile Island—occurred at a uranium reprocessing facility in Tokaimura, Japan, seriously exposing 69 people to radiation. In South Korea, local opposition stalled construction on two reactors, and the government is still unable to find radioactive-waste facilities.

TABLE 6.3

World net nuclear electric power generation, 1989–98
(billion kilowatt–hours)

Region Country	1989	1990	1991	1992	1993	1994	1995	1996	1997	1998[1]
North America										
Canada	75.4	69.2	80.7	76.6	90.1	102.4	93.0	88.1	77.9	67.5
Mexico	0.0	2.8	4.0	3.7	4.7	4.0	8.0	7.5	9.9	8.8
United States	529.4	577.0	612.6	618.8	610.4	640.5	673.4	674.7	628.6	673.7
Total	**604.8**	**649.0**	**697.4**	**699.1**	**705.1**	**747.0**	**774.4**	**770.3**	**716.4**	**750.0**
Central & South America										
Argentina	4.8	7.0	7.7	6.7	7.3	7.8	7.1	6.9	7.5	7.1
Brazil	1.5	1.9	1.4	1.7	0.4	0.1	2.4	2.3	3.0	3.1
Total	**6.3**	**9.0**	**9.1**	**8.4**	**7.7**	**7.9**	**9.5**	**9.2**	**10.5**	**10.3**
Western Europe										
Belgium	39.1	40.6	40.7	41.3	39.8	38.6	39.3	41.2	45.0	43.9
Finland	18.0	18.3	18.5	18.3	18.9	18.5	18.3	18.5	19.0	20.8
France	288.7	298.4	314.8	321.5	349.8	342.0	358.4	377.5	374.3	366.7
Germany	—	—	140.1	150.9	145.8	143.2	145.4	152.0	161.8	152.7
Germany, East	11.1	5.3	—	—	—	—	—	—	—	—
Germany, West	140.4	139.8	—	—	—	—	—	—	—	—
Netherlands	3.8	3.3	3.2	3.6	3.8	3.8	3.8	4.0	2.3	3.6
Spain	53.7	51.6	52.8	53.0	53.3	52.5	52.7	53.5	52.5	56.0
Sweden	62.8	64.8	72.9	60.4	58.3	69.5	66.4	69.6	66.7	70.8
Switzerland	21.5	22.4	21.7	22.3	22.2	23.1	23.7	23.9	24.0	24.5
United Kingdom	63.6	58.7	62.8	69.1	81.0	80.0	80.6	85.8	89.3	97.7
Former Yugoslavia	4.5	4.4	4.2	—	—	—	—	—	—	—
Slovenia	—	—	—	3.8	3.8	4.3	4.5	4.4	4.8	5.0
Total	**707.3**	**707.5**	**731.6**	**744.1**	**776.6**	**775.4**	**793.0**	**830.3**	**839.9**	**841.7**
Eastern Europe & Former U.S.S.R.										
Bulgaria	14.6	13.5	12.4	11.0	13.3	14.6	16.4	17.8	16.4	15.5
Former Czechoslovakia	23.2	23.4	22.5	23.3	—	—	—	—	—	—
Czech Republic	—	—	—	—	12.0	12.3	11.6	12.2	12.5	12.5
Slovakia	—	—	—	—	11.6	11.5	10.9	11.3	10.5	9.8
Hungary	13.1	13.0	13.0	13.3	13.1	13.3	13.3	13.5	13.3	13.3
Romania	0.0	0.0	0.0	0.0	0.0	0.0	0.0	0.9	5.1	4.9
Former U.S.S.R	212.7	201.3	201.5	—	—	—	—	—	—	—
Armenia	—	—	—	0.0	0.0	0.0	0.0	2.1	1.4	1.4
Kazakhstan	—	—	—	0.5	0.4	0.4	0.1	0.1	0.3	0.1
Lithuania	—	—	—	13.9	12.3	7.3	10.6	12.7	10.9	12.9
Russia	—	—	—	113.6	113.2	92.9	94.3	103.3	104.5	98.3
Ukraine	—	—	—	70.1	71.4	65.4	67.0	76.0	75.4	70.6
Total	**263.5**	**251.3**	**249.5**	**245.6**	**247.3**	**217.7**	**224.3**	**249.8**	**250.3**	**239.3**
Africa										
South Africa	11.1	8.4	9.1	9.3	7.3	9.7	11.3	11.8	12.6	13.6
Total	**11.1**	**8.4**	**9.1**	**9.3**	**7.3**	**9.7**	**11.3**	**11.8**	**12.6**	**13.6**
Far East & Oceania										
China	0.0	0.0	0.0	0.5	2.5	13.5	12.4	13.6	11.4	13.5
India	3.8	5.6	5.2	6.0	5.9	4.7	6.5	7.4	10.5	10.6
Japan	174.5	192.2	202.8	212.1	236.8	255.7	276.7	287.1	306.1	318.1
Korea, South	45.0	50.2	53.5	53.7	55.2	55.7	63.7	70.2	73.2	85.2
Pakistan	0.1	0.4	0.4	0.5	0.4	0.6	0.5	0.3	0.4	0.4
Taiwan	27.1	31.6	33.5	32.5	33.0	33.5	33.9	36.3	34.8	35.1
Total	**250.5**	**279.9**	**295.4**	**305.3**	**333.8**	**363.6**	**393.6**	**415.0**	**436.4**	**462.8**
World Total	**1,843.4**	**1,905.1**	**1,992.0**	**2,011.8**	**2,077.8**	**2,121.3**	**2,206.0**	**2,286.4**	**2,266.1**	**2,317.7**

[1] Preliminary.
— =Not applicable.
(s) = Value less than 50 million kilowatthours.
Notes: Sum of components may not equal total due to independent rounding.
Generation data consist of both utility and nonutility sources. Data are reported as net generation as opposed to gross. Net generation excludes the energy consumed by the generating unit.
No generation is reported for Middle East.

SOURCE: *Annual Energy Review 1999,* Energy Information Administration, Washington, D.C., 2000

Russian officials have announced their intention to fire up 26 new nuclear reactors by 2010 in an effort to overcome a growing energy shortage. This plan, if implemented, would more than double the number of reactors operating in Russia, although whether the Russian government will reach this goal is questionable. And China may pass South Korea as the world's leading builder of nuclear reactors: official plans call for an increase from the current 2,100 megawatts of capacity to 20,000 megawatts by 2010.

Worldwide, 81 reactors have already been taken out of service. On average, they were in use less than 17 years, far fewer years than had been estimated. Dozens of larger plants will likely be shut down over the next few years. Because of new construction, worldwide nuclear generating capacity is projected to increase worldwide until 2010, after which it is expected to decline.

International Agreement on Safety

In September 1994, 40 nations signed the International Convention on Nuclear Safety, an agreement that requires them to shut down nuclear power plants if necessary safety measures cannot be guaranteed. The agreement applies to land-based civil nuclear power plants and seeks to avert accidents like the 1986 explosion at Chernobyl, the world's worst nuclear disaster. Ukraine, which inherited the Chernobyl plant after the collapse of the Soviet Union, signed the agreement. Signers must immediately submit reports on atomic installations and, if necessary, make improvements to upgrade safety at the sites. Neighboring countries may call for an urgent study if they are concerned about a reactor's safety and the potential fallout that could affect their own population or crops.

NUCLEAR ENERGY AND GLOBAL WARMING

For some, concern about the greenhouse effect has been a consideration in the development of nuclear power plants. The increases in greenhouse gases in the atmosphere are mainly caused by burning fossil fuels and by worldwide deforestation. Nuclear fuel emits no greenhouse gases and can substitute for fossil fuel electricity. Some critics suggest that the threat of climate change could lead to a truce between the nuclear power industry and environmentalists, longtime bitter enemies. However, most environmentalists argue for increased development of conservation practices and renewable energy sources.

Despite concerns about greenhouse gases, no nation has yet considered nuclear power as the complete solution to its energy needs. Sweden has legislated a nuclear phaseout by 2010 (although it is not clear that this is an obtainable goal, since about half of Sweden's electricity comes from nuclear power), while the Netherlands and Germany have strong conservation policies and well-organized antinuclear movements. Although other countries are concerned about global warming, their governments claim that, at this time, the evidence of global warming is not yet strong enough to offset the cost and safety concerns of nuclear power. France's huge dependency on nuclear energy is based on a scarcity of other energy resources and has nothing to do with concerns about global warming.

PROBLEMS WITH NUCLEAR POWER FACILITIES

Aging Plants

The premature aging of nuclear power plants poses a critical problem. Some plants that originally were expected to last 40 years or more showed serious levels of deterioration after as few as 15 years. In many cases, pipes cracked or suffered "wasting," a situation in which pipe walls become thinner with use. Fixing aging nuclear plants is an expensive proposition. Backfits, which are modifications made to existing plants to conform to new safety requirements, have been estimated to cost $90 million for each of the United States's 35 oldest plants.

Dismantling Nuclear Power Plants

Eventually, the more than four hundred nuclear plants now operating worldwide will need to be retired. The oldest commercial nuclear generating plant, located in Shippingsport, Pennsylvania, started generating electricity in 1957. In 1986 dismantling procedures began there, costing an estimated $100 million. The dismantling was expected to take about five years to complete but is still not finished.

There are three methods of retiring ("decommissioning") a reactor: mothballing, entombment, and dismantling. Mothballing involves removing the fuel, monitoring the radiation, and guarding the structure (which will be contaminated for centuries) to prevent anyone from entering it. Entombment, which was used at Chernobyl, involves removing the fuel and permanently encasing the structure in thick concrete. This process exposes more workers to radiation than the mothballing method. Dismantling, the method used at Shippingsport, is initially more costly but removes the long-term costs of monitoring both the structure and the radiation levels. It also frees the site for other uses, including, possibly, even the construction of another nuclear power plant. Of the three methods, dismantling involves the highest worker exposure to radiation.

The decommissioning process can take 60 years or longer, at an estimated price of $124 million to $205 million. An additional cost of decommissioning is the storage of spent nuclear fuel. The U.S. Department of Energy does not have an interim facility to store the fuel, and such storage may not be available for another decade. Storing spent fuel on site beyond the expiration of a plant's operating license, and after revenues are no longer generated, is costly.

Paying for the closing, decontaminating, or dismantling of nuclear plants has become an issue of intense public debate. The early closing of several plants owing to safety concerns and poor economic performance has raised the question of how to pay for decommissioning when the utility industry cannot. Industry reports contend that the cost of decommissioning retired plants and handling radioactive wastes will continue to escalate, causing serious financial problems for electric utilities. Most electric utilities have concluded that nuclear power is no longer competitive with other power sources. Not only coal plants but also new technologies, such as wind turbines and geothermal energy, are less expensive than nuclear energy plants when all factors are considered.

RETROFITTING FOR ALTERNATIVE USES. As nuclear plants are dismantled, officials are seeking alternative uses for the nuclear shells, including the conversion of old nuclear plants to gas-fired plants. When the Shoreham Nuclear Power Station was decommissioned in 1991, research began on ways to adapt turbines, power lines, and the control room into a gas-fired plant. In 1993, in a controversial action, nuclear fuel from the defunct Shoreham plant began being shipped to France for reprocessing, at a cost of $74 million to utility customers.

NUCLEAR SAFETY PROBLEMS

Grave problems have plagued the nuclear industry almost from the beginning. Plant site selections have been considered questionable, especially those built near earthquake fault lines. Internal quality control has often been lax. The Nuclear Regulatory Commission (NRC) has come under fire for failure to adequately follow up on the observations of so-called "whistle-blowers," those who report problems in plant construction or operation.

Peach Bottom

In 1987 the NRC shut down the Peach Bottom nuclear plant in Delta, Pennsylvania, because control-room operators were found sleeping on duty. This was the first plant to shut down solely because of operator violations and misconduct. Many industry observers note that attempts to improve nuclear plant safety focus mainly on technical design improvements and do not recognize "people problems," which are just as serious as the mechanical problems. A nuclear analyst and spokesman for Public Citizen, Ralph Nader's consumer advocacy group, observed that, while the focus of the NRC is primarily technological, human error has been responsible for almost all nuclear accidents. Even supporters of nuclear power recognize that many of the jobs in question can be tedious, although they take a great deal of training to master.

Three Mile Island

On March 28, 1979, the Three Mile Island nuclear facility, near Harrisburg, Pennsylvania, was the site of the worst nuclear accident in American history. Information released several years after the accident revealed that the plant came much closer to meltdown than either the NRC or the industry had previously indicated. Temperatures inside the reactor were first said to have been 3,500 degrees Fahrenheit, but are now known to have reached at least 4,800 degrees. The temperature needed to melt uranium dioxide fuel is 5,080 degrees Fahrenheit. When meltdown occurs, an uncontrolled explosion may result, unlike the controlled nuclear reaction of normal operation.

The emergency core cooling system at Three Mile Island was designed to dump water on the hot core and spray water into the reactor building to stop the production of steam, but during the accident, the valves leading to the emergency water pumps closed. Another valve was stuck in the open position, drawing water away from the core, which then became partially uncovered and began to melt. The emergency core cooling system then began drawing water out of the basement supply and reusing it, contaminating the reactor pump, although limiting the radiation contamination to the interior of the building.

Chernobyl

On April 26, 1986, the most serious nuclear accident ever occurred at Chernobyl, a four-reactor nuclear plant complex located in the former Soviet Union (now Ukraine) near Kiev. At least 31 people died, and hundreds were injured in the explosion. About 500 people were hospitalized, and medical experts estimate that 6,000–24,000 cancer-related deaths occurred over the years as a result of the released radiation.

The cleanup was one of the biggest projects of its kind ever undertaken. Helicopters dropped 5,000 tons of limestone, sand, clay, lead, and boron on the smoldering reactor to stop the radiation leakage and reduce the heat. Workers built a giant steel and cement sarcophagus (or stone coffin) to enshrine the remains of the reactor and contain the radioactive waste. Approximately 135,000 people were evacuated from a 300-square-mile area around the power station. Topsoil had to be removed in a 19-mile area and buried as nuclear waste. Buildings were washed down, and the newly contaminated water and soil carted away and buried. Agricultural products from areas nearby were declared unmarketable throughout Europe. Many of the 600,000 people involved in the immediate cleanup have suffered long-term effects of radiation exposure.

Some people at Chernobyl received 400 rems of radiation immediately following the explosion. (A "rem" is a standard measure of the whole-body dose of radiation. Under normal conditions, a person receives about one-tenth of a rem annually.) With a dose of 25 rems, a person's blood begins to change. The DNA is damaged, preventing the red and white cells from reproducing. Sickness starts at 100 rems and severe sickness at 200

FIGURE 6.6

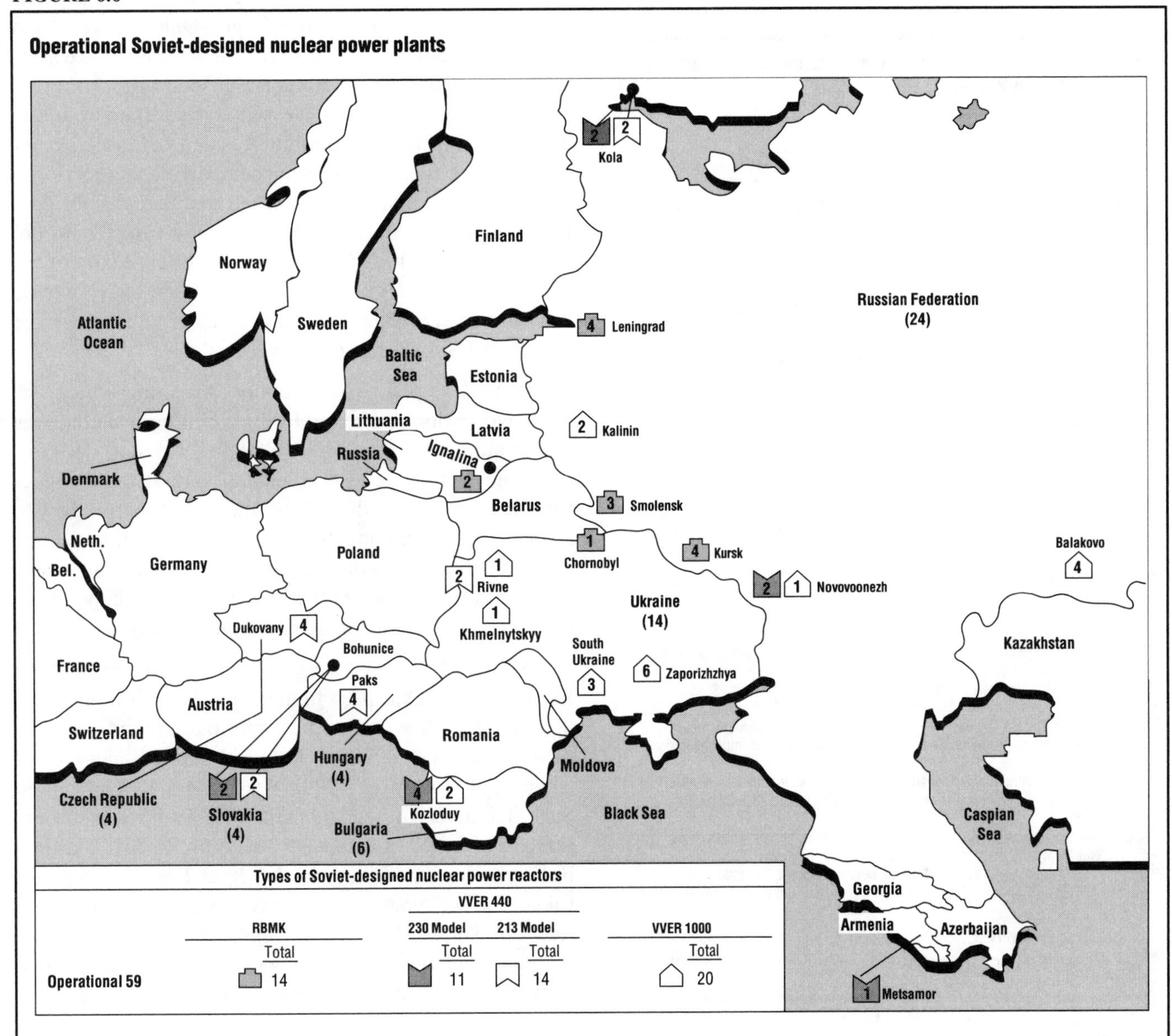

Notes:

1. Numbers in parentheses show the total number of reactors in each country, and numbers within symbols show the number of reactors of a specific type at a site.

2. DOE is providing assistance to five other nuclear power reactors in Russia. At one site, Bilibino, four small-scale RBMK reactors produce both steam and electricity. In addition, one fast-neutron reactor is located at Beloyarsk.

SOURCE: *Nuclear Safety: Concerns with the Continuing Operation of Soviet-Designed Nuclear Power Reactors*, U.S. General Accounting Office, Washington, D.C., April 2000

rems. Death of half the population occurs at 400 rems, and death to everyone within a week can be expected at 600 rems. One estimate is that 17.5 million people, including 2.5 million children under 7 years old, have had some significant exposure to radiation from Chernobyl.

In 1995 the United Nations reported that illnesses of all kinds were up 30 percent above normal in Ukraine; the incidences of depression, alcoholism, and divorce were also on the rise. Although the Chernobyl reactor is in Ukraine, Belarus, a neighboring nation of 10 million people, suffered more human and ecological damage from fallout because of prevailing winds. Thyroid cancers were 285 times pre-Chernobyl levels in Belarus, especially among children. Belarus's cabinet minister claimed that a quarter of his country's national income was being spent on alleviating the effects of the disaster. As many as 375,000 people in Belarus, Ukraine, and Russia have been displaced because of the accident. Contaminated forests spread radioactivity through fires, and seepage from the reactor is polluting waterways as far away as the Black Sea.

In April 1996 a fire in the woods near the Chernobyl reactor once more alarmed observers, who feared further

TABLE 6.4

Countries donating and receiving international nuclear safety assistance, as of November 1999
Dollars in millions

Pledged assistance			Received assistance	
Donor	**Amount**		**Recipient**	**Amount**
European Union	$753		Russia	$734
United States	532		Ukraine	629
Germany	168		Bulgaria	132
Japan	139		Lithuania	123
France	91		Regional/Unspecified[a]	58
United Kingdom	45		Slovak Republic	56
Sweden	44		Czech Republic	50
IAEA	43	→	Hungary	42
Canada	26		Armenia	32
Italy	25		NIS Regional[a]	26
Finland	18		Other[b]	35
Norway	14		Kazakhstan	15
Switzerland	13		**Total**	**$1,930**
Belgium	6			
Netherlands	5			
Denmark	4			
Spain	3			
EBRD	1			
Total	**$1,930**			

Notes:
1. Contributions to the Nuclear Safety Account are included in the amounts shown for each donor country.
2. The $532 million listed here as the U.S. contribution differs from the $545 million we identified because of the method used by the G-24 to classify projects and exchange rate variables. In addition, the G-24 data do not include amounts pledged by the United States for the Chornobyl Shelter Implementation Plan.
3. The cumulative contributions of Austria, the World Bank, and the Organization for Economic Cooperation and Development, which total less than $1 million, are not included in the figure.

[a] Nuclear Safety Account funds not yet allocated to specific recipients have been divided equally among these recipients.
[b] Other recipient countries include Azerbaijan, Belarus, Estonia, Georgia, Kyrgyzstan, Latvia, Moldova, Poland, Romania, Slovenia, and Uzbekistan.

SOURCE: *Nuclear Safety: Concerns with the Continuing Operation of Soviet-Designed Nuclear Power Reactors*, U.S. General Accounting Office, Washington, D.C., April 2000

danger to the entombed reactor and the release of radioactivity from soils and vegetation in the area. In addition, the sarcophagus built to contain the damaged reactor is reported to be crumbling. However, Ukraine, which suffers from 40 percent unemployment and other enormous economic woes, claims it needs the power from the remaining reactor and that completely shutting down Chernobyl would cost too much.

Some high technology companies have developed robots to clean up hazardous sites such as Chernobyl. Such efforts would, however, produce yet another problem: what to do with the nuclear waste once the Chernobyl sarcophagus is entered and cleaned up.

Problems with Soviet-Built Reactors

In April 2000 the U.S. General Accounting Office published a report (*Nuclear Safety: Concerns with the Continuing Operations of Soviet-designed Nuclear Power Reactors*) that underscored previous concerns about the danger of Soviet-built reactors. Of the 59 reactors in operation in the former Soviet Union, government officials stated that 25 reactors in Ukraine, Slovakia, Lithuania, Russia, and Bulgaria pose serious concerns because they fall below Western safety standards and can't be improved. Although Russian officials earlier claimed that the reactors were "comparable or superior to American designs," they admitted that, because of economic difficulty throughout the region, funds were not available to repair and staff some facilities. Many of the plants are old and scheduled for retirement, but economic pressures have kept them open. Figure 6.6 shows the location of Soviet-designed nuclear plants.

A nuclear plant in Metzamor, Armenia, which sits near two major and several minor geologic faults, was closed after an earthquake in 1988. In 1995, despite Western protests, the energy-starved nation restarted the reactor. In turning to the Metzamor reactor for power, Armenia is relying on a Soviet design that American scientists consider among the world's most dangerous. American scientists cite the lack of a dome-shaped containment vessel, which is standard on Western reactors.

Nuclear accidents are worldwide problems. Fallout from Chernobyl has been found in northern Europe and in the United States. The international community has pledged nearly two billion dollars in nuclear safety assistance as of November 1999. (See Table 6.4.) The United States has agreed to send more than $532 million in nuclear safety assistance to countries including Russia, Ukraine, Bulgaria, Hungary, Lithuania, the Czech Republic, and Slovakia. The U.S. goal for nuclear assistance is to increase the safety of old reactors without lengthening their operational lifetimes, and to find replacement sources of energy so that unsafe reactors can be shut down as quickly as possible.

Minami

On February 9, 1991, the Minami nuclear power plant in Mihama, Japan (220 miles west of Tokyo), almost suffered a meltdown when a pipe carrying superheated radioactive water broke in the 19-year-old plant. The water leaked, contaminating the water in the steam generator. A reported 20 tons of water were released before the plant shut down. After initially denying any radiation leak, the electric company said that an amount equal to about 8 percent of the plant's annual emissions had been released into the atmosphere. A sister plant was closed the next month because it had the same design flaw that existed at Minami. A storm of protest gave new life to the Japanese antinuclear movement.

The Japanese are in a difficult position concerning nuclear power. Japan has very few natural energy resources and is heavily dependent on Middle Eastern oil. The Japanese government foresaw nuclear power as providing increasing proportions of its electric energy, but there has been growing opposition from many Japanese people, including survivors of the nuclear bombings of Hiroshima and Nagasaki in World War II. With several of its biggest nuclear plants out of operation because of concern over safety issues, Japan's demand for electricity may challenge its generating capacity. In order to avoid that, Japan spends three times more than American industry on nuclear energy research.

A Retreat from Nuclear Power?

In December 1997 the U.S. Nuclear Regulatory Commission (NRC) fined Northeast Utilities $2.1 million for a host of violations of federal regulations at three nuclear reactors at the Millstone Nuclear Power Station in Waterford, Connecticut. The fine was nearly double the next-largest fine ever imposed by the NRC, and agency officials investigated criminal prosecution as well. (The previous largest fine of $1.25 million was assessed in 1988 against Philadelphia Electric Company's Peach Bottom plant.)

The commission had previously shut down the three Millstone reactors out of safety concerns, citing more than 50 violations from October 1995 to December 1996. Northeast Utilities will not be allowed to restart the reactors without a pending approval from an independent consultant and the NRC. Other plants have undergone reviews, but Millstone was the first in which the commission required the company to hire an independent consulting firm to examine it. Critics of the industry claimed the fine was not severe enough and that revocation of the license would have been a more appropriate penalty—and a message to the rest of the industry. Nevertheless, the fine was another blow to the already troubled nuclear industry.

Deregulation of the electric industry has raised questions about whether expensive nuclear power plants, which must comply with extensive government control, can compete. Since nuclear energy faces increased competition from cheaper energy sources, regulators worry that nuclear utilities may try to lower expenses by cutting corners on safety measures. The aging of the nation's reactors heightens the need for vigilance. Each year, maintenance problems grow more complex and expensive, raising supervision costs and requirements.

In 1998 Commonwealth Edison permanently closed 2 of its 12 nuclear reactors (in Zion, Illinois) because they were too expensive to operate under industry deregulation. The closing is estimated to cost $515 million. Spent fuel will be stored on site until 2014, when final decommissioning will begin. In addition to Zion, four other Commonwealth Edison plants are on the NRC's watch list. Industry experts predict Commonwealth Edison will be forced to close two of the other sites in order to focus on the remaining reactors.

THE PRICE-ANDERSON ACT—PAYING FOR A NUCLEAR ACCIDENT

Immediately after the Chernobyl incident in 1987, the Price-Anderson Act (PL95-256, amended), which limited how much a nuclear utility owner would have to pay to cover the costs of a nuclear accident, came up for renewal. At the time, the Price-Anderson Act required nuclear reactor owners to carry $650 million of off-site liability insurance, the amount at which the law limited their liability in the event of a nuclear accident. If damages from an accident exceeded $650 million, all nuclear plants could be charged up to $5 million per reactor to help cover the costs. After the Chernobyl accident, legislation was enacted to increase the liability cap to $7.4 billion.

NEW TECHNOLOGY?

Without new orders for nuclear plants, the U.S. nuclear industry could go out of business in the next few decades; the industry has had no new orders since 1978. Prompted by this grim future, suppliers have rethought plant design and reactor safety. In what government and industry officials hope will be a new era in nuclear power, they are now proposing smaller, standardized, and more simplified reactors that they claim will be 300 times safer than what current regulations require. These modular plants would take half as long to build, at barely a quarter of the cost of current designs. Designers claim that the plants will be serviced, in part, by robots and will rely on natural convection for emergency cooling, thus requiring fewer engineered safety features. The smaller plants will also use standardized parts that can be produced in commercial factories, eliminating the need for custom, on-site assembly.

NUCLEAR FUSION

On November 1, 1952, a thermonuclear bomb equivalent to 10 million tons of TNT was detonated at Eniwetok Atoll in the Pacific. That moment was both a nightmare and a dream. It showed the potential of even more destruction, as well as the possibility of generating cheap and abundant electricity by fusing hydrogen nuclei together. That dream of safe nuclear energy survives today—but just barely—as scientists have been experimenting with nuclear fusion as a potential major source of power for the 21st century.

Fusion is the process by which two atomic particles are joined together under certain conditions to release vast amounts of heat energy. In a fusion reaction, deuterium and tritium atoms fuse, or come together, creating helium and energetic neutrons. The neutrons escape through a chamber wall into a surrounding container, which absorbs heat that is used to create steam for generating electricity.

The closest natural example of fusion is the sun, where the temperature at the core is 14 million degrees centigrade. Because of extreme gravitational pressure, nuclei are driven so close together that they fuse and release vast energy. In a fusion reactor, the temperature must be even higher, around 200 million degrees centigrade, since there is less compression than at the center of the sun.

Fusion has several advantages over fossil fuels and nuclear fission power. Fusion does not create air pollutants that contribute to acid rain or global warming. Deuterium is available in essentially unlimited supply from seawater, and tritium can be generated on site as part of the fusion process. In theory, fusion could produce far more energy from the top two inches of Lake Erie than exists in all of Earth's known oil reserves.

Fusion experts believe the prospects for building fusion power plants that produce electricity in useful amounts are unlikely. In 1996 Congress slashed federal support for fusion research by one-third, to $244 million a year. The reductions cut into work at the Princeton Plasma Physics Laboratory, which operates the most powerful "magnetic confinement" fusion machine in the world: the Tokamak Fusion Test Reactor. (Tokamaks are donut-shaped vessels in which an element is compressed and heated to millions of degrees.)

No fusion reactor has, so far, produced more energy than it has used. The highest power ever reached, 10.6 million watts, was maintained for only a fraction of a second by the Princeton Tokamak in 1995. The Fusion Energy Advisory Committee, a panel of scientists that advises the DOE, recently concluded that the United States would have to join with the European Community, Japan, and Russia to jointly design an international experimental reactor to further test the concept, because only this collaboration would allow the U.S. to overcome the cost and complexity of the endeavor. Budget constraints could force the DOE to shut down the Tokamak Fusion Test Reactor in the future.

CHAPTER 7

NUCLEAR WASTE

Nuclear waste has sometimes been called the Achilles' heel of the nuclear power industry; much of the controversy over nuclear power centers on the lack of a disposal system for the highly radioactive spent fuel that must be regularly removed from operating reactors. As a result, progress on nuclear waste disposal is widely considered a prerequisite for any future growth of nuclear power.

— Congressional Research Service, August 1998.

A HISTORICAL PERSPECTIVE

In 1942 humanity's relationship with nature changed forever. Working in a laboratory in Chicago, Illinois, Italian physicist Enrico Fermi assembled enough uranium to cause a nuclear fission reaction. His discovery transformed both warfare and energy production. But his experiment also produced a small packet of radioactive waste materials that will remain dangerous for hundreds of thousands of years. That original waste lies buried under a foot of concrete and two feet of dirt on a hillside in Illinois. Scientists and governments have yet to find a way to dispose of this deadly residue or the many hundreds of thousands of tons that have since been generated.

Radioactive waste material is produced at all stages of the nuclear energy process, from the initial mining of the uranium to the final disposal of the spent fuel from the reactor. (Figure 7.1 shows the nuclear fuel cycle.) Nuclear waste is also a by-product of nuclear weapons plants, hospitals, and scientific research. Disposing of this waste is unquestionably one of the major problems associated with the development of nuclear power. Although federal policy is based on the assumption that nuclear waste can be disposed of safely, new storage and disposal facilities for all types of radioactive waste have frequently been delayed or blocked by concerns about safety, health, and the environment.

The highly toxic wastes must be isolated from the environment until the radioactivity decays to a safe level. In the case of plutonium, one of the most deadly substances known to man, the half-life (the time it takes for half of the atoms to disintegrate) is 26,000 years. At this rate, it will take at least 100,000 years before radioactive plutonium is no longer dangerous. Any facilities to store such materials must last virtually forever. The failure to handle deadly nuclear wastes properly has led to the pollution of water supplies and plant life in many areas. A massive cleanup operation will be required that may well test the capabilities of modern technology as well as the nation's financial resources.

From the 1940s, when the nation began to develop nuclear weapons, until the late 1980s, the predecessors of the U.S. Department of Energy (DOE) and the Nuclear Regulatory Commission (NRC) paid little attention to the environmental consequences of their activities. As a result, many DOE sites are now contaminated with radioactive and hazardous wastes, and the DOE faces the most complex cleanup task in the country, estimated to cost anywhere between $300 billion and $41 trillion. Congress is especially divided over this issue; nonetheless, when it comes to storing nuclear waste, NIMBY (not in my back yard) has been the position of most congressional representatives.

NUCLEAR WASTE DISPOSAL

Radioactive waste is a term that encompasses a broad range of material with widely varying characteristics. Some is barely radioactive and safe to handle, while other types are intensely hot in both temperature and radioactivity. Some waste decays to safe levels of radioactivity in a matter of days or weeks, while other types will remain dangerous for thousands of years. As illustrated in Figure 7.2, the DOE and the NRC define the major types of radioactive waste as follows:

- Uranium mill tailings are sand-like wastes produced in uranium refining operations. Although they emit

FIGURE 7.1

The nuclear fuel cycle

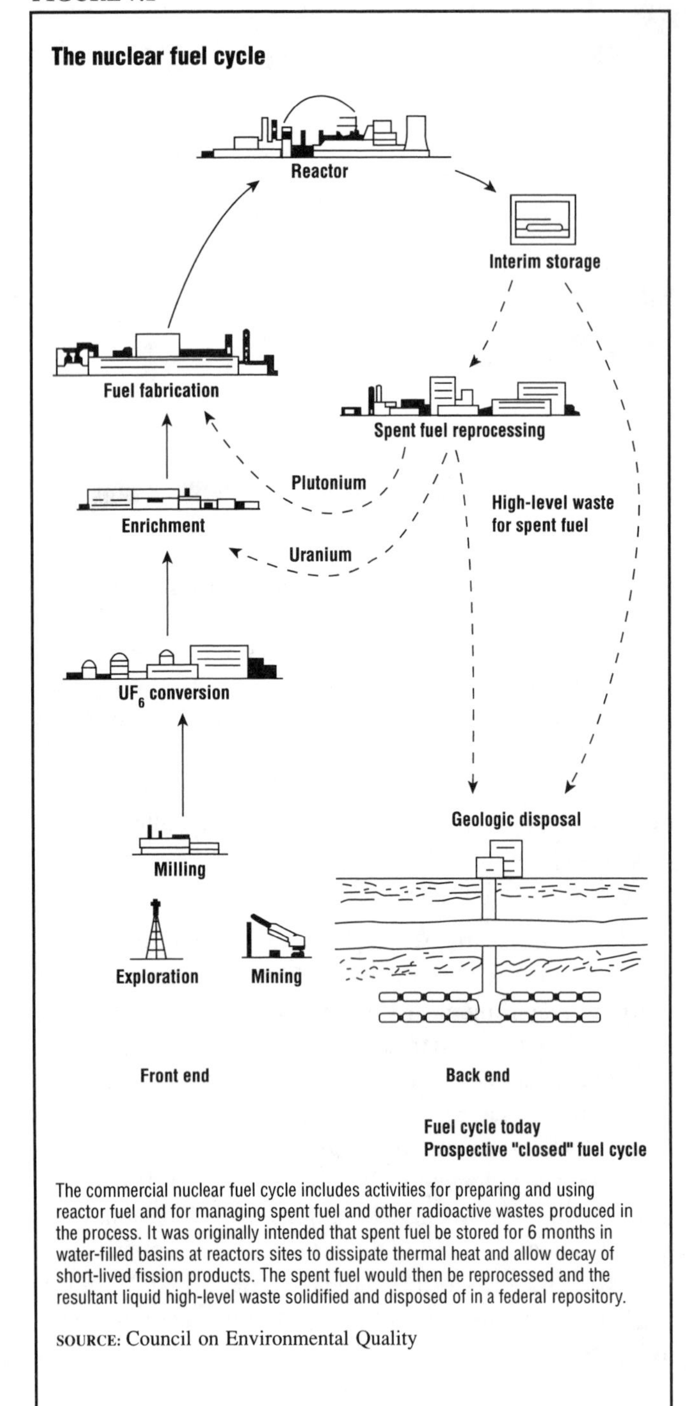

The commercial nuclear fuel cycle includes activities for preparing and using reactor fuel and for managing spent fuel and other radioactive wastes produced in the process. It was originally intended that spent fuel be stored for 6 months in water-filled basins at reactors sites to dissipate thermal heat and allow decay of short-lived fission products. The spent fuel would then be reprocessed and the resultant liquid high-level waste solidified and disposed of in a federal repository.

SOURCE: Council on Environmental Quality

low levels of radiation, their large volumes pose a hazard, particularly from radon emissions and groundwater contamination.

- Low-level waste contains varying lesser levels of radioactivity, and includes trash, contaminated clothing, and hardware. In general, low-level waste decays relatively quickly.
- Spent fuel is "used" reactor fuel that will be classified as waste if not reprocessed to recover the usable uranium and plutonium. In reprocessing, the used uranium and plutonium in spent reactor fuel can be removed for use again as nuclear reactor fuel. Spent fuel is the most radioactive type of civilian nuclear waste.
- High-level waste is the byproduct of reprocessing plants, containing highly toxic and extremely dangerous fission products. Although most of the uranium and plutonium has usually been removed for reuse, enough long-lived radioactive elements remain to require isolation for 10,000 years or more.
- Transuranic (TRU) wastes are 11 man-made radioactive elements with atomic numbers greater than that of uranium (92), with half-lives of thousands of years. They are found in refuse produced mainly by nuclear weapons plants and, therefore, are part of the nuclear waste problem but not directly the concern of nuclear power utilities.
- Mixed waste is high-level, low-level, or TRU waste that contains hazardous non-radioactive waste. Such waste poses serious institutional problems, because the radioactive portion is regulated by the DOE or NRC under the Atomic Energy Act, while the Environmental Protection Agency (EPA) regulates the non-radioactive elements under the Resource Conservation and Recovery Act (RCRA; PL 95-510).

Uranium Mill Tailings

These sand-like wastes emit low levels of radiation that can contaminate water and air. Most of these tailing sites are west of the Mississippi River, primarily in Utah, Colorado, New Mexico, and Arizona.

Prior to the early 1970s the tailings were believed to have such low levels of radiation that they were not harmful to humans. Miners, many of whom were Native Americans, received little protection from the radiation. Now many of these workers are reporting very high rates of cancer. Tailings were also left in scattered piles without warnings or safeguards, exposing anyone who came near. Some tailings were even used as landfill, and homes were built on top of them. Although too late for many, authorities now recognize that the handling and disposal of mill tailings must be properly managed to control radiation exposure.

Proper management of uranium mill tailings is particularly important, because they are generated in relatively large volumes—about 10 to 15 million tons annually. About 15 percent of the radioactivity is removed during the milling process, while the remainder (85 percent) stays in the tailings. Radium 226, the major radioactive waste product, retains its radioactivity for thousands of years and produces two potentially hazardous radiation components, gamma radiation and gaseous radon. There is a proven causal relationship between these radioactive elements and leukemia and lung cancer.

FIGURE 7.2

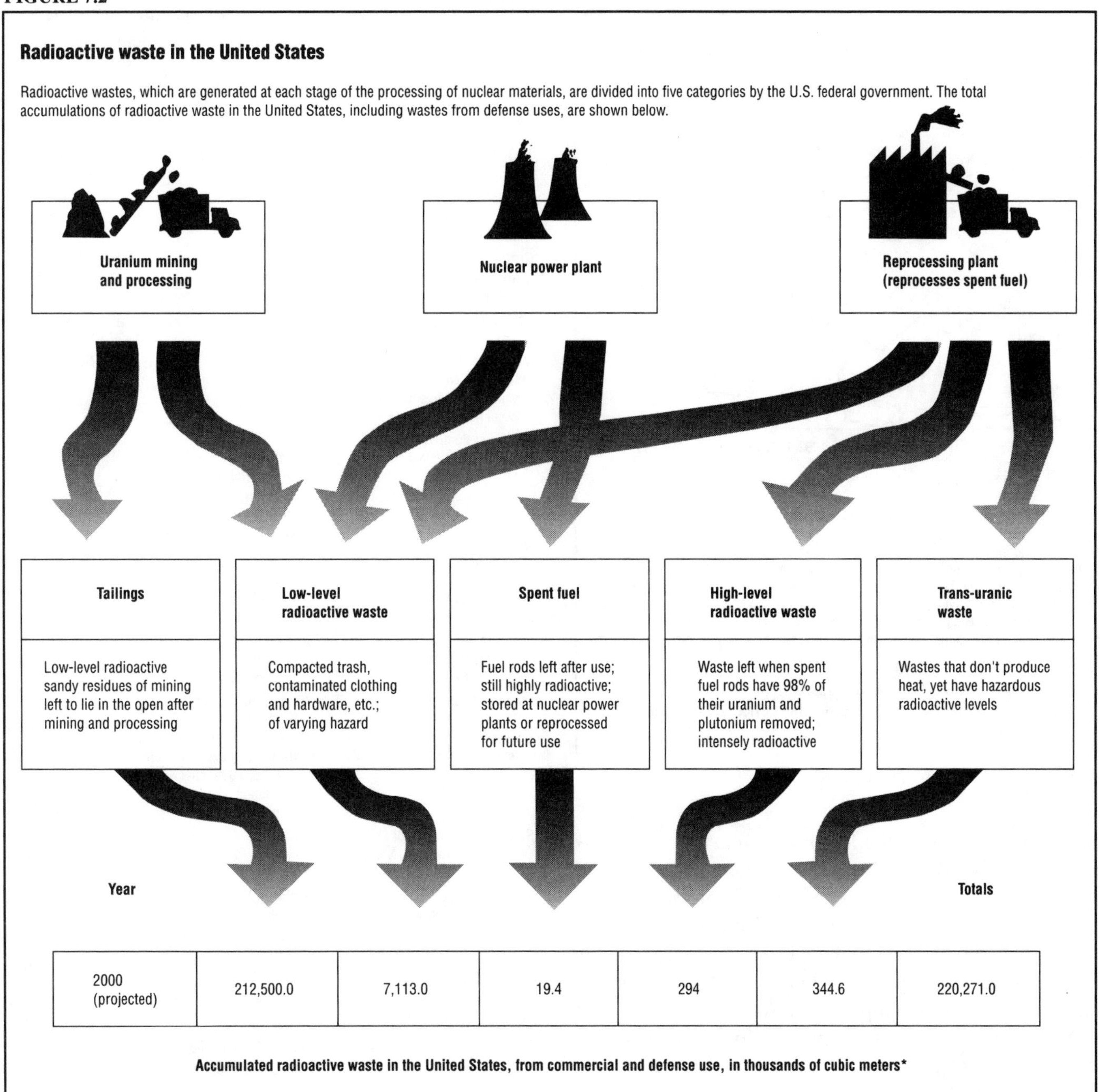

SOURCE: Natural Resources Defense Council; Worldwatch Paper 75: "Reassessing Nuclear Power: The Fallout from Chernobyl," by Christopher Revin, Worldwatch Institute, March 1987

In response to growing concern, Congress passed the Uranium Mill Tailing Radiation Control Act of 1978 (PL 95-604) to regulate mill tailing operations. The law called for the cleanup of abandoned mill sites, primarily at federal expense, although owners of still-active mines were financially responsible for their own cleanup. The legislation also required the EPA to prepare standards for the cleanup of both inactive and active sites, although it took the EPA until 1983 to issue such standards.

Low-Level Radioactive Waste

Low-level radioactive waste decays in 10 to 100 years. Until the 1960s the United States dumped low-level wastes into the ocean. The first commercial site to house such waste was opened in 1962, and six sites were licensed for disposal by 1971.

By 1979 only three commercial low-level waste sites were still operating—Hanford, Washington; Beatty, Nevada; and Barnwell, South Carolina. In response to the

FIGURE 7.3

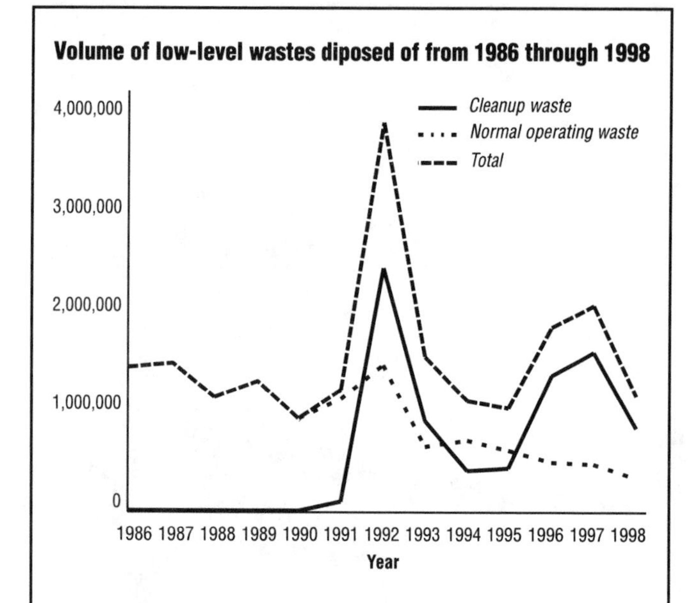

Note: The separation between operating wastes and cleanup wastes is not well defined. Operational wastes are generally defined as wastes that come from the clear power industry or from any entity that uses radioactive materials as part of an ongoing operation (even if that operation occurs only once every 2 years). Such wastes include materials like sludge and debris. In contrast, cleanup wastes are low-level radioactive wastes that have been contaminated by past activities; furthermore there is no longer any ongoing operation at the plant. Cleanup is a one-time event and, although the waste volume may be large, it is also very low in radioactivity.

SOURCE: *Low-Level Radioactive Wastes: States Are Not Developing Disposal Facilites,* U.S. General Accounting Office, Washington, D.C., September 1999

TABLE 7.1

Status of low-level radioactive waste disposal compacts and unaffiliated states

(dollars in millions)

State compacts (host state and state members)	Status of disposal siting efforts	Development costs
Appalachian compact (Pennsylvania, Delaware, Maryland, West Virginia)	Halted	$37.0
Central compact (Nebraska, Arkansas, Kansas, Louisiana, Oklahoma)	License application denied by Nebraska; Nebraska to withdraw from compact	95.6
Central Midwest compact (Illinois, Kentucky)	Halted	95.8
Midwest compact (No host state, Indiana, Iowa, Minnesota, Missouri, Ohio, Wisconsin)	Halted	Not available
Northeast compact (Dual hosts: Connecticut, New Jersey)	Connecticut: halted disposal facility siting, considering storage for 100 years or longer.	15.2
	New Jersey: halted siting effort	9.7
Northwest compact (Washington, Alaska, Hawaii, Idaho, Montana, Oregon, Utah, Wyoming)	Uses existing Richland disposal facility located on DOE's Hanford site	Not applicable
Rocky Mountain compact (No host state, Colorado, Nevada, New Mexico)	Contracted with Northwest compact to use the Richland facility	Not applicable
Southeast compact (North Carolina, Alabama, Florida, Georgia, Mississippi, Tennessee, Virginia	North Carolina halted licensing process for disposal facility, shut down its siting agency, and, on July 26, 1999, enacted legislation withdrawing from the compact	112.0
Southwestern compact (California, Arizona, North Dakota, South Dakota)	Halted	92.6
Texas compact (Texas, Maine, Vermont)	Halted, initial license application for original site denied by state's licensing authority	52.0
Unaffiliated states		
District of Columbia	No plans to site a facility	Not applicable
Massachusetts	Halted	Not available
Michigan	No efforts under way	12.6
New Hampshire	No plans to site a facility	Not applicable
New York	Halted	62.7
Puerto Rico	No plans to site a facility	Not applicable
Rhode Island	No plans to site a facility	Not applicable
South Carolina	Host state for Barnwell facility	Not applicable
Totals		**$585.2**

SOURCE: *Low-Level Radioactive Wastes: States Are Not Developing Disposal Facilities,* U.S. General Accounting Office, Washington, D.C., 1999

threatened closing of the South Carolina site, Congress passed the Low-Level Radioactive Waste Policy Act of 1980 (PL 96-573), calling for the establishment of a national system of such facilities. Each state would be responsible by 1986 for finding a low-level disposal site for wastes generated within its borders. The law also gave states the right to bar imports of low-level wastes if they were engaged in regional compacts for waste disposal. The disposal of high-level wastes, however, remains a federal responsibility.

The volume of low-level waste increased during the initial years (1963–80) of commercially generated waste disposal, until the Low-Level Radioactive Waste Policy Act. Since then, volume has decreased and the amount of radioactivity in the waste has remained steady. (See Figure 7.3.) The peak in the graph for 1992 reflects increased volumes due to cleanup efforts at nuclear facilities and sites.

COMPACTS. The 1980 waste law encouraged states to organize themselves into compacts to develop new low-level waste facilities. As of 1999, 10 compacts serving 44 states had been approved by Congress. (See Table 7.1.) Compacts and unaffiliated states have confronted significant barriers to developing disposal sites, including public health and environmental concerns, anti-nuclear sentiment, substantial financial requirements, political issues, and "not in my backyard" campaigns by some citizen activists.

No compact or state had successfully developed a new disposal facility for low-level wastes by 1999. California had planned a facility, but the land could not be obtained. Texas received federal approval for a site at

Sierra Blanca, which would receive waste from Maine and Vermont. The plan still requires the approval of the Texas Natural Resource Conservation Commission, however, and has been met with protests by Mexican officials. Certain conditions have led some states to remain uncommitted to disposal development and to consider other options. The reopening of the Barnwell, South Carolina, facility in 1995 eased some of the pressure on the states. Barnwell reopened for a 10-year period, taking waste from every state willing to pay sharply increased rates. The emergence in 1995 of new private sector nuclear waste handlers, Envirocare of Utah, Inc., has increased interest in the possibility of privately operated waste disposal facilities. Collectively the Barnwell, Hanford, and Envirocare facilities provided disposal capacity for almost all types of low-level wastes in 2000.

STORAGE PROBLEMS. Developing storage areas for low-level nuclear waste is difficult because regulatory requirements mandate a buffer zone of land surrounding each site. This acreage requires monitoring and limited land-use applications for at least a century. Although larger sites would collectively reduce the total number of acres required, some state officials believe that having more numerous local facilities would be safer by reducing the number of transportation accidents.

In addition, officials are concerned about degradation of the packages that contain stored waste. Depending on the environment, degradation of containers can occur from temperature fluctuations and corrosion. Some state officials worry that as the amounts of waste accumulate, with fewer sites for disposal, illegal dumping will increase. Finally, waste accumulation may lead to the reduction of nuclear health care and medical research to avoid adding to the waste problems.

Spent Fuel and High-Level Radioactive Waste

The most dangerous radioactive waste is irradiated uranium from commercial nuclear power plants. Approximately once a year, one-third of the nuclear fuel (uranium) inside a reactor is removed and replaced with fresh fuel. However, this leftover uranium is far from being completely "spent." It contains highly penetrating and toxic radioactivity and requires isolation from living things for thousands of years. It still contains significant amounts of uranium as well as plutonium created during the nuclear fission process. Spent fuel is a serious problem for nuclear power plants that will be decommissioned before a long-term, high-level waste disposal repository is available. (See Figure 7.4.) Unless a temporary site becomes available, decommissioned plants have the following options:

- Leave the fuel on site.
- Use on-site casks. This is not an option for hot fuel, which is fuel that is less than five years out of the core.
- Ship the spent fuel abroad for reprocessing. France, which is heavily dependent on nuclear power, developed the technology to reprocess spent fuel, something not available in the United States. In 1993 the British government also opened a nuclear fuel reprocessing plant that reprocesses spent fuel from nuclear power generators around the world. Nuclear watchgroups and some Americans fear that shipping spent fuel abroad will undermine efforts to halt the spread of nuclear arms because the process of transporting such materials increases the possibility for theft, in addition to increasing the risk of accidents.
- Ship the fuel to a monitored retrievable storage facility, if there is one available.
- Continue to operate the unit.

Only the natural decaying process, which can take hundreds of thousands of years, diminishes the radioactivity of high-level nuclear waste. The original "solution" was to bury the waste deep in the earth, but many scientists now believe that the deadly debris cannot be guaranteed to remain sealed off from the biosphere for hundreds of centuries. None of the options guarantee protection from radiation. Due to the scientific and political difficulties with geologic burial and other methods, above-ground "temporary" storage, despite the dangers, may remain the preferred option well into the twenty-first century.

The Vestiges of Nuclear Disarmament

Nuclear disarmament resulted in the dismantling of some of the United States' nuclear arsenal and created the need to store tons of plutonium. The federal government has proceeded to take apart as many as 15,000 warheads with intentions of eventually storing them at one of two former nuclear weapons-making plants: Pantex, near Amarillo, Texas, and Savannah River, South Carolina. More than 46,000 pounds of weapons-grade plutonium will be stored in above-ground bunkers by the time weapons disassembly is completed in 2003, although most experts believe the DOE is unlikely to meet this target year. After dismantling bombs, the government must then decontaminate buildings used at nuclear weapons facilities, dispose of millions of gallons of boiling radioactive water, and decontaminate hundreds of square miles at nuclear test sites.

The plutonium held by the DOE is in several forms: metals, oxides (fine powders), residues, solutions, and "pits," which are the cores of nuclear weapons. Six sites now hold the majority of the plutonium. (See Figure 7.5 and Table 7.2.) To date, government officials claim they can account for every ounce of plutonium that has been removed from nuclear weapons.

Plutonium will be disposed of in one of two ways: immobilization in glass or ceramic containers (called

FIGURE 7.4

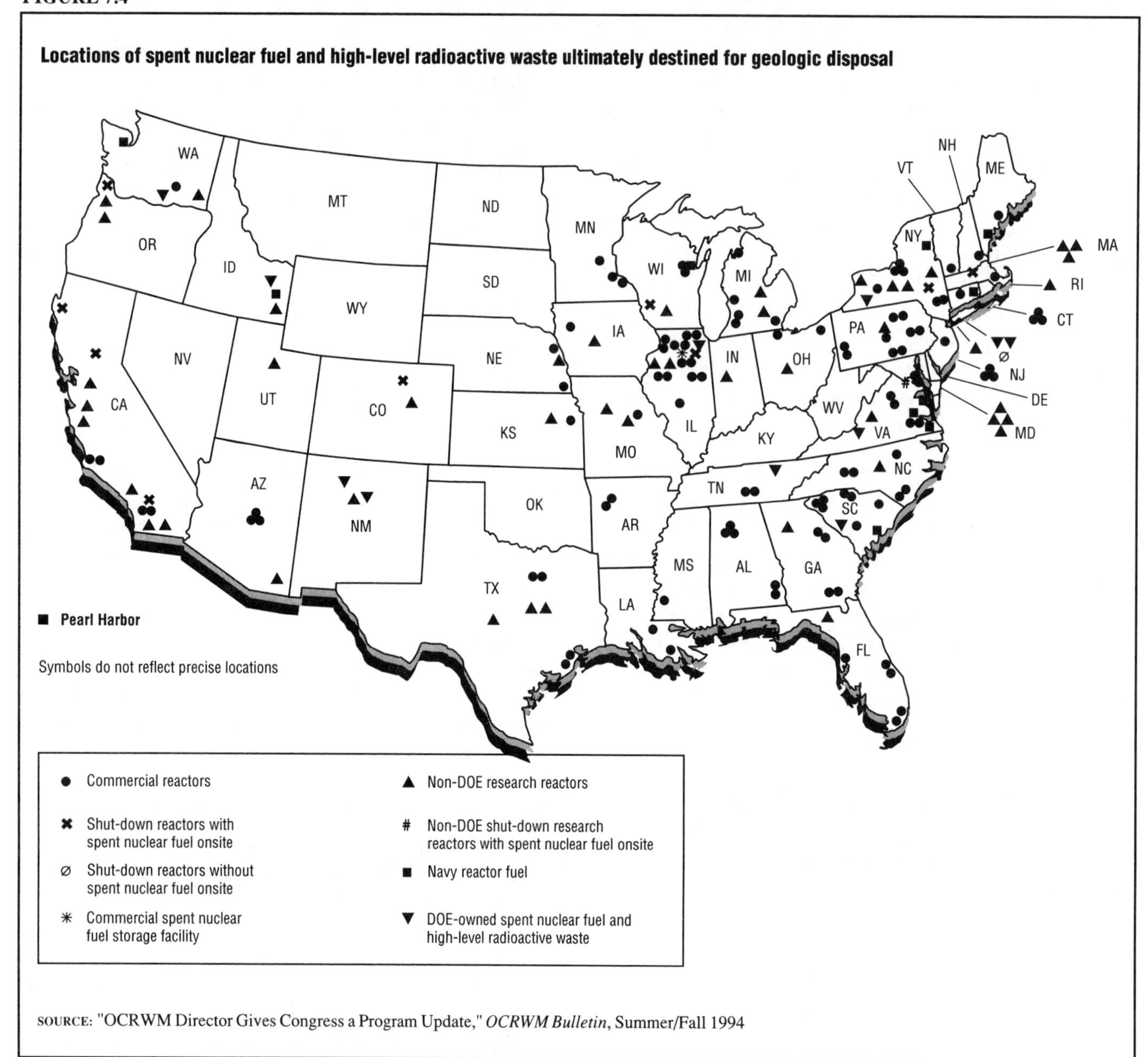

SOURCE: "OCRWM Director Gives Congress a Program Update," *OCRWM Bulletin*, Summer/Fall 1994

TABLE 7.2

Plutonium inventory, by form, at five DOE sites with plutonium stabilization activities
(metric tons)

Site	Total plutonium inventory[a]	Plutonium metals	Plutonium oxides	Plutonium residues	Plutonium solutions
Rocky Flats Environmental Technology Site	12.7	6.5	1.6	4.5	0.1
Hanford Site	3.5	0.8	2.0	0.4	0.3
Savannah River Site	1.8	0.7	0.7	0.1	0.3
Los Alamos National Laboratory	2.5	1.2	0.0	1.3	0.0
Lawrence Livermore National Laboratory	0.3	0.1	0.1	0.1	0.0
Total	**20.8**	**9.3**	**4.4**	**6.4**	**0.7**

[a]Inventory amounts are as of 1994; any updated amounts would be classified information. Amounts exclude spent nuclear fuel, reactor fuel, and special isotopes of plutonium.

SOURCE: *Department of Energy: Problems and Progress in Managing Plutonium,* U.S. General Accounting Office, Washington D.C., 1998

FIGURE 7.5

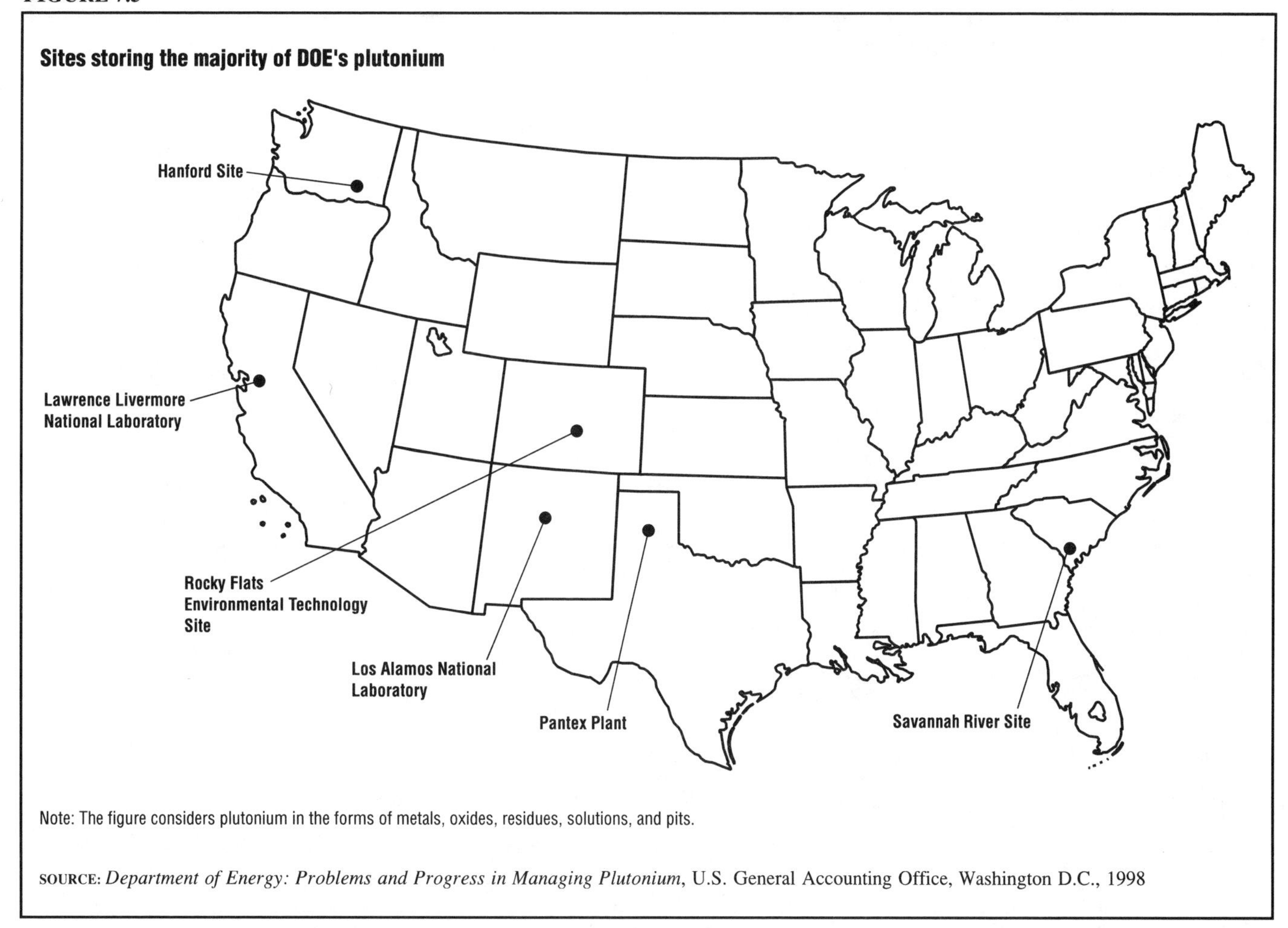

Sites storing the majority of DOE's plutonium

Note: The figure considers plutonium in the forms of metals, oxides, residues, solutions, and pits.

SOURCE: *Department of Energy: Problems and Progress in Managing Plutonium*, U.S. General Accounting Office, Washington D.C., 1998

"vitrification"), or burning as a mixed oxide fuel (MOX). Immobilization is the preferred method among nuclear safety watchdogs who fear increased nuclear waste through burning. MOX fuel could be used in existing reactors, a process used in Europe but never tried in the United States. The Clinton Administration preferred the dual-track approach, using both options.

NUCLEAR WASTE LEGISLATION

Nuclear Waste Policy Act of 1982

A major step toward shifting the responsibility for disposal of high-level nuclear wastes from the nuclear power industry to the federal government was taken in 1982. Congress passed the Nuclear Waste Policy Act (PL 97-425), which provided the first national, comprehensive policy and detailed timetable for the management and disposal of high-level nuclear waste. The major provisions of the act were to:

- Authorize construction of the first nuclear waste repository.
- Provide a schedule for the site selection and operation.
- Define the means of achieving cooperation with the host state.
- Assure funding for the program by charging 1 mill (0.1 cent) per kilowatt-hour of electricity generated by nuclear energy (this measure has been in effect since 1983, adding between 2 and 6 percent to the consumer's cost of electricity).
- Provide for the president to decide if civilian repositories may accept nuclear waste from military activities.
- Establish an interim storage program to ease the backlog of spent nuclear fuel at power plants.
- Direct the DOE to design a monitored retrievable storage program for long-term storage of spent fuel.

Congressional Amendments in 1987

In 1987 Congress amended the Nuclear Waste Policy Act as part of the Omnibus Budget Reconciliation Act (PL 100-203). Because costs of the program continued to escalate (from $23 to $30 billion), Congress directed the DOE to investigate only the Yucca Mountain site in Nevada (see below) as a potential site for the first repository. Congress limited the amount of waste that could be dis-

TABLE 7.3

TRU waste storage locations and pre-treatment volumes
(cubic meters)

Site	Location	Contact-handled TRU waste		Remote-handled TRU waste	
		Stored*	Projected through 2033**	Stored*	Projected through 2033**
Argonne National Laboratory-East	Argonne, IL	94	109	0	0
Hanford Reservation	Richland, WA	16,127	7,305	200	1,582
Idaho National Engineering and Environmental Laboratory	Idaho Falls, ID	64,575	15,009	86	53
Lawrence Livermore National Laboratory	Livermore, CA	297	835	0	0
Los Alamos National Laboratory	Los Alamos, NM	8,255	8,544	101	128
Mound Plant	Miamisburg, OH	241	6	0	0
Nevada Test Site	Nevada	618	19	0	0
Oak Ridge National Laboratory	Oak Ridge, TN	917	180	1,268	100
Rocky Flats Environmental Technology Site	Golden, CO	1,505	6,988	0	0
Savannah River Site	Aiken, SC	11,725	17,811	1	21
Small-Quantity Sites					
Ames Laboratory	Ames, IA	0	<1	0	0
ARCO Medical Products Company	West Chester, PA	<1	<1	0	0
Babcock & Wilcox - NES	Lynchburg, VA	20	0	0	0
Battelle Columbus Laboratories	Columbus, OH	0	0	0	369
Bettis Atomic Power Laboratory	West Mifflin, PA	0	114	0	2
Energy Technology Engineering Center	Santa Susana, CA	7	0	0	1
General Electric-Vallecitos Nuclear Center	Pleasanton, CA	6	3	8	5
Knolls Atomic Power Laboratory	Niskayuna, NY	0	0	<1	5
Lawrence Berkeley Laboratory	Berkeley, CA	<1	4	0	0
Missouri University Research Reactor	Columbia, MO	<1	1	0	0
Paducah Gaseous Diffusion Plant	Paducah, KY	2	0	0	0
Sandia National Laboratories	Albuquerque, NM	7	44	1	3
U.S. Army Material Command	Rock Island, IL	2.5	0	0	0
Total Waste Volumes* **		104,400	56,972	1,666	2,268

*Volumes prior to treatment and repackaging.
**Projected volumes include estimates from environmental restoration, decontamination and decommissioning, and future Departmental missions, for example, the disposition of weapons-useable plutonium at the Savannah River Site. Estimates will change based upon future compliance actions under environmental law.
***Totals reflect rounding of numbers.

SOURCE: *The National TRU Waste Management Plan,* U.S. Department of Energy, Carlsbad Area Office, Carlsbad, NM, December1997

posed of in the repository until a second repository was made ready.

The DOE was also authorized to develop a facility to temporarily receive and store waste until the second repository was built. In 1987 the DOE proposed developing a facility in Tennessee for temporary waste storage to begin accepting waste in 1998. Congress authorized the plan, but it cannot go into effect until the Nuclear Regulatory Commission (NRC) has authorized the construction of the Yucca Mountain repository, which is still pending.

FEDERAL NUCLEAR WASTE REPOSITORIES

The U.S. government is focusing on two locations as eventual long-term nuclear waste repositories: the Waste Isolation Pilot Plant (WIPP) in southeastern New Mexico for transuranic waste, and Nevada's Yucca Mountain for civilian waste.

The Waste Isolation Pilot Plant (WIPP)

The Waste Isolation Pilot Plant (WIPP) became the world's first deep depository for nuclear waste when it received its first shipment of waste on March 26, 1999. This large facility is located near Carlsbad, New Mexico. The WIPP is 655 meters below the earth's surface in the salt beds of the Salado Formation. It is intended to house up to 6.25 million cubic feet of transuranic waste for more than 10,000 years.

Under congressional mandate, the WIPP facility will only accept transuranic waste and not commercial or high-level waste. In 2000 more than 99 percent of existing transuranic waste was being temporarily stored in drums at nuclear defense sites in California, Colorado, Idaho, Illinois, Nevada, New Mexico, Ohio, Tennessee, South Carolina, and Washington. (See Table 7.3.) So far, the Energy Department has authorized the plant to receive waste from the Los Alamos National Laboratory in New Mexico, from the Idaho National Engineering and Environmental Laboratory, and from Rocky Flats, a former nuclear trigger factory near Denver. The waste is tracked by satellite and is moved only at night when traffic is lighter to avoid accidents. It can be transported only in good weather and must be routed around major cities.

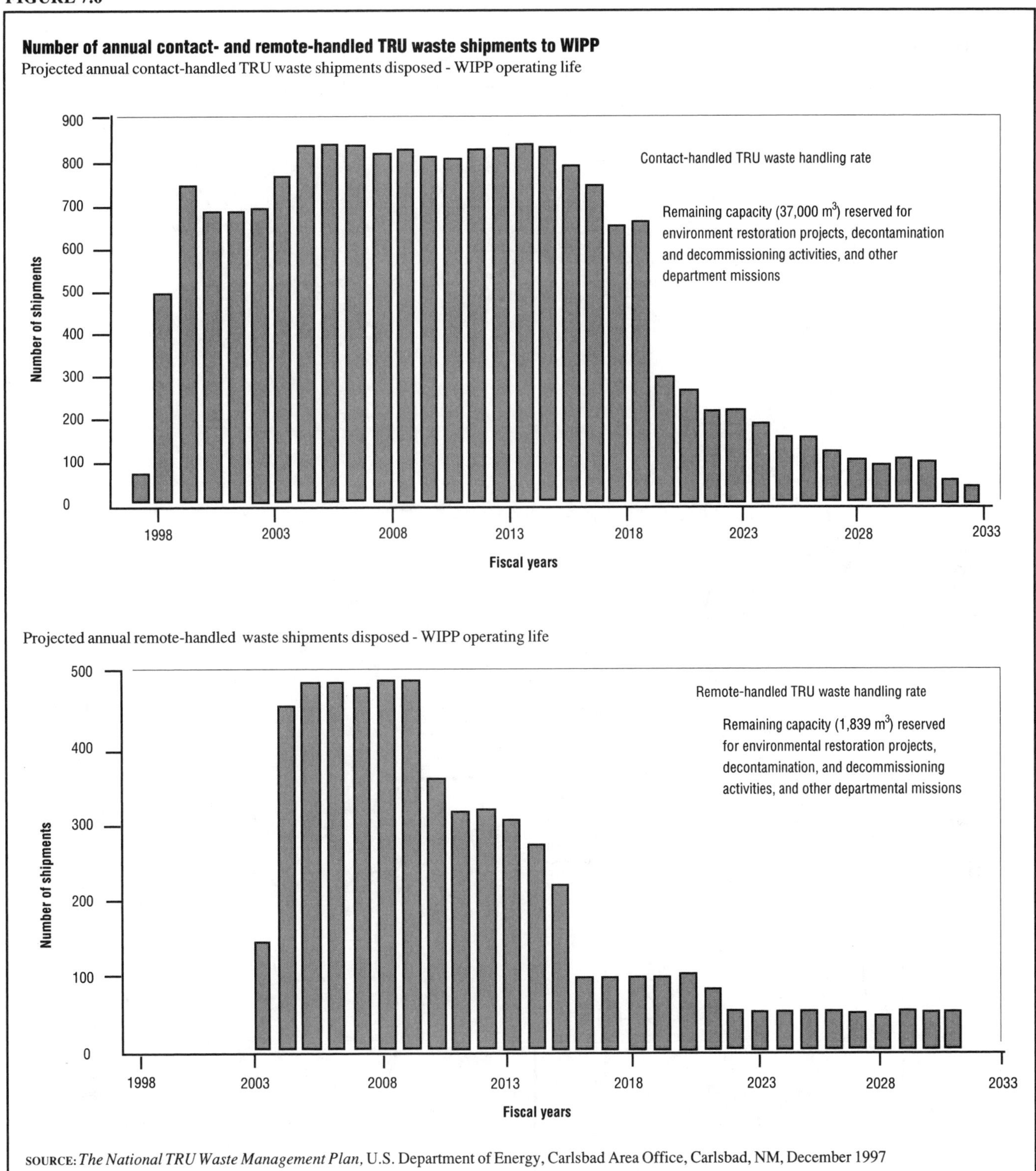

FIGURE 7.6

Number of annual contact- and remote-handled TRU waste shipments to WIPP

Projected annual contact-handled TRU waste shipments disposed - WIPP operating life

Projected annual remote-handled waste shipments disposed - WIPP operating life

SOURCE: *The National TRU Waste Management Plan,* U.S. Department of Energy, Carlsbad Area Office, Carlsbad, NM, December 1997

By the beginning of the twenty-first century about 61 million Americans lived within 50 miles of a military nuclear waste storage site. By the time the WIPP has been in operation for 10 years, the number should drop to 4 million. By 2035, barring court challenges, almost 40,000 truckloads of nuclear waste will be trucked across the country to the WIPP. Figure 7.6 shows the planned shipments to the WIPP by year.

Yucca Mountain

The centerpiece of the federal government's plan to dispose of highly radioactive waste is a proposed facility at Yucca Mountain in Nevada. The Nuclear Waste Policy Act of 1982 requires the secretary of energy to investigate the site and, if it is suitable, recommend to the president that the site be established. The investigation of Yucca Mountain is taking a long time. If scientists find Yucca

FIGURE 7.7

Yucca Mountain waste site

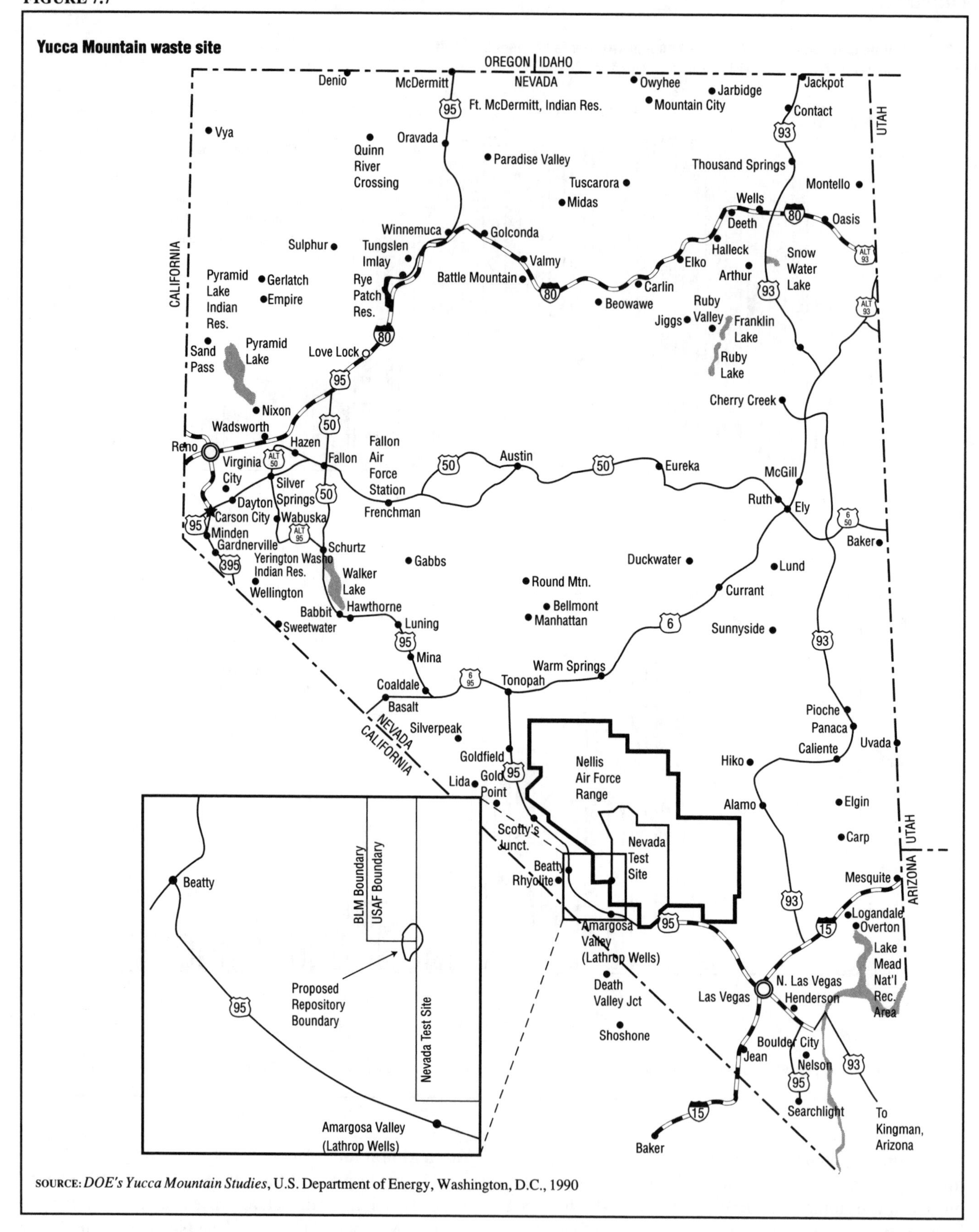

SOURCE: *DOE's Yucca Mountain Studies*, U.S. Department of Energy, Washington, D.C., 1990

Mountain is suitable as a deep geologic repository, the Department of Energy will apply for a repository license with the Nuclear Regulatory Commission. Repository construction cannot begin until the commission grants a construction authorization. The earliest date that the department could begin building the repository is 2005. It

FIGURE 7.8

Drawing of the proposed repository facilities on the surface and underground

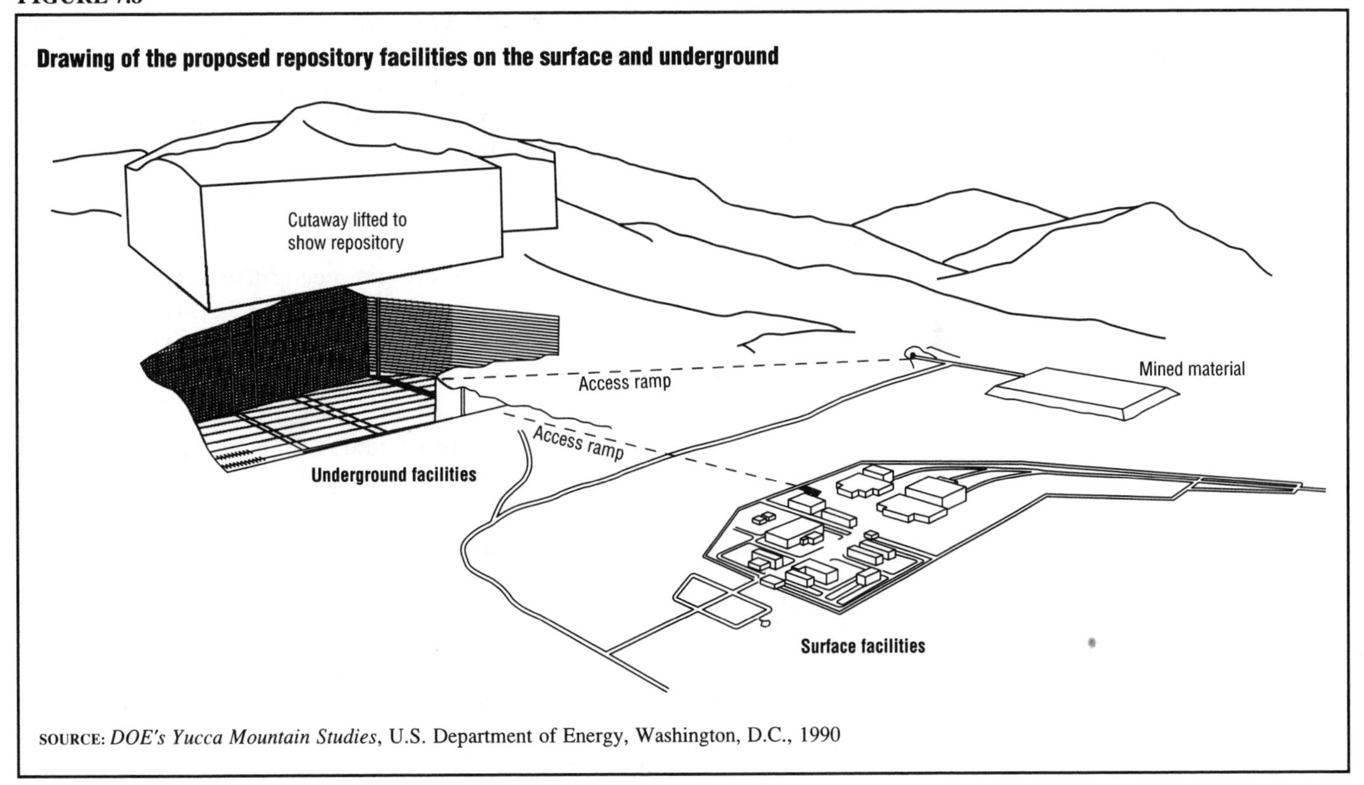

SOURCE: *DOE's Yucca Mountain Studies*, U.S. Department of Energy, Washington, D.C., 1990

would take another five years to complete construction and begin operations, 12 years later than originally expected.

Yucca Mountain is a flat-topped ridge, running six miles from north to south, that has changed little over the past million years. The site has a desert climate, important because water movement is the primary means by which radioactive waste could be transported from a repository into the environment. The repository would be built about 1,000 feet below the surface and 1,000 feet above the water table. (See Figure 7.7.)

It is not clear whether the Yucca Mountain site is suitable. It is located near volcanic and earthquake activity, and scientists are concerned by recent studies that found the area more geologically unstable than previously thought. Scientists have also discovered areas of "perched water" above the water table that could be affected by the facility. Although these areas may not ultimately prove to be a problem, researchers must determine their size and number in order to know for certain. In addition, local American Indian tribes are contesting the rights to the land, and some western states feel they have long been targeted for hazardous facilities. If the site is found to be acceptable, the president approves it, and a recommendation goes to Congress, the State of Nevada is expected to file a notice of disapproval.

The current schedule calls for the secretary of energy to decide in 2001 whether to recommend the site. If approval is granted at each step and there are no delays in the process, waste placement could begin in 2010. In December 1998 Secretary of Energy Bill Richardson submitted a Viability Assessment to the president and Congress. Secretary Richardson found no "show stoppers" at Yucca Mountain. The report concluded that scientific and technical work should proceed to support a decision by the secretary of energy in 2001 on whether to recommend the site to the president. The cost to license, construct, operate, monitor, and close the repository is estimated at $18.7 billion in 1998 dollars. This cost includes monitoring the repository for 100 years and disposing of 70,000 metric tons of spent nuclear fuel and high-level waste, currently the legal limit of what can be disposed.

The proposed repository would look like a large mining complex. It would have facilities on the surface for handling and packaging nuclear waste and a large mine about 1,000 feet underground. Plans call for the waste to be placed in sealed metal canisters placed vertically in the floor of underground tunnels. Aboveground facilities would cover approximately 400 acres and be surrounded by a three-mile buffer zone. Underground, about 1,400 acres would be mined, consisting of tunnels leading to the areas where waste containers would be placed and service areas near the shafts and ramps that provide access from the surface. (See Figure 7.8.)

STANDARDS FOR CONTAINMENT. In order for the Yucca Mountain Repository to be built, the DOE must satisfactorily demonstrate to the NRC that the combination of the site and the repository design complies with the standards set forth by the EPA. The EPA's standard is based on an approach of using numerical probabilities to

FIGURE 7.9

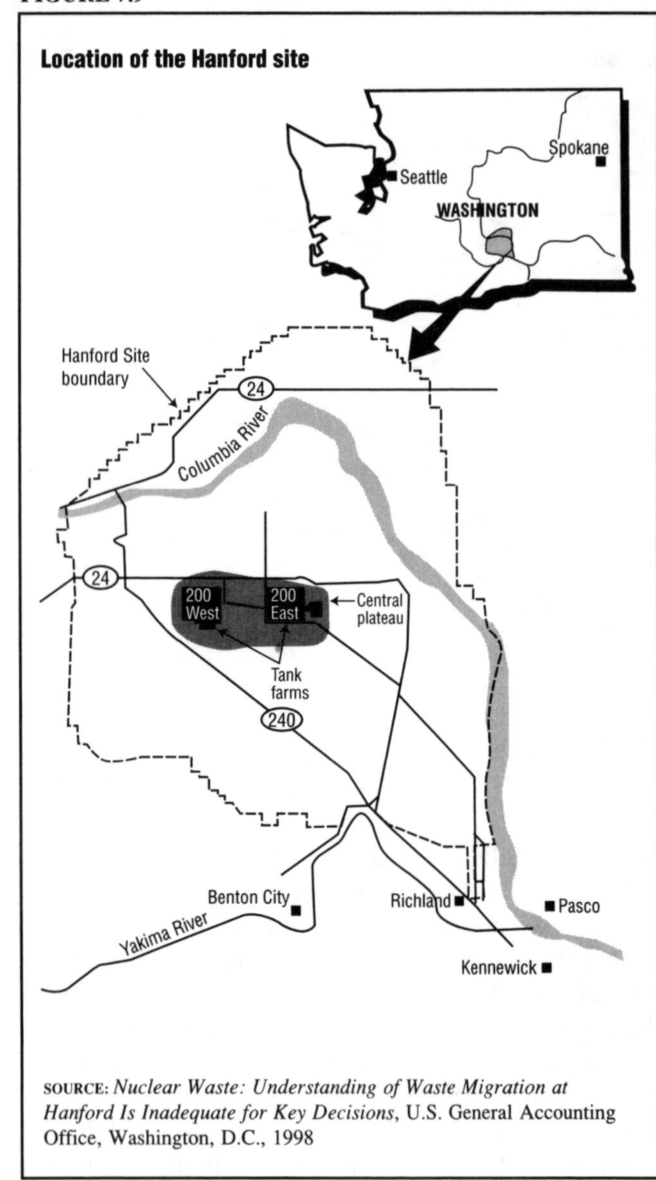

SOURCE: *Nuclear Waste: Understanding of Waste Migration at Hanford Is Inadequate for Key Decisions*, U.S. General Accounting Office, Washington, D.C., 1998

establish requirements for containing radioactivity within the repository. Their quantitative terms are:

- Cumulative releases of radioactivity from a repository must have a likelihood of less than one chance in ten of exceeding limits established in the standard, and a likelihood of less than one chance in 1,000 of exceeding ten times the limit, for a period of 10,000 years.
- Exposures of radiation to individual members of the public for 1,000 years must not exceed specified limits.
- Limits are placed on the concentration of radioactivity for 1,000 years after disposal from the repository to a nearby source of groundwater that 1) currently supplies drinking water for thousands of persons and 2) is irreplaceable.
- Prescribed technical or institutional procedures or steps must provide confidence that the containment requirements are likely to be met.

WHAT HAPPENS WHEN THE YUCCA MOUNTAIN REPOSITORY IS FULL? The repository would be designed to contain radioactive material by using layers of man-made and natural barriers. Regulations require that a repository isolate waste until the radiation levels decay to a level that is about the same as that from a natural underground uranium deposit. This decay time is estimated at about 10,000 years.

After the repository has been filled to capacity, regulations require the DOE to keep the facility open and to monitor it for at least 50 years from the fill date. This will allow experts to monitor conditions inside the repository and retrieve spent fuel if necessary. Eventually, the repository shafts will be filled with rock and earth and sealed. At the ground level, facilities will be removed, and to the extent possible, the DOE will take steps to return the site to its original condition.

Scientists assume that over thousands of years, some of the man-made barriers in a repository will break down. Once that happens, natural barriers will be counted on to stop or slow the movement of radiation particles. The most likely way for particles to reach humans and the environment would be through water, which is why water tables are of such concern to scientists. Yucca Mountain has certain chemical properties that act as another barrier to the movement of radioactive particles. Minerals in the rock called "zeolites" would stick to the particles and slow their movement throughout the environment.

Serious Leaks of Radioactive Waste at Government Facilities

HANFORD. In 1997 scientists discovered that about 900,000 gallons of radioactive waste had leaked into the soil from 68 of the 149 tanks at the nuclear weapons plant in Hanford, Washington. (See Figure 7.9.) Eventually all the tanks are expected to leak. The leak contaminated underground water moving toward the nearby Columbia River. (See Figure 7.10.) Managers at the plant maintained that the leaks were insignificant because radioactive materials would be trapped by the area above the water table called the "vadose zone." Furthermore, officials had been saying for decades that no waste from the tanks would reach the groundwater in the next 10,000 years.

Nonetheless, the groundwater under more than 85 square miles of the site is already contaminated. Washington's Governor Gary Locke called it a "Chernobyl waiting to happen." A threatened lawsuit by the State of Washington against the U.S. DOE over the leaks at the Hanford site resulted in an agreement to clean up the two indoor pools near the Columbia River by 2007.

BROOKHAVEN NATIONAL LABORATORY. Brookhaven National Laboratory (BNL) is a federally funded research facility, owned by the U.S. Department of Energy, on

FIGURE 7.10

Waste migration through the Vadose Zone at the Hanford site

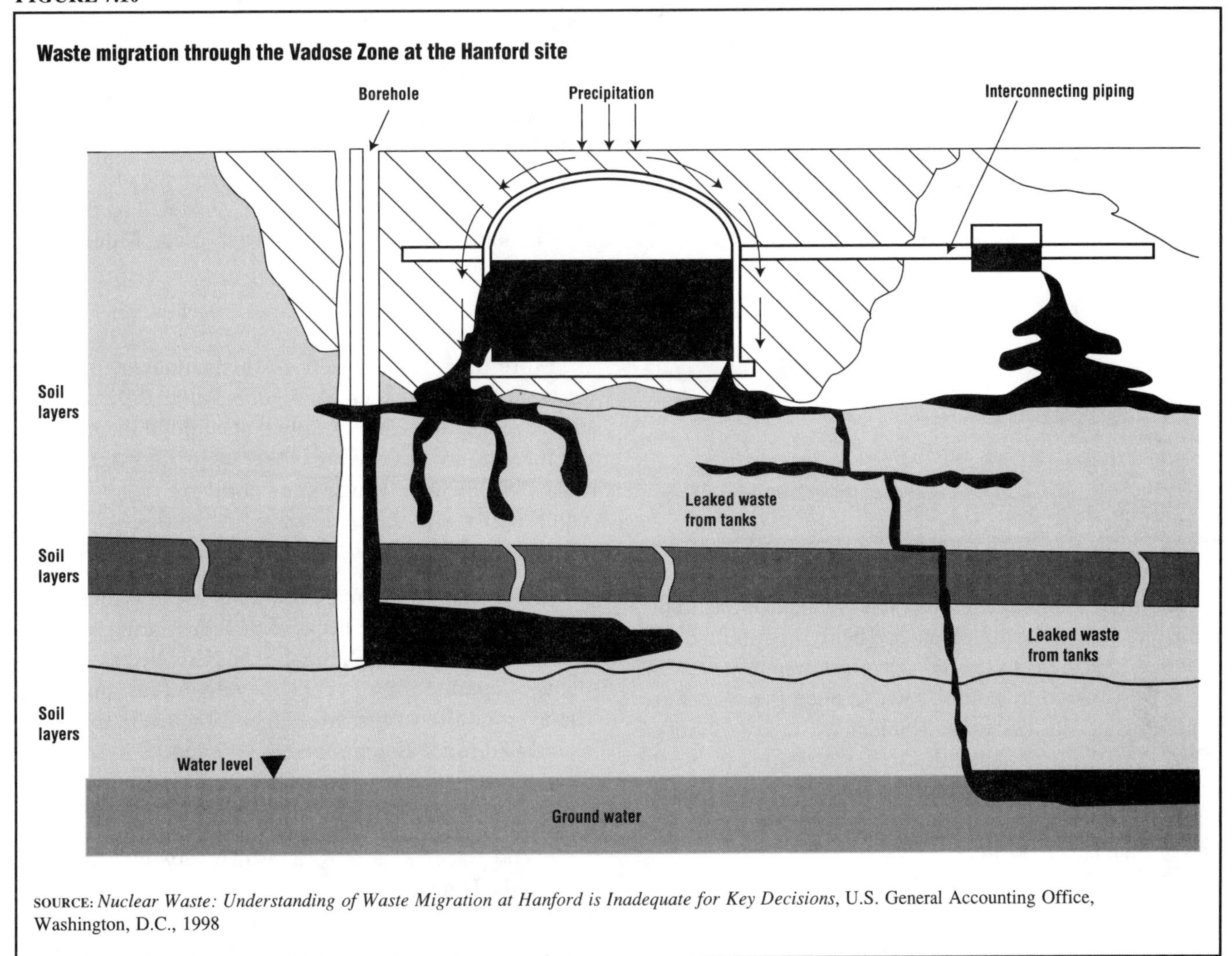

SOURCE: *Nuclear Waste: Understanding of Waste Migration at Hanford is Inadequate for Key Decisions*, U.S. General Accounting Office, Washington, D.C., 1998

Long Island, New York. BNL employs about 3,200 people, including 900 scientists and engineers. In January 1997 groundwater samples taken at BNL revealed concentrations of radioactive tritium that were twice the allowable federal drinking water standards. Some later samples found levels 32 times the standard.

Tritium was found to be leaking from the spent fuel pool of the laboratory's High Flux Beam Reactor into the aquifer that provides drinking water for local residents. An investigation revealed that the tritium had been leaking for as long as 12 years. Because Brookhaven did not monitor its reactor's spent-fuel pool for leaks, years passed before the tritium contamination was discovered. Built in the early 1960s, the spent-fuel pool is made of concrete but does not have a secondary container, such as a stainless steel liner, to protect against leaks. Newer reactors have secondary containment systems.

BNL's reactor will remain idle until improvements can be made in its operations. In the interim, engineers are draining a storage tank beneath the main reactor and redirecting the contaminated water 3,000 feet northward to an open basin. The water will then seep back into the ground. After 19 more years, the water should have undetectable levels of tritium.

Secret Government Activities

In 1993 then-Secretary of Energy Hazel O'Leary disclosed that over the past 45 years, the United States dumped tons of toxic waste across the nation, conducted hundreds of unannounced atomic tests (from 1948 to 1952), and experimented with the effects of plutonium on human subjects, often without their knowledge or approval. Secretary O'Leary revealed that of 925 nuclear tests, 204 were secret, and that the government was storing 33.5 metric tons of plutonium in six U.S. locations. In 1996 the U.S. government announced that compensation agreements had been reached with many of the victims who sought damages for their injuries from radiation.

CRISIS IN THE NUCLEAR POWER INDUSTRY

The long delay in providing disposal sites for nuclear wastes, coupled with the accelerated pace at which

TABLE 7.4

Total U.S. commercial spent nuclear fuel discharge, 1968–98

Reactor type	Stored at reactor sites	Stored at away-from-reactor facilities	Total
	Number of assemblies		
Boiling-water reactor	73,538	2,957	76,495
Pressurized-water reactor	56,778	491	57,269
Total	**130,316**	**3,448**	**133,764**
	Initial uranium content (MTU)		
Boiling-water reactor	13,230.3	554.0	13,784.2
Pressurized-water reactor	24,412.7	192.6	24,605.4
Total	**37,642.9**	**746.6**	**38,389.5**

Notes: A number of assemblies discharged prior to 1972 were reprocessed and are not included in this table.
Totals may not equal sum of components because of independent rounding.

SOURCE: *U.S. Commercial Spent Nuclear Fuel Annual 1995*, Energy Information Administration, Washington, D.C. 1996

nuclear plants are being retired, has created a crisis in the industry. Several aging plants are being maintained at a cost of $20 million a year for each reactor simply because there is no place to send the waste once the plants are decommissioned. Under the Nuclear Waste Policy Act of 1982, the DOE was scheduled to begin picking up waste on January 31, 1998. The utilities have been paying one-tenth of one cent per kilowatt-hour produced by the reactors to finance a repository.

Although the 1987 waste amendment designated Yucca Mountain as the site for a repository, little progress has been made in approving the project. In 1996 the U.S. Court of Appeals, in *Indiana Michigan Power Co. v. U.S. Department of Energy* (88 F3d 1272 [D.C. Ctr., 1996]), ruled that the Nuclear Waste Policy Act created an obligation for the DOE to start disposing of utilities' waste no later than January 31, 1998. Because the DOE missed the deadline, more than 20 utilities have sued.

In November 1998 the U.S. Court of Appeals for the District of Columbia, in *Northern States Power Company v. U.S. Department of Energy* (128 F3d 754) was unable to force the DOE to take possession of nuclear waste because there was no storage facility. The court ordered that utilities can postpone paying a portion of the fees owed to the Nuclear Waste Fund—an estimated $2.8 to $5 billion—and keep the earnings on the deferred fees. The utilities will be obligated to pay the withheld fees when the DOE begins to accept the spent fuel.

In February 1999 the DOE announced that, because it was unable to receive nuclear waste for permanent storage, it would take ownership of the waste and pay temporary storage costs with money the utilities have paid to develop the permanent repository. The waste will stay where it currently is being stored, and the DOE will pay the storage costs.

Both the Senate and the House passed legislation to build a temporary repository in Nevada. The Clinton Administration, however, vetoed the temporary site, claiming that it has not been proven safe and would deflect funds and engineering talent needed to build the permanent facility. Even without the expense of temporary storage, the nuclear waste fund (from the one cent per 10 kilowatt-hours of nuclear power generated by the utilities) is many billions short of what Yucca Mountain is expected to cost.

The Problem Is Now

More than 38,000 metric tons of nuclear uranium waste are sitting in spent fuel pools at the 109 operating and 20 closed nuclear energy plants around the country, and the amount is growing. (See Table 7.4.) According to the NRC, which licenses power plants, many of these plants will reach their capacity for storage within the next decade.

In addition many nuclear plants are shutting down well ahead of schedule because of skyrocketing maintenance and repair costs. Although the NRC licenses power plants to operate for 40 years, they do not last that long. The average life of the more than 20 reactors that have been shut down was approximately 13 years.

When a nuclear plant shuts down, the nuclear waste and the radioactive equipment stay on the premises because there is no place to put it. As a result, every nuclear power plant in the United States has become a temporary nuclear waste disposal site. Plants that close become mausoleums, largely untouched while they wait to be decommissioned or dismantled when a repository opens.

INTERNATIONAL APPROACH TO HIGH-LEVEL WASTE DISPOSAL

Governments around the world generally believe that deep geologic disposal offers the best option for isolating highly radioactive waste, although no country has yet built an operational facility. In fact, most do not plan to have a repository until 2020 or later. Table 7.5 shows estimates of repository opening dates in selected countries and the unique features of each nation's program, as of the most recent compilation. The United States faces a serious challenge because it has, by far, the largest civilian nuclear power program in the world. Other countries are under less pressure because they have temporary storage facilities that will be adequate for decades or, as in France, the countries are reprocessing their spent fuel.

Growing International Concern

In 1992 South Africa, Chile, Malaysia, and Indonesia barred Japanese ships from passing through their territories to transport plutonium from reprocessing centers in

TABLE 7.5

Comparison of waste programs

Country	Number of reactors	Nuclear-generated electricity in 1992 (approx.)	Earliest repository date	Likely geologic medium	Status	Unique features
Canada	22	15%	2025	Granite	Reviewing concept	Province of Ontario has 20 of Canada's 22 reactors
France	56	73%	2020	Granite or clay	Developing concept	Public opposition significantly slowed program
Germany	21	30%	2008	Salt	Constructing test facility	Opposition from state may affect licensing
Japan	43	27%	2030	Not selected	Searching for site	Government plans to increase use of nuclear power
Sweden	12	43%	2020	Crystalline rock	Searching for site	Waste managers plan to use long-lived copper canister
Switzerland	5	40%	2020	Crystalline rock or clay	Searching for site	Government would prefer to use an international repository
United Kingdom	37	23%	2040	Not selected	Delaying decision	Government plans lower-level waste repository
United States	109	22%	2010	Tuff	Constructing test facility	Federal law designated candidate site

SOURCE: Developed by GAO from data provided primarily by foreign officials. Data are as of June 1993.

Europe to civilian reactors in Japan. Concerns have also developed over the smuggling of radioactive nuclear materials from the former Soviet Union countries into Western Europe. The arrests of suspected smugglers have revealed the apparent failure of the Russian government to protect its radioactive materials.

Russian scientists have disclosed that, for the past three decades, the former USSR secretly pumped billions of gallons of atomic waste directly into the earth. They claim that this practice of injecting the waste, which violates generally accepted global standards for waste disposal, continues today. The Russians report that about half of all nuclear waste the former Soviet Union ever generated has been pumped into the ground at three sites near several major rivers — the Volga, the Ob, and the Yenisei Rivers. Russian scientists contend the practice is safe because the wastes are pumped under layers of clay and shale. Nonetheless, the wastes at one site have already leaked beyond the expected range.

In 1994 the United States and Russia agreed to allow each other access to nuclear sites where weapons are being dismantled. The United States and several newly independent countries of the former Soviet Union also pledged to reduce their nuclear arsenals as part of post-cold-war agreements.

Scientific expeditions are underway in the Arctic fishing grounds near Norway to map illegal nuclear dumping by the Soviet Union. Russian authorities acknowledge that the area was used for three decades as a dumping ground for radioactive wastes. The radioactive wastes include 18 nuclear reactors and 80 nuclear submarines.

As recently as 1993, Russia also dumped hundreds of tons of nuclear waste into the Sea of Japan. Russian officials claim that their countries lack the technology and funds to safely remove spent uranium from the reactors or storage sites. Although radiation levels in area waters are currently within normal levels, Norway considers the region a time bomb and has committed $35 million since 1994 to Russian nuclear safety. The United States also continues to provide aid to Russia to upgrade its aging nuclear reactors to make them safer.

Other countries are struggling with waste disposal as well. In 1998, with 30,000 German police guarding the route against protesters and threats of violence, 60 tons of atomic waste were moved to a temporary waste site in Ahaus, in southern Germany. Germany has no site for permanent storage of nuclear waste. Germany's Chancellor Gerhard Schroeder and major power companies agreed to close Germany's nuclear power plants within 20 years.

CHAPTER 8
ELECTRICITY

Since 1879, when Thomas Edison flipped the first switch to light Menlo Park, New Jersey, the use of electrical power has become nearly universal in the United States.

ELECTRICITY DEFINED

Electricity is a property of all matter. Electrons, negatively charged particles, and protons, positively charged particles, can be separated using energy from friction, magnetism, or chemical change, which produces electric current. In 1882 Sir W. Siemens proposed the name "watt," after the Scottish engineer James Watt, as the standard measure of electrical power. The term "wattage" refers to the amount of electrical power required to operate a particular appliance or device. A "kilowatt" is a unit of electrical power equal to 1,000 watts, while a "kilowatt-hour" is a unit of electrical work equal to that done by one kilowatt acting for one hour.

Electrical Capacity

The generating capacity of an electrical plant, measured in watts, indicates its ability to produce electrical energy. A 1,000-kilowatt generator running at full capacity for one hour supplies 1,000 kilowatt-hours of power. That generator operating continuously for an entire year could produce 8,760,000 kilowatt-hours of electricity (1,000 kilowatts x 24 hours/day x 365 days a year). However, no generator can operate at 100 percent capacity during an entire year because of "down-time" for routine maintenance, outages, or legal restrictions. On the average, about one-fourth of the potential generating capacity of an electrical plant is not available at any given time.

Electricity demands vary daily and seasonally, so the continuous operation of electrical generators is not necessary. Utilities depend on steam, nuclear, and large hydroelectric plants to meet routine demand. Auxiliary gas, turbine, internal combustion, and smaller hydroelectric plants are normally used during short periods of high demand.

An Electric Power System

An electric power system has several components. Figure 8.1 illustrates a simple electric system. Generating units produce electricity, transmission lines carry electricity over long distances, and distribution lines deliver the electricity to customers. Substations connect the pieces of the system together, while energy control centers coordinate the operation of all the components.

DOMESTIC ELECTRICITY USAGE

Domestic Production

In the United States coal has been and continues to be the largest source for electricity production, accounting for over half the electricity generated in 1999. (See Figures 8.2 and 8.3.) Nuclear power was the second largest source of electricity, followed by natural gas, petroleum, and hydroelectric power.

A record 3.7 trillion kilowatt-hours of electricity was generated in 1999. Table 8.1 shows that electricity use, or retail sales, in the United States has increased almost every year since 1950, by an average annual rate of approximately 5.6 percent. Conventional steam plants, run by fossil fuels, were responsible for most of the production of electricity. According to Energy Information Administration figures, coal, petroleum, and natural gas accounted for almost two-thirds of the total electricity generated in 1999, or 2.6 trillion kilowatt-hours. Coal alone accounted for a record 1.89 trillion kilowatt-hours, while petroleum and natural gas contributed to the production of 661 billion kilowatt-hours. Nuclear power accounted for 728 billion kilowatt-hours, while hydroelectric generation totaled 312 billion kilowatt-hours. All other renewable energy sources, including geothermal, wood, municipal waste,

FIGURE 8.1

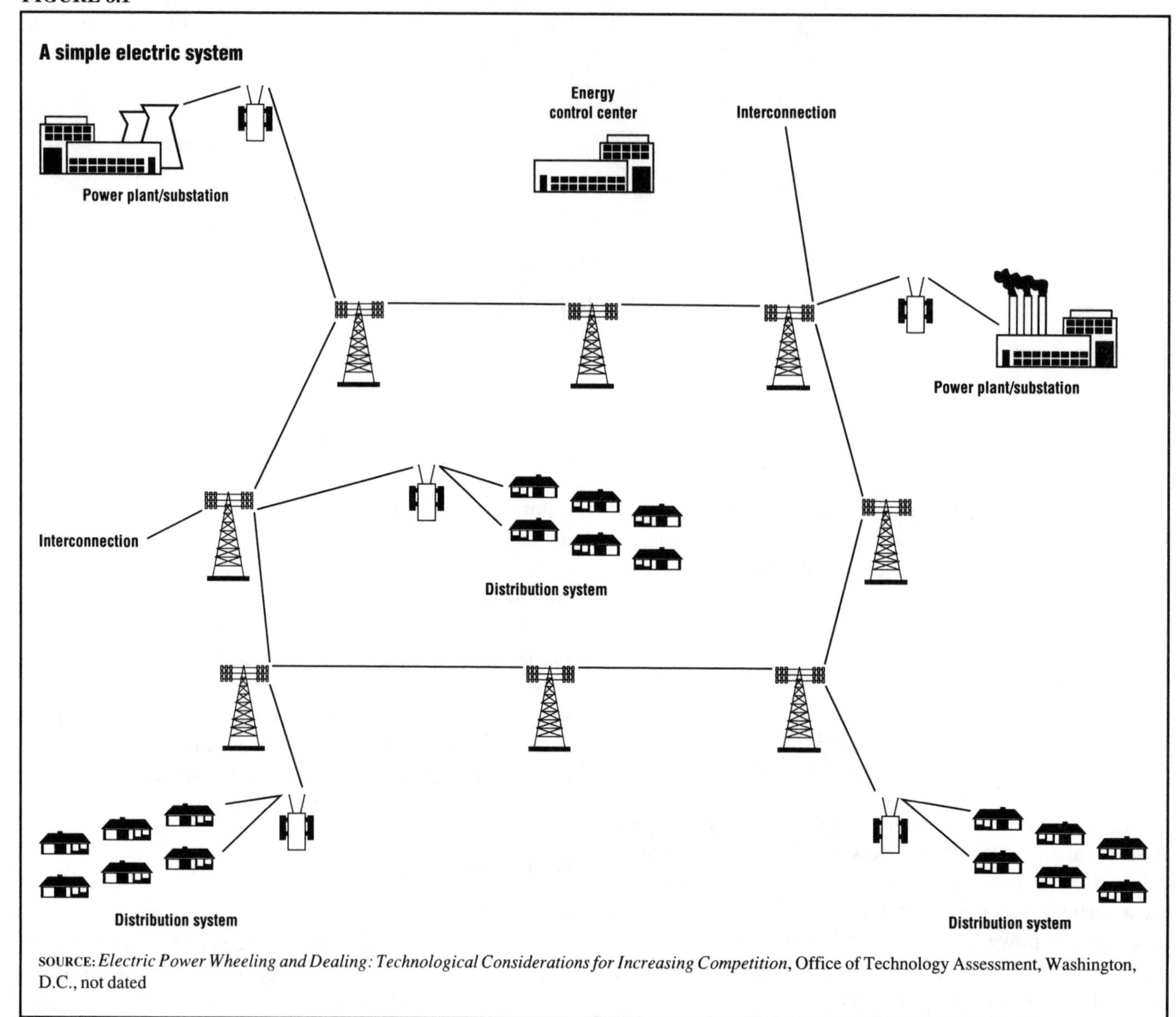

SOURCE: *Electric Power Wheeling and Dealing: Technological Considerations for Increasing Competition*, Office of Technology Assessment, Washington, D.C., not dated

wind, and solar energy, produced 83 billion kilowatt-hours in 1999.

The structure of the electric power industry is evolving, moving away from traditional, government-regulated electric utilities and toward an environment marked by less regulation and increased competition from non-utility power producers. In 1999, 13 percent of the total net generation of electricity came from non-utility producers, such as independent power producers and cogenerators. (See Figure 8.4.)

Domestic Consumption

From 1949 through 1990 the industrial sector was the largest consumer of electricity in the U.S., but since then sales to the residential sector have been higher due to changing economic factors. (See Table 8.2.) In 1999, 1,139 billion kilowatt-hours went to residential use; 1,050 billion kilowatt-hours to industrial customers; and 975 billion kilowatt-hours to commercial users.

Consumption of electricity is growing because electricity has increasingly taken over the tasks formerly done with coal, natural gas, or human muscle: manufacturing steel, assembling cars, or milking cows. Electricity is being used extensively in rapidly growing technology fields such as the computer industry. Residential and commercial use of electricity is increasing with new appliances, air conditioners, computers, and many other developing applications.

THE ELECTRIC BILL

The price paid by a consumer for electricity includes the cost of converting energy into electricity from its original form, such as coal, as well as the cost of delivering it. In 1997, according to the latest figures

FIGURE 8.2

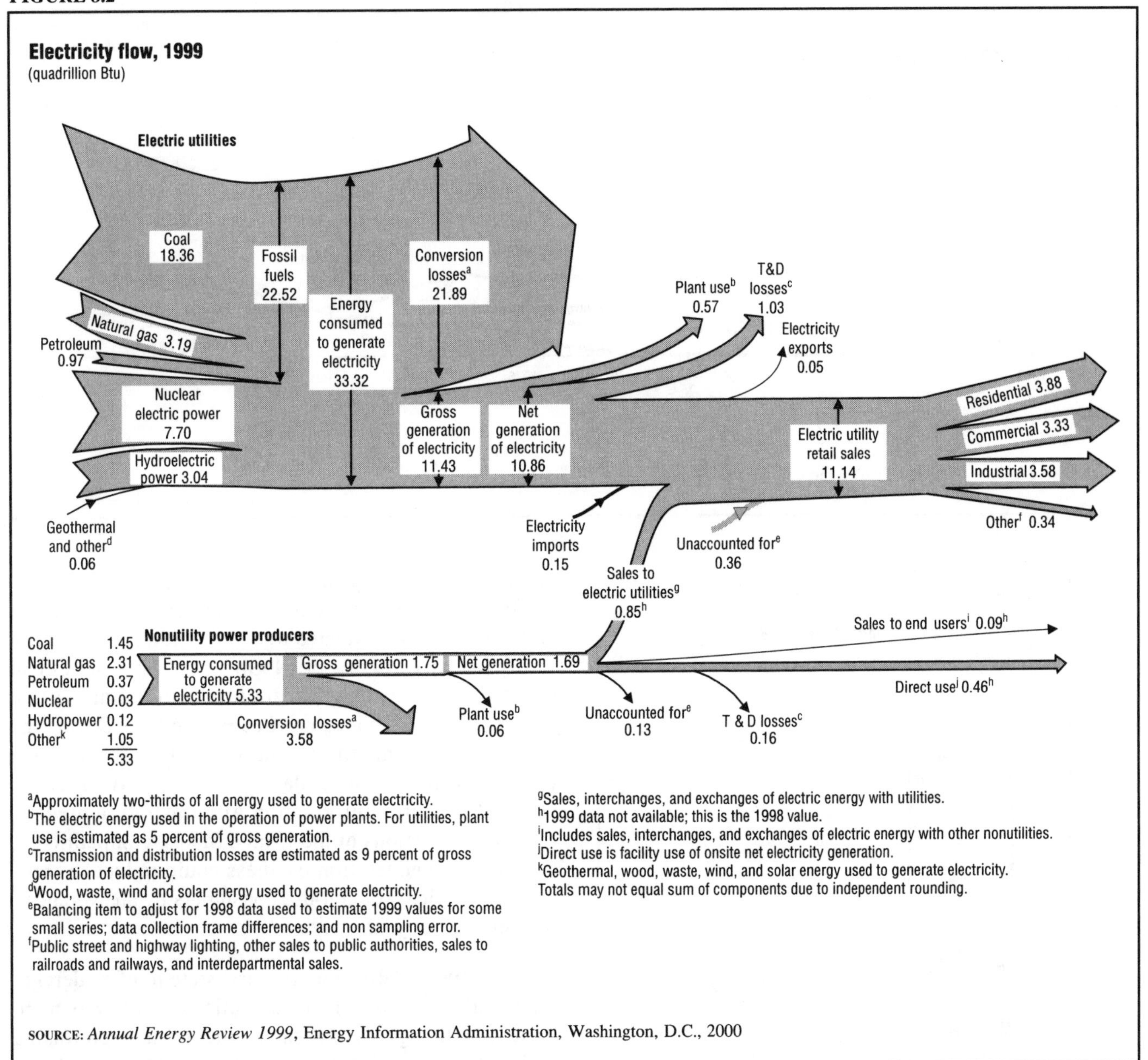

[a]Approximately two-thirds of all energy used to generate electricity.
[b]The electric energy used in the operation of power plants. For utilities, plant use is estimated as 5 percent of gross generation.
[c]Transmission and distribution losses are estimated as 9 percent of gross generation of electricity.
[d]Wood, waste, wind and solar energy used to generate electricity.
[e]Balancing item to adjust for 1998 data used to estimate 1999 values for some small series; data collection frame differences; and non sampling error.
[f]Public street and highway lighting, other sales to public authorities, sales to railroads and railways, and interdepartmental sales.
[g]Sales, interchanges, and exchanges of electric energy with utilities.
[h]1999 data not available; this is the 1998 value.
[i]Includes sales, interchanges, and exchanges of electric energy with other nonutilities.
[j]Direct use is facility use of onsite net electricity generation.
[k]Geothermal, wood, waste, wind, and solar energy used to generate electricity.
Totals may not equal sum of components due to independent rounding.

SOURCE: *Annual Energy Review 1999*, Energy Information Administration, Washington, D.C., 2000

from the Energy Information Administration, consumers paid an average of $20.15 per million Btu for the electric power delivered to their residences, compared to only $4.62 per million Btu for natural gas and $9.73 per million Btu for gasoline.

The unit cost of electricity is high because of the amounts of energy expended in creating the electricity and moving it to the point of use. In 1999 for example, about 33 quadrillion Btu of energy were consumed to generate electricity in the United States, but only about 11 quadrillion Btu were actually used. (See Figure 8.2.) Most of the remaining 22 quadrillion Btu was lost during the energy conversion process, and some was lost during the transmission and distribution process (about 1 quadrillion Btu). In the end, for every three units of energy that are converted to create electricity, slightly less than one unit actually reaches the end user.

Between 1935 and 1970, the price of electricity declined, but it began to increase during the 1970s due to the OPEC oil embargo. Since 1985 the price of electricity has been dropping again because of the decline in energy resource prices. (See Figure 8.5.) Prices vary depending upon the location. In 1999 electricity was most expensive in the Middle Atlantic and New England states. The average real price (adjusted for inflation) of electricity sold to the residential sector was 7.8 cents per kilowatt-hour in 1999, while the commercial sector paid 6.9 cents per kilowatt-hour. Industrial users paid less per kilowatt-hour, 4.2 cents in 1999, because they use huge amounts of electricity and receive volume discounts.

FIGURE 8.3

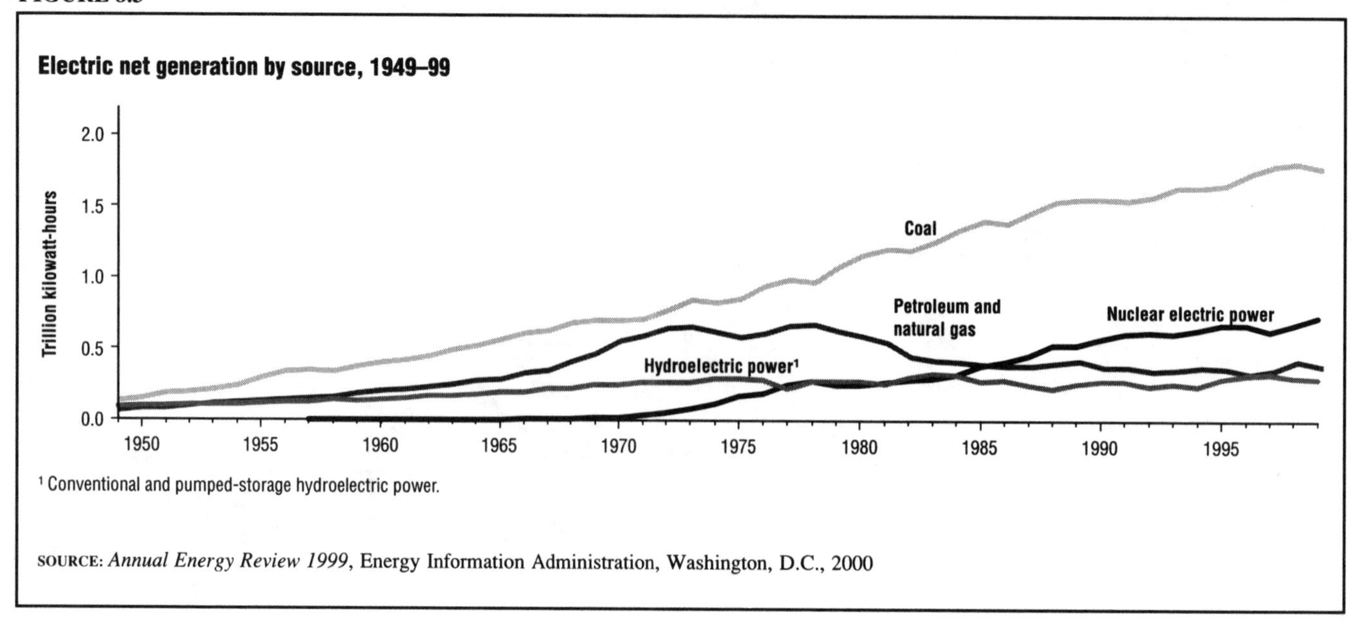

[1] Conventional and pumped-storage hydroelectric power.

SOURCE: *Annual Energy Review 1999*, Energy Information Administration, Washington, D.C., 2000

FIGURE 8.4

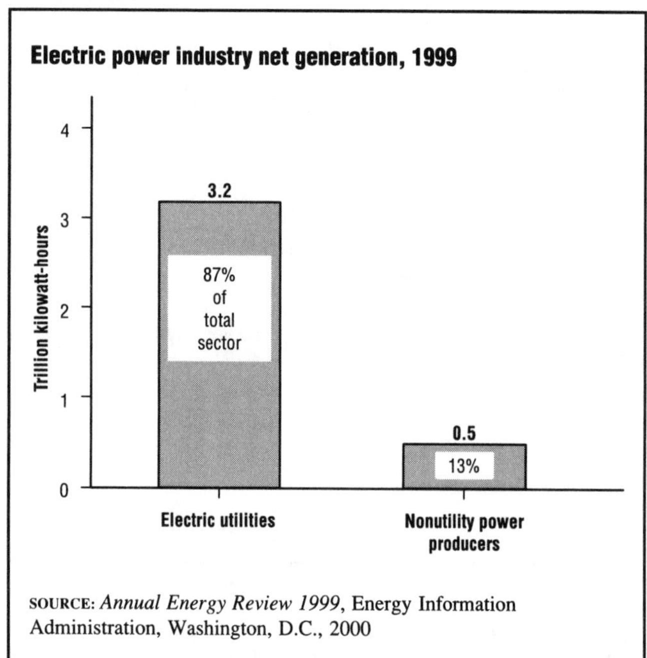

SOURCE: *Annual Energy Review 1999*, Energy Information Administration, Washington, D.C., 2000

DEREGULATION OF ELECTRIC UTILITIES

Regulated for decades as "natural monopolies," much like the railroad and telecommunications industries, electric utilities are in the midst of a radical shift toward unregulated markets and increased competition. In 1978 Congress passed the Public Utilities Regulatory Policy Act (PURPA; PL 95-617), which required that utilities buy electricity from private companies when that was a lower-cost alternative to building their own power plants. The Energy Policy Act of 1992 (PL 102-486) gave greater access to the market for other generators, resulting in a flurry of activity in state and federal legislatures as a host of interest groups debated regulatory, economic, energy, and environmental policies. State public utility commissions are conducting proceedings and designing rules related to competition in the electric utility industry. As of April 1998 all but three states had enacted or planned significant restructuring activity in the electric power industry. (See Figure 8.6.) During the transition to competition, many electric utilities have been downsizing, consolidating, and merging to reduce costs. The Energy Information Administration predicts that in most regions of the United States, competition in the electrical generation business could reduce average electricity prices for end-use consumers by an average of 0.6 percent per year.

In 1998 in California, the first state to offer deregulated options to consumers, a cautious public was bombarded with aggressive sales pitches and confusing claims by many electricity marketers. Most residents, however, remained with their original utility providers. During peak periods in 2000, many parts of California experienced severe shortages of electricity, causing the public to question whether deregulation was going to work at all. This occurred because several of California's largest utility companies were facing bankruptcy. Therefore they could not afford to buy wholesale electricity from other producers with which to meet all of their customer's demands.

The utilities blamed the slow process of deregulation for the problems, saying that it had forced them into paying high market prices for wholesale electricity even though they were not allowed to pass those costs on to their customers. The result was that the utilities had no money left with which to get more electricity. The California state government eventually had to step in and help the utilities get affordable wholesale electricity.

TABLE 8.1

Electricity overview, 1950–99
(billion kilowatt-hours)

Year	Net generation			Imports[1]	Exports[1]	Losses and unaccounted for[2]	End use			
								Nonutility power producers		
	Electric utilities	Nonutility power producers	Total				Electric utility retail sales	Direct use[3]	Sales to end users	Total
1950	329	NA	329	2	(s)	NA	291	NA	NA	NA
1960	756	NA	756	5	1	NA	688	NA	NA	NA
1970	1,532	NA	1,532	6	4	NA	1,392	NA	NA	NA
1975	1,918	NA	1,918	11	5	NA	1,747	NA	NA	NA
1980	2,286	NA	2,286	25	4	NA	2,094	NA	NA	NA
1981	2,295	NA	2,295	36	3	NA	2,147	NA	NA	NA
1982	2,241	NA	2,241	33	4	NA	2,086	NA	NA	NA
1983	2,310	NA	2,310	39	3	NA	2,151	NA	NA	NA
1984	2,416	NA	2,416	42	3	NA	2,286	NA	NA	NA
1985	2,470	NA	2,470	46	5	NA	2,324	NA	NA	NA
1986	2,487	NA	2,487	41	5	NA	2,369	NA	NA	NA
1987	2,572	NA	2,572	52	6	NA	2,457	NA	NA	NA
1988	2,704	NA	2,704	39	7	NA	2,578	NA	NA	NA
1989	2,784	[R,4]188	[R]2,972	26	15	[R]236	2,647	[4]83	[4]18	2,747
1990	2,808	[R,4]217	[R]3,025	[R]18	[R]16	[R]210	2,713	[4]84	[4]20	2,817
1991	2,825	[R,4]246	[R]3,071	[R]22	[R]2	218	2,762	[4]100	[4]11	2,873
1992	2,797	286	3,083	[R]28	[R]3	[R]224	2,763	111	11	2,885
1993	2,883	314	3,197	[R]31	[R]4	[R]236	2,861	111	16	2,988
1994	2,911	343	3,254	[R]47	[R]2	223	2,935	123	18	3,075
1995	2,995	363	3,358	[R]43	[R]4	[R]235	3,013	134	16	3,162
1996	3,077	370	3,447	[R]43	[R]3	[R]241	3,098	135	14	3,247
1997	3,123	372	3,494	[R]43	[R]9	[R]240	3,140	131	18	3,289
1998	3,212	[R]406	[R]3,618	[R]40	[R]13	[R]245	[R]3,240	[R]134	[R]26	[R]3,400
1999[P]	3,183	495	3,678	43	14	NA	3,265	NA	NA	NA

[1] Electricity transmitted across U.S. borders with Canada and Mexico.

[2] Energy losses that occur between the point of generation and delivery to the customer, and data collection frame differences and nonsampling error.

[3] Facility use of onsite net electricity generation.

[4] Data for 1989-1991 were collected for facilities with capacities of 5 megawatts or more. In 1992, the threshold was lowered to include facilities with capacities of 1 megawatt or more. Estimates of the 1-to-5 megawatt range for 1989-1991 were derived from historical data. The estimation did not include retirements that occurred prior to 1992 and included only the capacity of facilities that came on line before 1992.

R=Revised. P=Preliminary. NA=Not available. (s)=Less than 0.5 billion kilowatt-hours.

Totals may not equal sum of components due to independent rounding.

SOURCE: *Annual Energy Review 1999*, Energy Information Administration, Washington, D.C., 2000

FIGURE 8.5

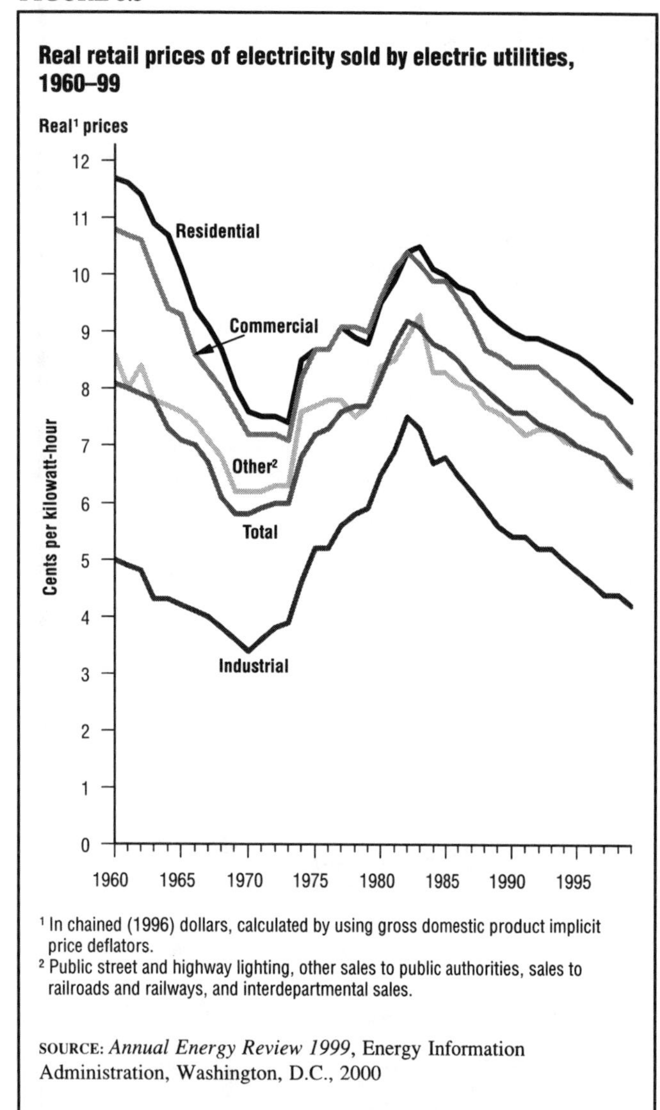

¹ In chained (1996) dollars, calculated by using gross domestic product implicit price deflators.
² Public street and highway lighting, other sales to public authorities, sales to railroads and railways, and interdepartmental sales.

SOURCE: *Annual Energy Review 1999*, Energy Information Administration, Washington, D.C., 2000

On April 15, 1999, the Clinton Administration submitted to Congress its proposed Comprehensive Electricity Competition Act (CECA), which further guides and empowers the states to increase competition in the electricity market and ultimately reduce consumer prices. The act also has provisions to stimulate research and investment in renewable energy and efficiency, give consumers choice and protection, and reduce emissions. As of late 2000, the CECA was still pending ratification by Congress.

INTERNATIONAL ELECTRICITY USAGE

World Production

In 1998 over 13.6 trillion kilowatt-hours of electricity were generated around the world. (See Table 8.3.) North America accounted for 32 percent; the Far East and Oceania, 26 percent; Western Europe, 20 percent; and Eastern Europe and the former USSR, 11 percent. In 1998 fossil fuel electricity production (coal, gas, and oil) accounted for 63 percent of all electricity generated; hydroelectric, 19 percent; nuclear, 17 percent; and geothermal and other sources, less than 1 percent of the world's total net electricity production.

World Consumption

World total electricity consumption continues to increase, rising from 10.3 trillion kilowatt-hours in 1989 to 12.6 trillion kilowatt-hours in 1998. (See Table 8.4.) North America accounted for 32 percent of the world's total consumption; the Far East and Oceania, 26 percent; Western Europe, 20 percent; and Eastern Europe and the former USSR, 11 percent in 1998.

FUTURE TRENDS IN THE U.S. ELECTRIC INDUSTRY

The Energy Information Administration of the U.S. Department of Energy, in its *Annual Energy Outlook 2000* (1999), predicts relatively slower growth for U.S. electricity consumption (sales) in the next 20 years, at a rate of 1.4 percent annually. Efficiency gains are predicted to be offset by increased use of electricity. Residential electricity demand is projected to grow by 1.5 percent per year, while commercial demand grows by 1.2 percent and industrial demand by 1.3 percent per year.

Historically, the demand for electricity has been related to economic growth. This positive relationship will continue, but at a slower rate. During the 1960s, electricity demand grew by more than 7 percent per year. In the 1970s electricity demand growth was only 1.5 percent per year, and this dropped to 1.0 percent per year in the 1980s. Several factors led to this occurrence, including increased market saturation of electric appliances and improvements in efficiency. (See Figure 8.7.) Electricity demand growth has been especially high in the West and South, reflecting the higher population growth in those parts of the country.

To meet increased demand, the EIA estimates that the U.S. will need 300 gigawatts of new generating capacity by 2020. Nuclear energy is expected to lose 40 gigawatts of its current 97 gigawatt capacity by 2020, due to the retirement of plants; so coal, natural gas, and renewable sources will be required to increase production. Electricity provided from natural gas is expected to increase from 14 percent of total generation in 1998 to 31 percent in 2020, as producers take advantage of natural gas's efficiency, lower plant costs, and lower emissions. Renewable sources, such as solar, wind, and biomass, are projected to increase their electricity production by 0.5 percent per year to 2020.

Utility companies plan to keep up with demand with new construction and improvements in existing plants, as well as increasing their use of renewable sources of power and importing more power from Canada and Mexico. The

TABLE 8.2

Electricity end use, 1949–99

(billion kilowatt-hours)

Year	Electric utility retail sales					Nonutility power producers		Total
	Residential	Commercial	Industrial	Other [1]	Total	Direct Use [2]	Sales to end users	
1949	67	45	123	20	255	NA	NA	NA
1950	72	51	146	22	291	NA	NA	NA
1960	201	131	324	32	688	NA	NA	NA
1970	466	307	571	48	1,392	NA	NA	NA
1975	588	403	688	68	1,747	NA	NA	NA
1980	717	488	815	74	2,094	NA	NA	NA
1981	722	514	826	85	2,147	NA	NA	NA
1982	730	526	745	86	2,086	NA	NA	NA
1983	751	544	776	80	2,151	NA	NA	NA
1984	780	583	838	85	2,286	NA	NA	NA
1985	794	606	837	87	2,324	NA	NA	NA
1986	819	631	831	89	2,369	NA	NA	NA
1987	850	660	858	88	2,457	NA	NA	NA
1988	893	699	896	90	2,578	NA	NA	NA
1989	906	726	926	90	2,647	[3]83	[3]18	2,747
1990	924	751	946	92	2,713	[3]84	[3]20	2,817
1991	955	766	947	94	2,762	[3]100	[3]11	2,873
1992	936	761	973	93	2,763	111	11	2,885
1993	995	795	977	95	2,861	111	16	2,988
1994	1,008	820	1,008	98	2,935	123	18	3,075
1995	1,043	863	1,013	95	3,013	134	16	3,162
1996	1,082	887	1,030	98	3,098	135	14	3,247
1997	1,076	928	1,033	103	3,140	131	18	3,289
1998	R1,128	R969	R1,040	R104	R3,240	R134	R26	R3,400
1999P	1,139	975	1,050	100	3,265	NA	NA	NA

[1] Public street and highway lighting, other sales to public authorities, sales to railroads and railways, and interdepartmental sales.

[2] Facility use of onsite net electricity generation.

[3] Data for 1989-1991 were collected for facilities with capacities of 5 megawatts or more. In 1992, the threshold was lowered to include facilities with capacities of 1 megawatt or more. Estimates of the 1-to-5 megawatt range for 1989-1991 were derived from historical data. The estimation did not include retirements that occurred prior to 1992 and included only the capacity of facilities that came on line before 1992.

R=Revised. P=Preliminary. NA=Not available.

Totals may not equal sum of components due to independent rounding.

SOURCE: *Annual Energy Review 1999,* Energy Information Administration, Washington, D.C., 2000

FIGURE 8.6

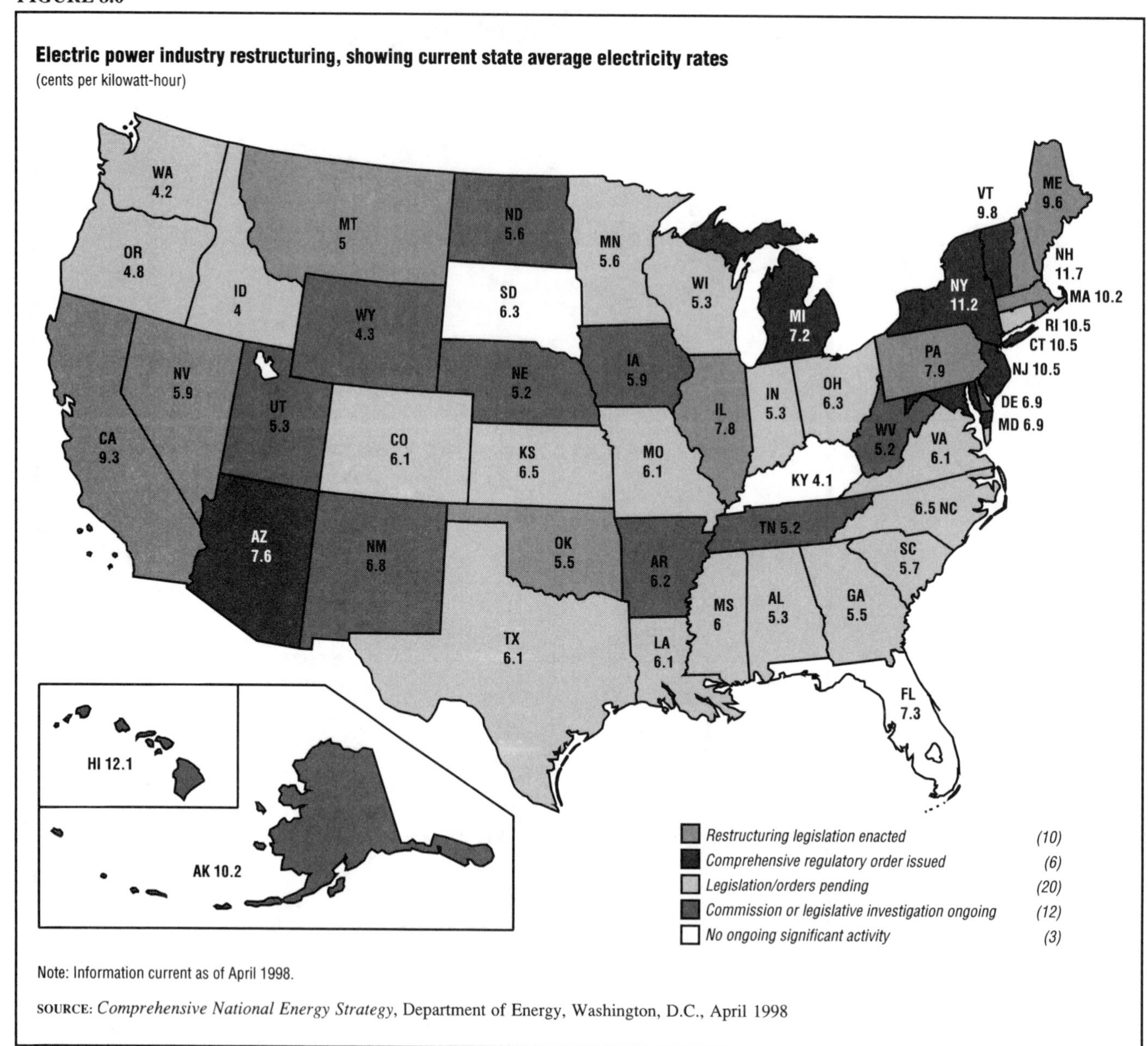

Note: Information current as of April 1998.

SOURCE: *Comprehensive National Energy Strategy*, Department of Energy, Washington, D.C., April 1998

EIA estimates that up to 1000 new plants may be required to meet 2020 demand levels. The issue of electric growth is important, and carries financial risks for electric companies. If the industry underestimates future needs for electricity, it could mean power shortages or losses. On the other hand, excessive projection of the nation's needs could mean billions of dollars spent on unneeded equipment.

Electricity prices are expected to decrease from 6.7 cents per kilowatt-hour in 1998 to 5.8 cents in 2020, a decrease of 0.6 percent per year, due to efficiency gains and increased competition brought about by new legislation.

Growing concerns about acid rain and global warming could result in tightened environmental emission standards, which may have an impact on electrical utility expansion decisions, prices, and supply. Continued advances in solar and wind turbine technology could make renewable sources of electrical power more economical in the future. Some energy experts claim that increased efficiency and conservation efforts are the most sensible alternatives to new construction or to burning more fossil fuels in existing plants.

TABLE 8.3

World net generation of electricity by type, 1980, 1997, 1998

(billion kilowatt-hours)

Region and country	Fossil fuel			Nuclear electric power			Hydroelectric power[1]			Total[2]		
	1980	1997	1998[P]	1980	1997	1998[P]	1980	1997	1998[P]	1980	1997	1998[P]
North America	**1,880.3**	**[R]2,680.9**	**2,829.5**	**287.0**	**716.4**	**750.0**	**546.9**	**[R]728.7**	**672.5**	**[R]2.721.8**	**[R]4,223.2**	**4,345.5**
Canada	79.8	[R]132.8	149.7	35.9	77.9	67.5	251.0	[R]347.6	329.3	[R]367.9	[R]562.2	550.9
Mexico	46.0	[R]124.7	137.5	0.0	9.9	8.8	16.7	26.2	24.3	63.6	[R]166.1	176.1
United States	1,754.0	[R]2,422.6	2,541.5	251.1	628.6	673.7	279.2	354.9	318.9	2,289.8	[R]3,494.2	3,617.9
Other	0.5	0.7	0.7	0.0	0.0	0.0	0.0	0.0	0.0	0.5	0.7	0.7
Central and South America	**99.8**	**[R]159.7**	**165.2**	**2.2**	**10.5**	**10.3**	**201.5**	**[R]512.4**	**523.1**	**[R]308.4**	**[R]696.4**	**712.8**
Argentina	22.2	[R]33.5	32.1	2.2	7.5	7.1	17.3	[R]34.8	35.8	[R]42.0	[R]76.0	75.2
Brazil	7.5	[R]14.7	15.6	0.0	3.0	3.1	128.4	[R]276.2	288.5	138.3	[R]303.5	316.9
Colombia	5.1	[R]12.2	13.6	0.0	0.0	0.0	14.3	30.9	31.2	[R]19.5	[R]43.4	45.0
Venezuela	17.6	[R]16.7	17.9	0.0	0.0	0.0	14.4	[R]56.6	52.5	32.0	[R]73.2	70.4
Other	47.3	[R]82.5	86.0	0.0	0.0	0.0	27.0	[R]113.9	115.2	[R]76.6	[R]200.2	205.2
Western Europe	**1,180.1**	**[R]1,267.1**	**1,316.6**	**219.2**	**839.9**	**841.7**	**431.7**	**[R]502.1**	**519.6**	**[R]1,844.5**	**[R]2,659.3**	**2,731.9**
Austria	11.9	18.4	17.6	0.0	0.0	0.0	28.5	[R]35.6	37.0	[R]40.7	[R]55.5	56.1
Finland	22.0	[R]33.9	31.3	6.6	19.0	20.8	10.1	[R]12.1	14.8	38.7	[R]72.8	75.3
France	118.0	[R]38.2	51.8	63.4	374.3	366.7	68.3	[R]61.6	59.9	[R]250.8	[R]476.6	481.0
Germany	390.3	[R]335.8	345.5	55.6	161.8	152.7	18.8	[R]17.2	16.8	[R]469.9	[R]524.7	525.4
Italy	125.5	[R]188.2	195.0	2.1	0.0	0.0	45.0	[R]41.2	42.0	[R]176.4	234.6	243.0
Netherlands	58.0	[R]78.5	81.0	3.9	2.3	3.6	0.0	0.1	0.1	[R]62.9	[R]84.7	88.7
Norway	0.1	0.7	0.7	0.0	0.0	0.0	82.7	108.7	114.5	82.9	[R]109.6	115.5
Spain	74.5	[R]89.4	86.6	5.2	52.5	56.0	29.2	[R]34.3	34.4	[R]109.2	[R]178.5	179.5
Sweden	10.1	[R]9.6	9.6	25.3	66.7	70.8	58.1	[R]68.3	72.9	[R]94.3	[R]148.2	156.8
Switzerland	0.9	[R]2.0	2.3	12.9	24.0	24.5	32.5	[R]33.7	33.2	[R]46.4	[R]60.8	61.1
Turkey	12.0	59.6	64.6	0.0	0.0	0.0	11.2	39.4	41.8	[R]23.3	[R]99.4	106.7
United Kingdom	228.9	[R]226.5	234.1	32.3	89.3	97.7	3.9	4.1	5.1	265.1	[R]325.9	343.1
Other	127.8	[R]186.4	196.5	11.9	49.8	48.9	43.5	[R]45.9	47.2	[R]183.8	[R]288.0	299.8
Eastern Europe and Former U.S.S.R.	**1,309.3**	**[R]1,019.4**	**1,003.7**	**83.2**	**[R]250.3**	**239.3**	**211.3**	**[R]249.1**	**245.8**	**[R]1,604.1**	**[R]1,520.2**	**1,490.2**
Czech Republic	—	[R]46.6	46.4	—	12.5	12.5	—	[R]1.7	1.6	—	[R]61.7	61.5
Kazakhstan	—	[R]42.8	43.2	—	[R]0.3	0.1	—	[R]6.4	6.0	—	49.5	49.3
Poland	111.1	[R]130.6	130.1	0.0	0.0	0.0	3.2	[R]3.8	4.3	[R]114.7	[R]134.8	134.9
Romania	51.4	[R]32.2	31.0	0.0	5.1	4.9	12.5	[R]17.3	16.6	63.9	[R]54.6	52.5
Russia	—	[R]527.0	523.1	—	104.5	98.3	—	[R]152.5	150.5	—	[R]784.0	771.9
Ukraine	—	[R]83.2	76.0	—	75.4	70.6	—	[R]9.9	11.3	—	[R]168.6	157.9
Other	1,146.8	[R]157.0	153.9	83.2	**[R]52.4**	52.8	195.5	[R]57.5	55.5	1,425.6	[R]267.0	262.2
Middle East	**82.8**	**[R]333.7**	**354.4**	**0.0**	**0.0**	**0.0**	**9.6**	**[R]18.8**	**18.9**	**92.4**	**[R]352.5**	**373.3**
Iran	15.7	[R]83.1	88.0	0.0	0.0	0.0	5.6	[R]7.3	7.3	[R]21.3	[R]90.4	95.3
Saudi Arabia	20.5	[R]103.8	110.1	0.0	0.0	0.0	0.0	0.0	0.0	20.5	[R]103.8	110.1
Other	46.6	[R]146.8	156.3	0.0	0.0	0.0	4.1	[R]11.5	11.6	[R]50.7	[R]158.3	167.9
Africa	**129.1**	**[R]304.3**	**305.3**	**0.0**	**12.6**	**13.6**	**60.6**	**[R]62.7**	**63.1**	**189.7**	**[R]380.0**	**382.4**
Egypt	8.6	[R]42.9	45.5	0.0	0.0	0.0	9.7	[R]11.9	12.3	18.3	54.8	57.8
South Africa	92.1	[R]181.4	176.8	0.0	12.6	13.6	1.0	2.1	1.6	93.1	[R]196.2	192.0
Other	28.4	[R]79.9	82.9	0.0	0.0	0.0	49.9	[R]48.7	49.2	78.4	[R]129.0	132.6

TABLE 8.3

World net generation of electricity by type, 1980, 1997, 1998 [CONTINUED]

(billion kilowatt-hours)

	Fossil fuel			Nuclear electric power			Hydroelectric power[1]			Total[2]		
Region and country	**1980**	**1997**	**1998[P]**	**1980**	**1997**	**1998[P]**	**1980**	**1997**	**1998[P]**	**1980**	**1997**	**1998[P]**
Far East and Oceania	**907.7**	**[R]2,531.1**	**2,558.4**	**92.7**	**436.4**	**462.8**	**[R]275.2**	**[R]500.4**	**515.7**	**[R]1,280.5**	**[R]3,508.3**	**3,577.7**
Australia	74.5	[R]155.8	167.5	0.0	0.0	0.0	12.8	[R]16.6	15.6	[R]87.7	[R]175.5	186.4
China	227.9	850.0	882.5 0.0		11.4	13.5	57.6	[R]193.1	202.9	285.5	[R]1,054.5	1,098.8
India	69.7	[R]355.8	358.4	3.0	10.5	10.6	46.5	[R]74.0	76.2	119.3	[R]441.1	446.1
Indonesia	10.6	[R]62.3	64.5	0.0	0.0	0.0	3.0	[R]5.9	6.1	13.5	[R]70.7	73.1
Japan	381.6	[R]579.9	564.5	78.6	306.1	318.1	87.8	[R]88.9	89.5	549.1	[R]999.3	996.0
South Korea	29.8	[R]154.3	131.8	3.3	73.2	85.2	[R]1.5	[R]2.8	4.2	[R]34.6	[R]230.3	221.3
Taiwan	31.3	[R]79.7	88.1	7.8	34.8	35.1	2.9	[R]9.4	10.5	42.0	[R]124.0	133.6
Thailand	12.3	78.9	75.7	0.0	0.0	0.0	1.3	[R]7.1	7.1	[R]13.6	[R]86.0	82.8
Other	70.1	[R]214.6	225.5	0.0	0.4	0.4	61.8	[R]102.5	103.6	[R]135.3	[R]327.0	339.6
World	**5,589.0**	**[R]8,296.1**	**8,533.1**	**684.4**	**[R]2,266.1**	**2,317.7**	**[R]1,736.8**	**[R]2,574.2**	**2,558.7**	**[R]8,041.5**	**[R]13,339.9**	**13,613.9**

[1]Excludes pumped storage, except for the United States.
[2]Geothermal, wood, other biomass, waste, solar, wind, hydrogen, sulfur, batteries, and chemicals are included in total.

R=Revised. P=Preliminary. — = Not applicable.

Data include both electric utility and non-electric utility sources.

SOURCE: *Annual Energy Review 1999*, Energy Information Administration, Washington, D.C., 2000

FIGURE 8.7

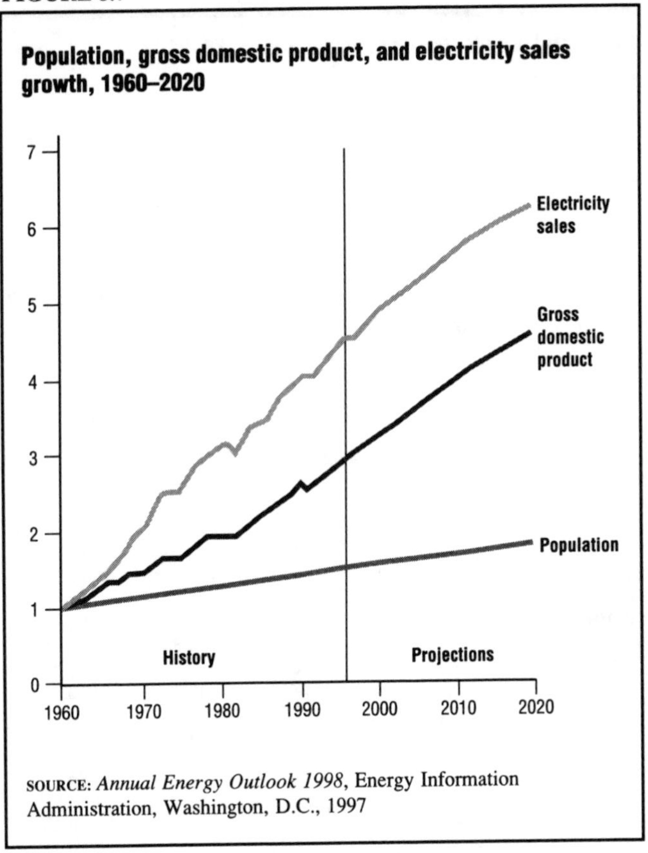

SOURCE: *Annual Energy Outlook 1998*, Energy Information Administration, Washington, D.C., 1997

World total net electricity consumption, 1989–98

(billion kilowatt-hours)

Region/country	1989	1990	1991	1992	1993	1994	1995	1996	1997	1998
North America										
Canada	441.9	437.1	443.6	450.1	454.5	462.8	473.2	484.8	487.3	484.5
Mexico	108.1	107.1	110.5	114.1	119.2	128.9	133.7	143.7	155.9	164.8
United States	2,747.3	2,816.8	2,873.1	2,885.1	2,988.4	3,075.5	3,162.4	3,246.7	3,288.5	3,367.4
Other	0.6	0.7	0.7	0.7	0.7	0.7	0.7	0.7	0.7	0.7
Total	**3,297.9**	**3,361.7**	**3,427.9**	**3,450.0**	**3,562.7**	**3,667.9**	**3,770.0**	**3,876.0**	**3,932.4**	**4,017.4**
Central & South America										
Argentina	47.7	45.9	49.7	56.8	60.7	63.2	66.8	70.7	75.7	75.6
Bolivia	1.9	2.0	2.2	2.3	2.3	2.4	2.5	2.4	2.2	2.4
Brazil	225.1	228.6	242.1	246.3	259.4	271.7	288.2	307.2	322.7	336.2
Chile	16.5	19.3	18.1	20.3	21.8	22.9	25.4	27.7	30.1	26.7
Colombia	31.9	33.3	33.9	31.0	35.3	38.4	42.1	40.2	40.5	42.0
Costa Rica	3.2	3.4	3.6	3.8	4.1	4.7	4.3	4.7	5.5	5.3
Cuba	14.4	13.6	11.6	10.3	9.4	10.3	10.9	11.6	12.3	14.2
Dominican Republic	3.1	3.3	3.5	5.2	5.4	5.7	6.0	6.8	7.3	7.9
Ecuador	5.2	5.8	6.3	6.5	6.8	7.5	7.7	8.4	8.7	9.0
El Salvador	1.9	1.9	1.8	1.9	2.3	2.7	2.8	2.4	3.1	3.3
Guatemala	2.2	2.2	2.4	2.3	2.6	2.7	2.7	3.1	2.8	2.9
Honduras	1.9	2.1	2.1	2.2	2.3	2.5	2.5	2.6	2.1	2.7
Jamaica	2.0	2.3	1.9	2.0	3.3	4.2	5.1	5.3	5.5	5.9
Nicaragua	1.4	1.8	1.6	1.7	1.8	1.7	2.0	2.2	2.4	2.5
Panama	2.6	2.7	3.3	3.6	3.9	4.2	3.9	4.0	4.1	4.3
Peru	12.5	12.3	12.9	11.9	13.5	13.5	16.1	15.4	15.8	17.0
Puerto Rico	12.5	13.4	13.8	14.4	14.5	15.6	16.6	16.7	16.7	16.5
Suriname	1.2	1.4	1.6	1.6	1.6	1.6	1.7	1.7	1.9	1.9
Trinidad and Tobago	3.0	3.1	3.3	3.5	3.3	3.6	3.8	4.0	4.4	4.4
Uruguay	4.1	5.6	4.6	4.8	5.1	5.4	5.8	6.0	6.4	6.5
Venezuela	52.0	53.6	57.2	60.8	62.5	64.3	66.5	68.2	68.1	65.5
Virgin Islands, U.S.	0.8	0.9	0.9	0.9	0.9	0.9	0.9	0.9	0.9	0.9
Other	5.5	5.5	5.9	6.2	6.9	7.0	7.7	7.7	8.4	9.2
Total	**452.7**	**463.7**	**484.2**	**500.2**	**529.6**	**556.6**	**591.8**	**619.7**	**647.6**	**662.9**
Western Europe										
Austria	43.1	45.2	47.9	47.8	47.0	47.9	49.2	50.8	50.9	51.9
Belgium	56.9	58.7	61.5	63.8	64.7	67.7	69.8	71.3	72.9	74.5
Denmark	29.2	29.4	29.9	30.9	31.2	30.9	32.2	32.5	32.6	33.0
Finland	56.5	59.0	58.8	64.1	67.1	70.4	69.5	70.5	75.4	79.3
France	316.2	325.8	348.3	354.8	356.0	359.2	366.6	383.9	377.8	389.3
Germany	0.0	0.0	474.8	468.4	465.2	469.6	479.3	485.4	485.6	488.0
Germany, East	110.4	97.7	0.0	0.0	0.0	0.0	0.0	0.0	0.0	0.0
Germany, West	384.1	391.4	0.0	0.0	0.0	0.0	0.0	0.0	0.0	0.0
Greece	29.6	31.2	32.0	33.4	34.3	35.9	37.2	38.7	40.4	42.2
Iceland	4.2	4.1	4.1	4.2	4.3	4.4	4.6	4.7	5.1	5.8
Ireland	11.8	12.5	13.1	13.8	14.2	14.8	15.4	16.4	17.3	18.4
Italy	216.5	222.4	228.2	232.1	233.7	239.9	247.1	249.2	257.0	266.7
Luxembourg	4.4	4.5	4.8	4.6	4.7	5.1	5.5	5.3	5.6	5.9
Netherlands	69.6	71.9	75.0	77.5	78.9	81.6	83.6	86.8	91.4	94.3
Norway	94.8	96.2	99.2	99.3	102.6	103.4	106.0	105.4	105.7	111.0
Portugal	24.4	25.9	27.2	28.3	28.5	29.7	31.2	32.7	34.3	36.2
Spain	128.7	134.0	137.1	139.6	139.4	145.1	152.0	156.3	162.9	170.3
Sweden	130.9	131.2	132.4	131.3	132.4	130.5	133.9	133.2	135.1	135.1
Switzerland	46.4	47.5	48.6	48.4	48.0	48.5	50.1	50.3	49.8	50.8
Turkey	43.2	50.6	54.0	60.0	65.8	69.4	76.6	84.9	94.6	102.2
United Kingdom	287.2	287.4	294.5	295.0	299.3	301.1	309.1	321.1	319.6	331.5
Former Yugoslavia	73.0	72.8	65.5	0.0	0.0	0.0	0.0	0.0	0.0	0.0
Bosnia and Herzegovina	0.0	0.0	0.0	10.9	10.9	1.9	2.2	2.2	2.3	2.1
Croatia	0.0	0.0	0.0	10.9	10.7	11.1	11.5	11.5	12.7	12.9
Macedonia, TFYR	0.0	0.0	0.0	5.6	5.2	5.0	5.4	5.7	5.9	6.2
Serbia and Montenegro	0.0	0.0	0.0	33.6	29.7	30.7	33.0	33.7	35.8	36.1
Slovenia	0.0	0.0	0.0	8.9	9.0	11.4	9.5	9.7	10.0	10.7
Other	1.2	1.2	1.5	1.6	1.5	1.6	1.7	1.7	1.7	1.8
Total	**2,162.2**	**2,200.7**	**2,238.4**	**2,268.6**	**2,284.3**	**2,316.7**	**2,382.1**	**2,444.0**	**2,482.8**	**2,556.2**

TABLE 8.4

World total net electricity consumption, 1989–98 [CONTINUED]

(billion kilowatt-hours)

Region/country	1989	1990	1991	1992	1993	1994	1995	1996	1997	1998
Eastern Europe & Former U.S.S.R.										
Albania	3.2	3.1	2.7	2.6	3.1	3.6	4.1	5.6	5.4	5.3
Bulgaria	44.4	40.5	36.3	33.9	33.6	33.5	36.7	37.7	33.8	35.5
Former Czechoslovakia	81.1	80.7	76.4	70.9	0.0	0.0	0.0	0.0	0.0	0.0
Czech Republic	0.0	0.0	0.0	0.0	49.6	51.2	54.0	56.3	56.2	54.7
Slovakia	0.0	0.0	0.0	0.0	23.9	24.3	24.8	26.5	26.0	21.1
Hungary	37.0	36.3	33.7	31.2	31.4	31.5	32.3	33.0	33.2	33.3
Poland	129.3	118.5	115.7	112.3	115.1	116.2	119.3	122.6	123.2	121.9
Romania	74.7	65.8	55.2	52.0	50.9	49.6	52.9	54.8	51.0	49.6
Former U.S.S.R.	1,491.0	1,488.4	1,475.3	0.0	0.0	0.0	0.0	0.0	0.0	0.0
Armenia	0.0	0.0	0.0	8.3	5.7	5.1	4.7	5.4	5.4	5.4
Azerbaijan	0.0	0.0	0.0	16.7	16.9	15.7	15.4	15.2	15.8	15.5
Belarus	0.0	0.0	0.0	39.3	35.2	31.4	29.4	28.9	30.6	28.7
Estonia	0.0	0.0	0.0	7.1	6.6	6.9	6.8	7.1	7.1	7.6
Georgia	0.0	0.0	0.0	11.3	9.9	7.1	7.0	6.7	6.7	6.1
Kazakhstan	0.0	0.0	0.0	86.2	83.4	64.9	64.3	56.8	49.5	48.8
Kyrgyzstan	0.0	0.0	0.0	8.8	9.3	9.4	12.6	10.6	9.9	11.1
Latvia	0.0	0.0	0.0	7.5	6.1	5.9	5.9	6.0	5.9	4.9
Lithuania	0.0	0.0	0.0	11.2	10.3	10.1	10.0	8.5	8.8	7.8
Moldova	0.0	0.0	0.0	9.8	6.9	7.8	7.2	6.9	6.6	7.1
Russia	0.0	0.0	0.0	879.9	830.6	730.6	738.2	726.3	709.4	702.7
Tajikistan	0.0	0.0	0.0	16.3	15.1	14.8	14.3	14.1	13.9	12.6
Turkmenistan	0.0	0.0	0.0	8.6	7.9	6.6	6.5	6.0	6.5	5.5
Ukraine	0.0	0.0	0.0	216.7	200.6	178.1	168.8	159.5	156.7	144.0
Uzbekistan	0.0	0.0	0.0	44.2	40.4	43.7	40.5	44.0	42.5	41.3
Total	**1,860.7**	**1,833.3**	**1,795.3**	**1,674.7**	**1,592.5**	**1,448.0**	**1,455.5**	**1,438.7**	**1,404.2**	**1,370.5**
Middle East										
Bahrain	3.1	3.1	3.1	3.4	3.7	4.0	4.2	4.2	4.3	4.4
Cyprus	1.6	1.7	1.8	2.1	2.3	2.3	2.2	2.3	2.4	2.5
Iran	42.6	51.9	56.4	60.4	66.9	72.0	74.6	79.8	84.1	88.6
Iraq	25.3	19.3	17.7	22.1	23.0	24.5	25.4	25.4	25.9	26.4
Israel	17.5	17.9	18.4	21.2	22.4	24.4	25.7	27.4	28.3	31.8
Jordan	3.0	3.2	3.4	4.0	4.3	4.6	5.5	5.6	5.9	6.1
Kuwait	18.8	18.0	9.4	14.7	17.6	19.9	20.7	22.3	23.7	25.1
Lebanon	2.7	1.7	2.7	3.3	4.2	4.7	5.0	6.8	8.1	9.6
Oman	4.1	4.7	4.8	4.5	5.1	5.4	5.6	5.9	6.4	6.8
Qatar	4.1	4.2	4.1	4.5	4.8	5.1	5.2	5.7	6.0	6.2
Saudi Arabia	57.3	60.4	64.4	68.8	76.4	84.6	87.3	91.0	96.5	102.4
Syria	9.2	10.4	11.0	11.4	11.4	13.7	13.7	15.1	16.2	16.3
United Arab Emirates	13.6	14.9	15.1	15.3	15.4	16.5	16.7	17.3	18.0	18.7
Yemen	1.5	1.5	1.6	1.7	1.8	1.9	2.1	2.0	2.1	2.1
Total	**204.4**	**212.8**	**213.9**	**237.5**	**259.5**	**283.7**	**293.8**	**310.8**	**327.8**	**347.2**
Africa										
Algeria	13.4	14.0	14.5	15.1	15.8	16.3	17.0	17.9	19.0	19.9
Angola	1.7	1.7	1.7	1.7	1.7	1.7	1.7	1.7	1.7	1.8
Cameroon	2.5	2.5	2.5	2.5	2.5	2.5	2.6	2.7	2.9	3.1
Congo (Kinshasa)	6.3	5.0	4.7	5.4	4.1	4.1	5.5	5.9	5.7	5.5
Cote d'Ivoire (Ivory Coast)	2.9	2.1	1.7	1.5	2.0	2.0	2.6	2.8	2.9	3.2
Egypt	35.1	38.5	39.6	40.5	44.4	46.6	48.4	48.1	51.0	53.8
Ghana	4.6	5.2	5.4	5.4	5.4	5.3	5.2	5.0	5.3	5.4
Kenya	2.8	2.9	3.1	3.1	3.4	3.5	3.6	3.8	4.0	4.1
Libya	14.1	14.7	14.8	14.8	14.9	15.6	15.7	16.0	15.9	15.7
Morocco	8.0	8.6	8.7	9.5	10.0	10.8	10.7	11.1	11.7	12.4
Nigeria	11.3	11.1	12.6	13.1	12.8	13.7	12.9	13.4	13.5	13.7
South Africa	142.7	143.8	146.1	144.6	149.4	156.2	161.1	168.6	175.8	174.5
Tunisia	4.4	4.8	5.0	5.4	5.5	5.9	6.5	6.8	6.9	7.5
Zambia	4.8	5.7	5.7	5.7	5.7	5.7	5.7	5.9	6.2	6.4
Zimbabwe	8.4	10.0	9.7	8.7	8.9	7.9	7.9	8.8	9.9	8.4
Other	14.4	16.5	17.3	17.5	19.3	19.2	19.8	20.9	21.0	20.3
Total	**277.1**	**287.1**	**293.2**	**294.5**	**305.7**	**316.9**	**327.0**	**339.4**	**353.4**	**355.6**

TABLE 8.4

World total net electricity consumption, 1989–98 [CONTINUED]

(billion kilowatt-hours)

Region/country	1989	1990	1991	1992	1993	1994	1995	1996	1997	1998
Far East & Oceania										
Afghanistan	1.0	1.0	1.0	0.8	0.7	0.8	0.7	0.6	0.6	0.5
American Samoa	0.1	0.1	0.1	0.1	0.1	0.1	0.1	0.1	0.1	0.1
Australia	129.5	136.1	138.2	142.3	146.1	149.5	154.5	158.4	163.2	173.3
Bangladesh	6.5	7.1	7.8	8.4	8.7	9.3	10.5	10.9	11.2	11.0
Bhutan	0.0	0.0	0.1	0.1	0.1	0.1	0.1	0.3	0.4	0.3
Brunei	1.0	1.1	1.2	1.2	1.4	1.5	1.7	1.9	2.1	2.4
Burma	2.2	2.2	2.4	2.7	3.0	3.2	3.6	3.5	3.8	4.0
Cambodia	0.1	0.1	0.1	0.2	0.2	0.2	0.2	0.2	0.2	0.2
China	518.2	550.9	600.9	670.6	744.1	816.2	880.9	926.0	972.7	1,014.0
Fiji	0.4	0.4	0.4	0.4	0.4	0.5	0.5	0.5	0.5	0.5
French Polynesia	0.2	0.3	0.3	0.3	0.3	0.3	0.3	0.3	0.3	0.3
Guam	0.7	0.7	0.7	0.7	0.7	0.7	0.7	0.7	0.7	0.7
Hong Kong	22.2	23.5	24.7	25.6	27.3	25.4	28.8	31.9	32.6	34.6
India	238.1	257.1	280.6	295.1	316.9	341.9	369.7	385.4	411.7	416.3
Indonesia	35.9	43.2	45.4	46.0	41.2	46.0	49.2	59.1	65.7	68.0
Japan	712.5	764.6	791.3	797.9	806.4	857.9	881.4	899.0	929.4	926.3
Korea, North	48.2	48.1	48.2	34.3	34.3	33.4	32.5	31.7	30.7	29.7
Korea, South	81.8	93.4	103.0	113.6	125.4	159.4	176.3	196.3	214.2	205.8
Laos	0.2	0.3	0.2	0.2	0.3	0.5	0.3	0.4	0.4	0.5
Macau	0.7	0.8	0.9	1.0	1.1	1.2	1.5	1.4	1.4	1.4
Malaysia	19.2	22.3	24.9	25.8	30.6	34.5	40.1	45.2	50.8	53.4
Mongolia	3.3	3.1	2.9	2.7	2.3	2.5	2.7	2.7	2.7	2.8
Nepal	0.6	0.7	0.8	0.9	0.8	0.9	0.9	1.1	1.1	1.2
New Caledonia	1.1	1.0	1.0	1.0	1.0	1.1	1.5	1.4	1.4	1.4
New Zealand	28.6	29.9	30.9	29.5	31.3	32.8	33.0	34.2	33.9	33.3
Pakistan	36.0	33.8	36.8	40.7	43.5	45.3	47.9	50.8	52.7	55.1
Papua New Guinea	1.6	1.6	1.6	1.6	1.6	1.6	1.6	1.6	1.6	1.6
Philippines	22.9	22.7	22.0	21.4	21.9	26.9	29.7	32.5	35.2	36.8
Samoa	0.0	0.0	0.0	0.1	0.1	0.1	0.1	0.1	0.1	0.1
Singapore	12.3	13.7	14.8	15.5	16.5	18.1	19.4	20.5	22.9	24.7
Sri Lanka	2.6	2.9	3.1	2.9	3.7	4.0	4.4	4.1	4.7	5.1
Taiwan	74.6	77.5	84.6	90.2	98.3	106.2	113.3	109.1	115.3	124.2
Thailand	35.1	41.2	44.7	50.7	56.1	63.0	70.6	77.2	80.7	77.6
U.S. Pacific Islands	0.2	0.2	0.2	0.2	0.2	0.2	0.2	0.2	0.2	0.2
Vietnam	7.0	7.9	8.4	8.9	9.7	11.2	12.5	15.5	17.5	19.2
Other	0.1	0.1	0.1	0.1	0.2	0.2	0.2	0.2	0.2	0.2
Total	**2,044.8**	**2,189.7**	**2,324.6**	**2,433.6**	**2,576.5**	**2,796.4**	**2,971.6**	**3,104.7**	**3,262.8**	**3,327.3**
World Total	**10,299.8**	**10,549.0**	**10,777.4**	**10,859.1**	**11,110.7**	**11,386.2**	**11,791.8**	**12,133.2**	**12,411.0**	**12,636.9**

Last Updated on 01/11/2000

SOURCE: *International Energy Annual 1998,* Energy Information Administration, Washington, D.C., 2000

CHAPTER 9

RENEWABLE ENERGY

RENEWABLE ENERGY DEFINED

Imagine an energy source that uses no oil, produces no pollution, cannot be affected by political events and cartels, creates no radioactive waste, and yet is economical. Although it sounds impossible, some experts claim that technological advances could make a renewable energy possible within a few decades. Renewable energy is from sources that are naturally regenerated and are therefore virtually unlimited. These energy sources include the sun (solar), wind, water (hydropower), vegetation (biomass), and the heat of the earth (geothermal).

Solar energy, wind energy, hydropower, and geothermal power are all renewable, cheap, and clean sources of energy. Each of these alternative energy sources has advantages and disadvantages, and many observers hope that one or more of them may provide a substantially better energy source than conventional, fossil fuel methods. As the United States and the rest of the world continue to expand their energy needs, putting a strain on the environment, alternative sources of energy continue to be explored.

A HISTORICAL PERSPECTIVE

Before the nineteenth century, most energy came from renewable sources. People burned wood for heat, used sails to harness the wind and propel boats, and installed water wheels on streams to grind grain. The large-scale shift to nonrenewable energy sources began in the 1700s with the Industrial Revolution, a period marked by the rise of factories first in Europe and then in North America. As the demand for energy grew, coal replaced wood as the main fuel source. Coal was the most efficient fuel for the steam engine, one of the most important inventions of the Industrial Revolution.

Until the early 1970s most Americans were unconcerned about the sources of the nation's energy. Supplies of coal and oil, which together provided more than 90 percent of U.S. energy, were believed to be plentiful. The decades preceding the 1970s were characterized by cheap gasoline and little public discussion of energy conservation.

That carefree approach to energy consumption ended in the 1970s. A fuel oil crisis made Americans more aware of the importance of developing alternative sources of energy to supplement and perhaps even replace fossil fuels. In major cities throughout the United States, gasoline rationing became commonplace, lower heat settings for offices and living quarters were encouraged, and people waited in line to fill their gas tanks. In a country where mobility and personal transportation are highly valued, the oil crisis was a shocking reality for many Americans. As a result, the administration of President Jimmy Carter (1977–81) encouraged federal funding for research into alternative energy sources.

In 1978 the U.S. Congress passed the Public Utilities Regulatory Policies Act (PURPA; PL 95-617), which was designed to help the struggling alternative energy industry. The act exempted small alternative producers from state and federal utility regulations and required existing local utilities to buy electricity from them. The renewable energy industries responded by growing rapidly, gaining experience, improving technologies, and lowering costs. This act was the single most important factor in the development of the commercial renewable energy market.

In the 1980s President Ronald Reagan decided that private sector financing for the short-term development of alternative energy sources was best. As a result, he proposed the reduction or elimination of federal expenditures for alternative energy sources. Although funds were severely cut, the U.S. Department of Energy (DOE) continued to support some research and development to explore alternate sources of energy. The Clinton Administration reemphasized the importance of renewable energy, and increased funding in several areas.

TABLE 9.1

Energy consumption by source, 1949–99

(quadrillion Btu)

Year	Fossil fuels: Coal	Fossil fuels: Coal coke net imports	Fossil fuels: Natural gas[1]	Fossil fuels: Petroleum[2]	Fossil fuels: Total fossil fuels	Nuclear electric power	Hydroelectric pumped storage[3]	Renewable energy: Conventional hydroelectric power[4]	Renewable energy: Geothermal[5]	Renewable energy: Wood and waste[6]	Renewable energy: Solar	Renewable energy: Wind	Renewable energy: Total renewable energy	Total[7]
1949	11.981	-0.007	5.145	11.883	29.002	0	([8])	1.449	0	1.549	0	0	2.998	32.000
1950	12.347	0.001	5.968	13.315	31.632	0	([8])	1.440	0	1.562	0	0	3.003	34.635
1951	12.553	-0.021	7.049	14.428	34.008	0	([8])	1.454	0	1.535	0	0	2.988	36.996
1952	11.306	-0.012	7.550	14.956	33.800	0	([8])	1.496	0	1.474	0	0	2.970	36.770
1953	11.373	-0.009	7.907	15.556	34.826	0	([8])	1.439	0	1.419	0	0	2.857	37.684
1954	9.715	-0.007	8.330	15.839	33.877	0	([8])	1.388	0	1.394	0	0	2.783	36.660
1955	11.167	-0.010	8.998	17.255	37.410	0	([8])	1.407	0	1.424	0	0	2.832	40.242
1956	11.350	-0.013	9.614	17.937	38.888	0	([8])	1.487	0	1.416	0	0	2.903	41.791
1957	10.821	-0.017	10.191	17.932	38.926	(s)	([8])	1.557	0	1.334	0	0	2.890	41.816
1958	9.533	-0.007	10.663	18.527	38.717	0.002	([8])	1.629	0	1.323	0	0	2.952	41.670
1959	9.518	-0.008	11.717	19.323	40.550	0.002	([8])	1.587	0	1.353	0	0	2.940	43.493
1960	9.838	-0.006	12.385	19.919	42.137	0.006	([8])	1.657	0.001	1.320	0	NA	2.977	45.120
1961	9.623	-0.008	12.926	20.216	42.758	0.020	([8])	1.680	0.002	1.295	0	NA	2.977	45.755
1962	9.906	-0.006	13.731	21.049	44.681	0.026	([8])	1.822	0.002	1.300	0	NA	3.124	47.832
1963	10.413	-0.007	14.403	21.701	46.509	0.038	([8])	1.772	0.004	1.323	0	NA	3.099	49.647
1964	10.964	-0.010	15.288	22.301	48.543	0.040	([8])	1.907	0.005	1.337	0	NA	3.248	51.831
1965	11.581	-0.018	15.769	23.246	50.577	0.043	([8])	2.058	0.004	1.335	0	NA	3.397	54.016
1966	12.143	-0.025	16.995	24.401	53.514	0.064	([8])	2.073	0.004	1.369	0	NA	3.446	57.024
1967	11.914	-0.015	17.945	25.284	55.127	0.088	([8])	2.344	0.007	1.340	0	NA	3.691	58.906
1968	12.331	-0.017	19.210	26.979	58.502	0.142	([8])	2.342	0.009	1.419	0	NA	3.771	62.415
1969	12.382	-0.036	20.678	28.338	61.362	0.154	([8])	2.659	0.013	1.440	0	NA	4.113	65.628
1970	12.265	-0.058	21.795	29.521	63.522	0.239	([8])	2.654	0.011	[R]1.429	0	NA	[R]4.094	[R]67.856
1971	11.598	-0.033	22.469	30.561	64.596	0.413	([8])	2.861	0.012	[R]1.430	0	NA	[R]4.303	[R]69.312
1972	12.077	-0.026	22.698	32.947	67.696	0.584	([8])	2.944	0.031	[R]1.501	0	NA	[R]4.476	[R]72.756
1973	12.971	-0.007	22.512	34.840	70.316	0.910	([8])	3.010	0.043	[R]1.527	0	NA	[R]4.579	[R]75.806
1974	12.663	0.056	21.732	33.455	67.906	1.272	([8])	3.309	0.053	[R]1.538	0	NA	[R]4.900	[R]74.078
1975	12.663	0.014	19.948	32.731	65.355	1.900	([8])	3.219	0.070	[R]1.497	0	NA	[R]4.786	[R]72.041
1976	13.584	(s)	20.345	35.175	69.104	2.111	([8])	3.066	0.078	[R]1.711	0	NA	[R]4.855	[R]76.070
1977	13.922	0.015	19.931	37.122	70.989	2.702	([8])	2.515	0.077	[R]1.837	0	NA	[R]4.429	[R]78.120
1978	13.766	0.125	20.000	37.965	71.856	3.024	([8])	3.141	0.064	[R]2.036	0	NA	[R]5.242	[R]80.122
1979	15.040	0.063	20.666	37.123	72.892	2.776	([8])	3.141	0.084	[R]2.150	0	NA	[R]5.375	[R]81.042
1980	15.423	-0.035	20.394	34.202	69.984	2.739	([8])	3.118	0.110	[R]2.483	0	NA	[R]5.710	[R]78.434
1981	15.908	-0.016	19.928	31.931	67.750	3.008	([8])	3.105	0.123	2.590	0	NA	5.818	76.569
1982	15.322	-0.022	18.505	30.232	64.037	3.131	([8])	3.572	0.105	[R]2.615	0	NA	[R]6.292	[R]73.441
1983	15.894	-0.016	17.357	30.054	63.290	3.203	([8])	3.899	0.129	2.831	0	(s)	6.860	73.317
1984	17.071	-0.011	18.507	31.051	66.617	3.553	([8])	3.800	0.165	2.880	0	(s)	6.845	76.972
1985	17.478	-0.013	17.834	30.922	66.221	4.149	([8])	3.398	0.198	[R,9]2.862	0	(s)	[R,9]6.458	[R,9]76.777
1986	17.260	-0.017	16.708	32.196	66.148	4.471	([8])	3.446	0.219	[R,9]2.840	0	(s)	[R,9]6.506	[R,9]77.065
1987	18.008	0.009	17.744	32.865	68.626	4.906	([8])	3.117	0.229	[R]2.822	0	(s)	[R]6.169	[R]79.633
1988	18.846	0.040	18.552	34.222	71.660	5.661	([8])	2.662	0.217	[R,9]2.940	0	(s)	[R,9]5.819	[R,9]83.071
1989	18.926	0.030	19.384	34.211	72.551	5.677	([8])	[R,10]2.999	[R,10]0.338	[R,10]3.050	[R,10]0.059	[R,10]0.024	[R,10]6.470	[R,10]84.593
1990	19.101	0.005	19.296	33.553	71.955	[R]6.162	-0.036	[R,11]3.140	[R]0.359	[R]2.665	0.063	[R]0.032	[R]6.260	[R]84.186
1991	18.770	[R]0.010	19.606	32.845	[R]71.231	[R]6.580	-0.047	[R]3.222	[R]0.368	[R]2.679	0.066	[R]0.032	[R]6.367	[R]84.063
1992	[12]19.158	[R]0.035	20.131	33.527	[R,12]72.850	[R]6.608	-0.043	2.863	0.379	[R]2.826	0.068	0.030	[R]6.167	[R,12]85.512
1993	19.776	[R]0.027	20.827	33.841	[R]74.471	[R]6.520	-0.042	3.147	0.393	[R]2.782	0.071	0.031	[R]6.424	87.309
1994	19.960	[R]0.058	21.288	[R]34.670	[R]75.976	[R]6.838	-0.035	2.971	0.395	[R]2.914	0.072	0.036	[R]6.387	[R]89.234
1995	20.024	[R]0.061	22.163	[R]34.553	[R]76.802	7.177	-0.028	3.474	0.339	[R]3.044	0.073	0.033	[R]6.963	[R]90.940
1996	20.940	[R]0.023	[R]22.559	[R]35.757	[R]79.279	7.168	-0.032	[R]3.915	0.352	[R]3.104	0.075	0.035	[R]7.482	[R]93.911
1997	21.444	[R]0.046	[R]22.530	[R]36.266	[R]80.286	6.678	-0.042	[R]3.940	[R]0.328	[R]2.982	0.074	[R]0.034	[R]7.358	[R]94.316
1998	[R]21.593	[R]0.067	[R]21.921	[R]36.934	[R]80.515	7.157	-0.046	[R]3.552	[R]0.335	[R]2.991	0.074	[R]0.031	[R]6.984	[R]94.570
1999[P]	21.698	0.058	22.096	37.706	81.557	7.733	-0.063	3.417	0.327	3.514	0.076	0.038	7.373	96.596

[1] Includes supplemental gaseous fuels.

[2] Petroleum products supplied, including natural gas plant liquids and crude oil burned as fuel.

[3] Represents total pumped storage facility production minus energy used for pumping.

[4] Through 1988, includes all net imports of electricity. From 1989, includes only the portion of net imports of electricity that is derived from hydroelectric power.

[5] Includes electricity imports from Mexico that are derived from geothermal energy.

[6] Values are estimated. For all years, includes wood consumption in all sectors. Beginning in 1970, includes electric utility waste consumption. Beginning in 1981, includes industrial sector waste consumption, and transportation sector use of ethanol blended into motor gasoline. Beginning in 1989, includes expanded coverage of nonutility wood and waste consumption.

[7] From 1989, includes net imported electricity from nonrenewable sources and removes ethanol blended into motor gasoline, which would otherwise be double counted in both petroleum and renewable energy.

[8] Through 1989, pumped storage is included in conventional hydroelectric power.

[9] Not all data were available; therefore, values were interpolated.

[10] There is a discontinuity in this time series between 1988 and 1989 due to the expanded coverage of renewable energy beginning in 1989.

[11] There is a discontinuity in this time series between 1989 and 1990; beginning in 1990, pumped storage is removed and expanded coverage of use of hydroelectric power is included.

[12] Independent power producers' use of coal is included beginning in 1992.

R=Revised. P=Preliminary. (s)=Less than 0.0005 and greater than -0.0005 quadrillion Btu. NA=Not available.

Note: Totals may not equal sum of components due to independent rounding.

SOURCE: *Annual Energy Review 1999,* Energy Information Administration, Washington, D.C., 2000

DOMESTIC RENEWABLE ENERGY USAGE

Renewable energy currently contributes only a small portion of the nation's energy supply, although its importance is expected to grow in the future. In 1999 the United States consumed an estimated 7.4 quadrillion Btu of renewable energy, almost 8 percent of the nation's total energy consumption. (See Table 9.1 and Figure 9.1.) Biomass sources (wood and waste) contributed 3.5 quadrillion Btu, while hydropower provided 3.4 quadrillion Btu. Together, biomass and hydropower provided 94 percent of renewable energy, or 7 percent of all energy. Geothermal energy was the third largest source, with .33 quadrillion Btu. Solar power contributed .08 quadrillion Btu, and wind provided .04 quadrillion Btu. Electric utilities and the industrial sector were the biggest consumers of renewable energy.

FIGURE 9.1

Renewable energy as a share of total energy, 1999

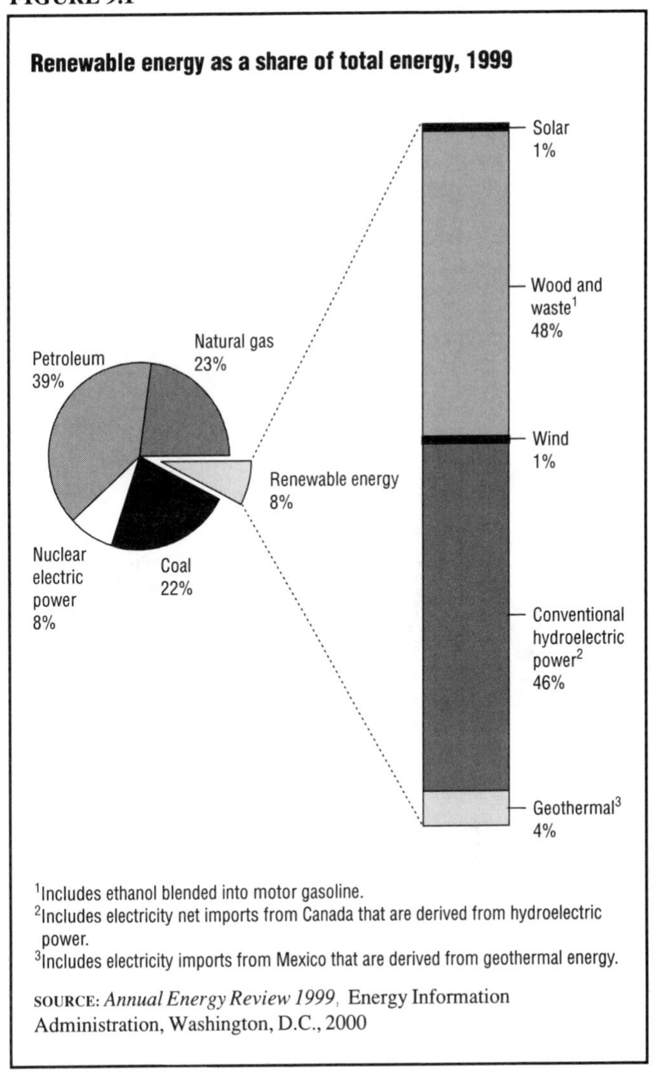

[1]Includes ethanol blended into motor gasoline.
[2]Includes electricity net imports from Canada that are derived from hydroelectric power.
[3]Includes electricity imports from Mexico that are derived from geothermal energy.

SOURCE: *Annual Energy Review 1999*, Energy Information Administration, Washington, D.C., 2000

BIOMASS ENERGY

Biomass refers to organic material such as plant and animal waste, wood, seaweed and algae, and garbage. These raw materials can be converted into liquid or gaseous fuels or used directly to provide heat and electricity. A biofuel is the product of biomass conversion. The byproducts of biomass conversion can be used for fertilizers and chemicals. Wood, the most commonly used biofuel, is used to heat millions of homes every year. In 2000 wood and other biomass resources provided the largest source of renewable electricity produced in the United States.

Some uses of biomass are not without environmental problems. When wood is widely used as a fuel in an area, deforestation can occur, resulting in the possibility of soil erosion and mudslides. Burning wood, like fossil fuels, also pollutes the environment.

There are two types of biofuel energy (bioenergy) conversion processes: thermochemical conversion and biochemical conversion.

Thermochemical Conversion

Thermochemical conversion uses heat to produce chemical reactions in biomass. Direct combustion is the easiest and most commonly used method. Materials such as dry wood or agricultural wastes are chopped and burned to produce steam, electricity, or heat for industries, utilities, and homes. Industrial-size wood boilers are operating throughout the country, and the Department of Energy projects that many more will be built during the next decade. The burning of agricultural wastes is also becoming more widespread. In Florida, sugar cane producers use the residue from the cane to generate much of their energy.

Wood burning in stoves and fireplaces is another example of thermochemical conversion. In the United States, the number of homes burning wood for fuel—20 million—has remained relatively unchanged since 1980, although these homes are using less fuel now. Homes in the South consume far greater amounts of wood energy than in other parts of the country.

Pyrolysis, also called gasification or carbonization, is a thermochemical process that uses heat to break down biomass and yields liquid, gaseous, and solid substances. Charcoal is an example of this process.

Biochemical Conversion

Biochemical conversion uses enzymes, fungi, or other microorganisms to convert high-moisture biomass into either liquid or gaseous fuels. Bacteria convert manure, agricultural wastes, paper, and algae into methane, which is used as fuel. Sewage treatment plants have used anaerobic (without oxygen) digestion for many years to generate methane gas. Small-scale digesters have been used on farms, primarily in Europe and Asia, for hundreds of years. The DOE estimates that many thousands of these small biofuel plants are in use today in Korea and perhaps half a million plants operate in China.

Another type of biochemical conversion process, fermentation, uses yeast to decompose carbohydrates to

FIGURE 9.2

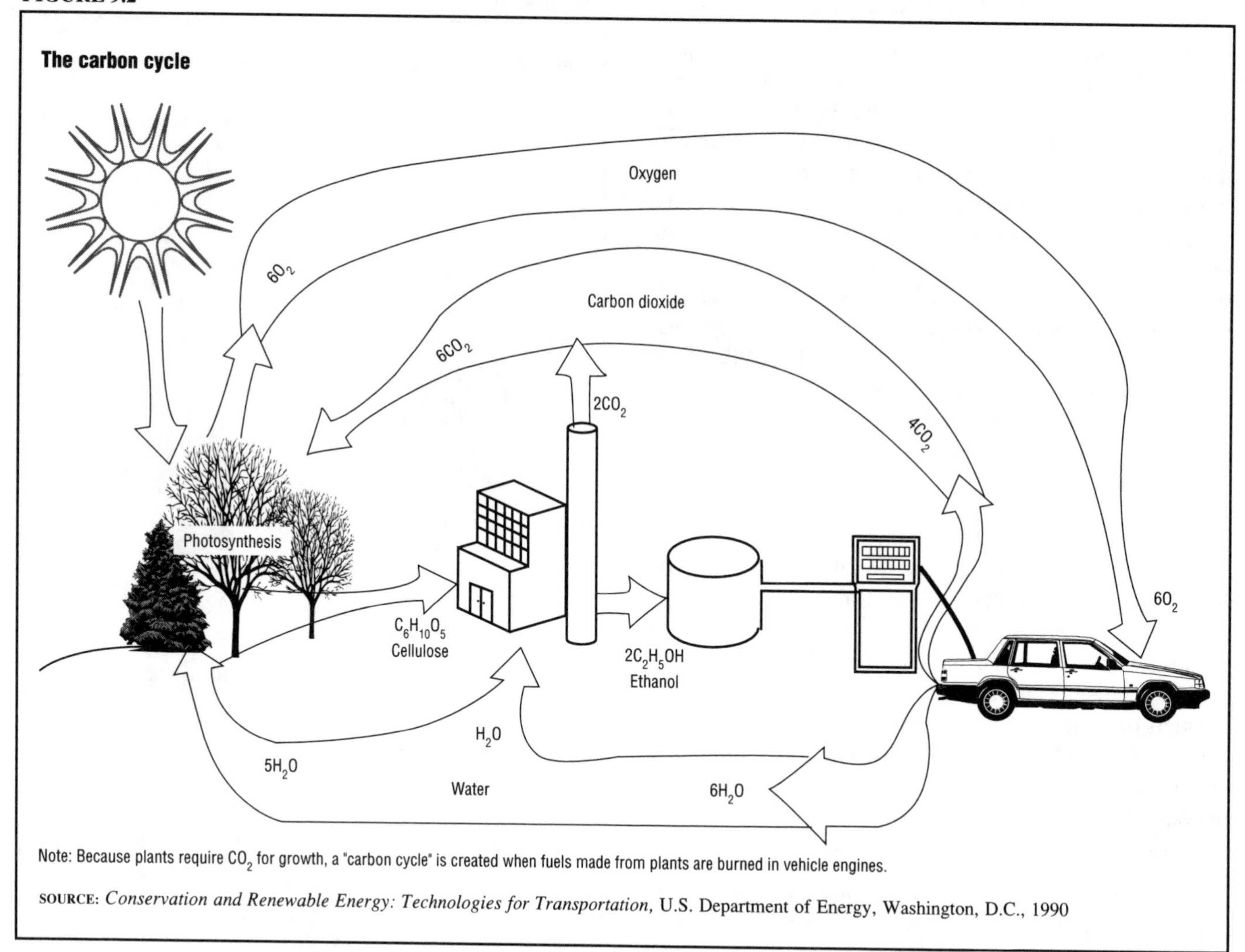

Note: Because plants require CO_2 for growth, a "carbon cycle" is created when fuels made from plants are burned in vehicle engines.

SOURCE: *Conservation and Renewable Energy: Technologies for Transportation,* U.S. Department of Energy, Washington, D.C., 1990

yield ethyl alcohol (ethanol) and carbon dioxide. Sugar crops, grains (corn, in particular), potatoes, and other starchy crops are common yields that supply the sugar for ethanol production.

Ethanol and Methanol

Ethanol (ethyl alcohol) is a colorless, nearly odorless, flammable liquid derived from fermenting plant material that contains carbohydrates in the form of sugar. Most of the ethanol manufactured for use as fuel is derived from corn, wood, and sugar. A mixture of 10 percent ethanol and 90 percent gasoline is usable in any internal combustion engine. Ethanol is difficult and expensive to produce in bulk. The development of ethanol as a fuel source may depend more upon the political support of legislators from farming states than upon real economic benefits.

Some scientists believe ethanol made from wood, sawdust, corncobs, or rice hulls could liberate the alcohol fuel industry from its dependence on food crops, such as corn and sugar cane. Worldwide, there are enough corncobs and rice hulls left over from annual crop production to produce more than 40 billion gallons of ethanol.

In just a decade, research has cut the cost of wood-derived ethanol from $4.00 per gallon to $1.35 per gallon. Advocates of wood-derived ethanol believe that it could eventually be a sustainable liquid fuel industry that does not rely on pollution-generating fossil fuels. For instance, if new trees were planted to replace those that were cut for fuel, they would be available for later harvesting while alleviating global warming by continuing their carbon-dioxide-processing function. Other scientists warn that an increased demand for wood for transportation fuels might accelerate the destruction of old-growth forests and endanger ecosystems.

Methanol (methyl alcohol) fuels have also been tested successfully. Using methanol instead of diesel fuel virtually eliminates sulfur emissions and reduces other environmental pollutants usually emitted from trucks and buses. Producing methanol from biofuels, however, is costly.

Burning biofuels in vehicle engines creates a "carbon cycle" in which the earth's vegetation can in turn make use of the products of combustion and therefore reduce net greenhouse gases. (See Figure 9.2.)

Municipal Waste Recovery

Each year millions of tons of garbage are buried in landfills and city dumps. This method of disposal is becoming increasingly costly as many landfills across the nation near capacity. Many communities discovered that they could solve both problems at once by constructing waste-to-energy plants. Not only is the garbage burned and reduced in volume by 90 percent, but energy in the form of steam or electricity is generated in a cost-effective way. The potential energy benefit is significant: the solid waste generated by the nation's households is equal in energy to more than 200 million barrels of oil per year. Use of municipal waste as fuel has increased steadily over the past two decades. In 1981 municipal waste, including landfill gas, generated 88 trillion Btu of energy, which grew to 571 trillion Btu by 1999.

The two most common waste-to-energy plant designs are the mass burn (also called direct combustion) system and the refuse derived fuel (RDF) system.

MASS BURN SYSTEMS. Most waste-to-energy plants in the United States use the mass burn system. This system's advantage is that the waste does not have to be sorted or prepared before burning, except for removing obviously noncombustible, oversized objects. The mass burn eliminates expensive sorting, shredding, and transportation machinery that may be prone to break down.

Waste is carried to the plant in trash trucks and dropped into a storage pit. Large overhead cranes lift the garbage into a furnace feed hopper that controls the amount and rate of waste that is fed into the furnace. Next the garbage is moved through a combustion zone so that it burns to the greatest extent possible. The burning garbage produces heat, and that heat is used to produce steam. The steam can be used directly for industrial needs or can be sent through a turbine to power a generator to produce electricity.

REFUSE DERIVED FUEL (RDF). RDF systems process waste to remove noncombustible objects and to create homogeneous and uniformly sized fuel. Large items such as bedsprings, dangerous materials, and flammable liquids are removed by hand. The trash is then shredded and carried to a screen to remove glass, rocks, and other material that cannot be burned. The remaining material is usually sifted a second time with an air separator to yield fluff. The fluff is sent to storage bins before being burned, or it can be compressed into pellets or briquettes for long-term storage. This fuel can be used as an energy source by itself in a variety of systems, or it can be used with other fuels, such as coal or wood.

PERFORMANCE OF WASTE-TO-ENERGY SYSTEMS. Most waste-to-energy systems can produce two to four pounds of steam for every pound of garbage burned. A 1,000-ton-per-day mass burn system will burn an average of 310,250 tons of trash each year and will recover 2 trillion Btu of energy. In addition, the plant will emit 96,000 tons of ash (32 percent of waste input) for landfill disposal. An RDF plant produces less ash but sends almost the same amount of waste to the landfill because of the noncombustibles that accumulate in the separation process before burning.

DISADVANTAGES OF WASTE-TO-ENERGY PLANTS. The major problem with increasing the use of municipal waste-to-energy plants is their effect on the environment. The emission of particles into the air is partially controlled by electrostatic precipitators, and many gases can be eliminated by proper combustion techniques. There is concern, however, over the amounts of dioxin (a very dangerous air pollutant) and other toxins that are often emitted from these plants. Noise from trucks, fans, and processing equipment at these plants can be unpleasant for nearby residents.

Landfill Gas Recovery

Landfills contain a large amount of biodegradable matter. Gas is created because a lack of oxygen helps the growth of methagens, which are types of bacteria that produce methane gas and carbon dioxide. In the past, as landfills aged, these gases built up and leaked out. This gas leakage prompted some communities to drill holes and burn off the dangerous methane.

The energy crisis of the 1970s made this methane gas an energy resource too valuable to waste, and efforts were made to find an inexpensive way to tap the gas. The first landfill gas-recovery site was finished in 1975 at the Palos Verdes Landfill in Rolling Hills Estates, California.

In a typical operation, garbage is allowed to decompose for several months. When a sufficient amount of methane gas has developed, it is piped out to a generating plant where it is turned into electricity. In its purest form, methane gas is equivalent to natural gas and can be used in exactly the same way. Depending on the extraction rates, most sites can produce gas for about 20 years. The advantages of tapping gas from a landfill go beyond the energy provided by the methane, as extraction reduces landfill odors and the chances of explosions.

HYDROPOWER

Hydropower is the world's largest renewable energy source. Hydropower is the energy that comes from the natural flow of water. The energy of falling water or flowing water is converted into mechanical energy and then to electrical energy. In the past, water's energy was harnessed by waterwheels to grind grain or turn saws, but today water is used to turn modern turbines. Hydropower is a renewable, nonpolluting, and reliable energy source.

Advantages and Disadvantages

At present hydropower is the only means of storing large quantities of electrical energy for almost

FIGURE 9.3

Cross-section of the earth showing source of geothermal energy

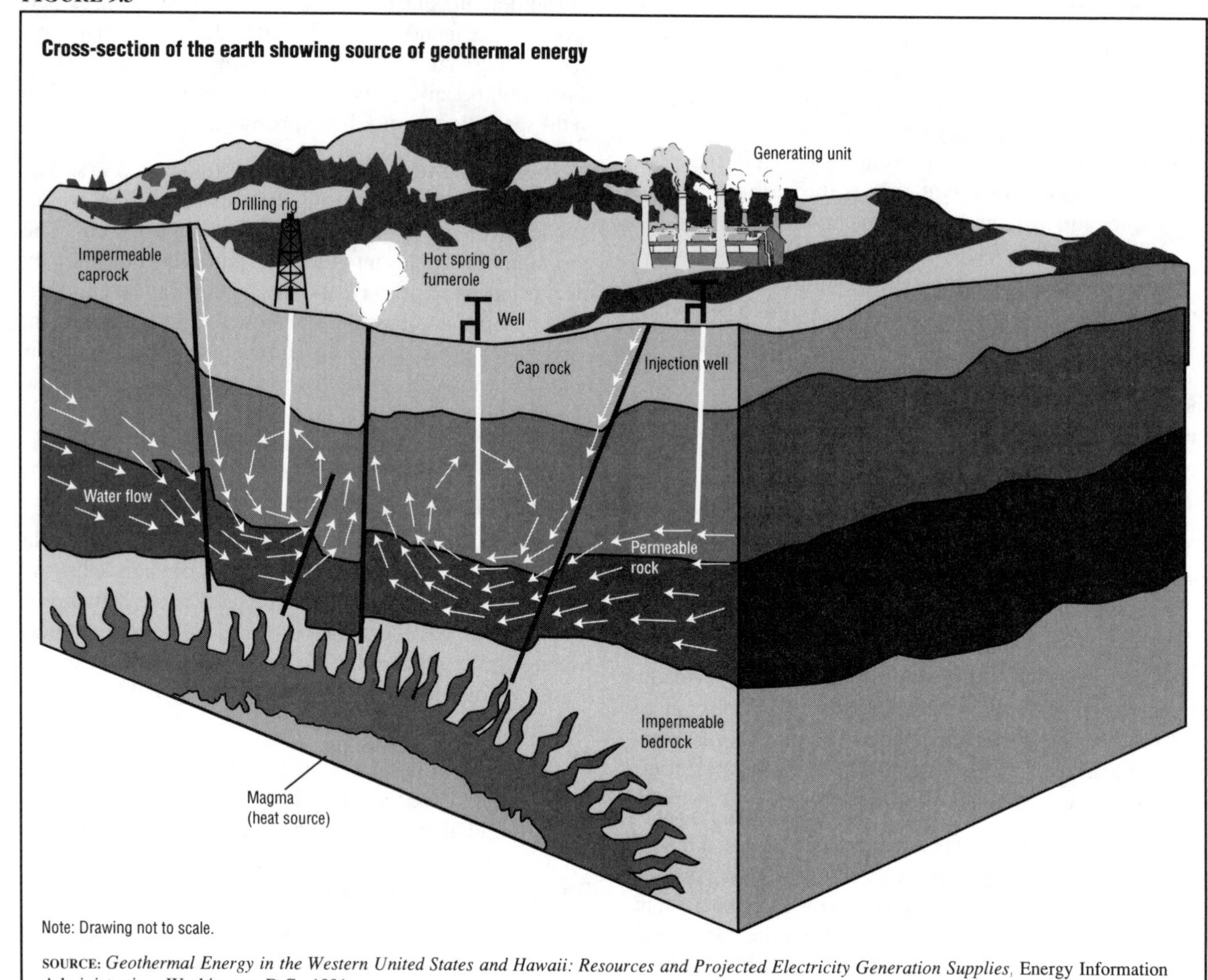

Note: Drawing not to scale.

SOURCE: *Geothermal Energy in the Western United States and Hawaii: Resources and Projected Electricity Generation Supplies*, Energy Information Administration, Washington, D.C., 1991

instant use. This is done by holding water in a large reservoir behind a dam, with a hydroelectric power plant below. The dam creates a height from which water flows. The fast-moving water pushes the turbine blades that turn the rotor part of the electric generator. Whenever power is needed at peak times, the valves are opened and in a short amount of time, turbine generators produce power.

On the other hand, nearly all the best sites for large hydropower plants have already been used in the United States. Small hydropower plants in the U.S. are costly to build, but become cost-efficient because of their low operating costs. One of the disadvantages of small hydropower generators is their reliance on rain and melting snow to fill reservoirs, as some years bring drought conditions. There is strong protest by environmental groups over constructing new dams in America. Ecologists express concern that dams ruin streams, dry up waterfalls, and interfere with marine life habitats.

New Directions in Hydropower Energy

The United States and Europe have developed a major proportion of their hydroelectric potential. Large-scale hydropower development has virtually stopped in the United States, with not one new dam being approved for federal funding in more than a decade. Until recently in the United States, dams were usually funded entirely with federal monies. Since 1986, however, any new dam proposed in the United States must be built with at least half the money being put up by local governments. Hydropower's contribution to U.S. energy generation should remain relatively constant, although existing sites can become more efficient as new generators are added. Any new major supplies of hydroelectric power for the United States will likely come from Canada.

Most of the new development in hydropower is occurring in the Third World as developing nations see it as an effective method of supplying power to growing populations. Most hydropower development programs are massive public works projects requiring huge amounts of money, which is mostly borrowed from the developed world. Third World leaders believe that, despite threats to the environment, hydroelectric dams will pay for themselves by bringing cheap electric power to their people.

GEOTHERMAL ENERGY

Since ancient times, humans have exploited Earth's natural hot water sources. Although bubbling hot springs became public baths in ancient Rome, using hot water and underground steam to produce power is a relatively recent development. Electricity was first generated from natural steam in Italy in 1904. The world's first steam power plant was built in 1958 in a volcanic region of New Zealand. A field of 28 geothermal power plants covering 30 square miles in northern California was completed in 1960.

What Is Geothermal Energy?

Geothermal energy is the natural, internal heat of the earth trapped in rock formations deep underground. Only a fraction of this vast storehouse of energy can be extracted, usually through large fractures in the earth's crust. Hot springs, geysers, and fumaroles (holes in or near volcanoes from which vapor escapes) are the most easily exploitable sources of geothermal energy. (See Figure 9.3.) Geothermal reservoirs provide hot water or steam that can be used for heating buildings, processing food, and generating electricity.

To produce power from a geothermal energy source, pressurized steam or hot water is extracted from the earth and directed toward turbines. The electricity produced by turbines is then fed into a utility grid and distributed to residential and commercial customers. Today, electricity production accounts for almost two-thirds of the world's geothermal energy commercial use.

Types of Geothermal Energy

Like most natural energy sources, geothermal energy is usable only when it is concentrated in one spot in what is called a "thermal reservoir." The four basic categories of thermal reservoirs are hydrothermal reservoirs, dry rock, geopressurized reservoirs, and magma resources. Most of the known areas for geothermal power in the United States are located west of the Mississippi River. (See Figure 9.4.)

Hydrothermal reservoirs consist of a heat source covered by a permeable formation through which water circulates. Steam is produced when hot water boils underground and some of the steam escapes to the surface under pressure. Once at the surface, impurities and tiny rock particles are removed, and the steam is then piped directly to the electrical generating station. These systems are the cheapest and simplest form of geothermal energy. The Geysers, 90 miles north of San Francisco, California, are the most famous example of this type. The Geysers produce enough electricity to meet the needs of about 1.3 million people.

Dry rock formations are the most common geothermal source, especially in the West. To tap this source of energy, water is injected into hot rock formations that have been fractured, to produce steam or water for collection.

Geopressurized reservoirs are sedimentary formations containing hot water and methane gas. Supplies of geopressurized energy remain uncertain, and drilling is expensive. Scientists hope that advanced technology will eventually permit the commercial exploitation of the methane content in these reservoirs.

Magma resources are found from 10,000 to 33,000 feet below the earth's surface where molten or partially liquefied rock is located. Because magma is so hot, ranging from 1,650 degrees to 2,200 degrees Fahrenheit, it is a good geothermal resource. The process for extracting energy from magma is still in the experimental stages.

Disadvantages of Geothermal Energy

Geothermal plants are expensive because they must be built near the source. Other drawbacks include low efficiency, bad odors from sulfur released in processing, noise, lack of access for most states, potentially harmful pollutants (hydrogen sulfide, ammonia, and radon), and poisonous arsenic or boron often found in geothermal waters. Serious environmental concerns have been raised over the release of chemical compounds, potential water contamination, the collapse of land surface around the area from which the water is being drained, and potential water shortages resulting from massive withdrawals of water.

Domestic Production of Geothermal Energy

Geothermal energy ranks third in renewable energy production in the United States, after biomass and hydroelectric energy. Public sector involvement in the geothermal industry began with the passage of the Geothermal Steam Act of 1970 (PL 91-581), which authorized the U.S. Department of Interior to lease geothermal resources on federal lands.

Although the United States is the greatest producer of geothermal power, with 44 percent of the world's capacity, the geothermal industry in the United States is currently static. Most of the easily exploited geothermal reserves have already been developed. In addition, utility companies and independent power producers are arguing over who should build additional generating capacity and what prices should be paid for the power. Continued growth in the American market depends on the regulatory environment, oil price trends, and the success of unproven technologies for economically exploiting some of the presently inaccessible geothermal reserves.

FIGURE 9.4

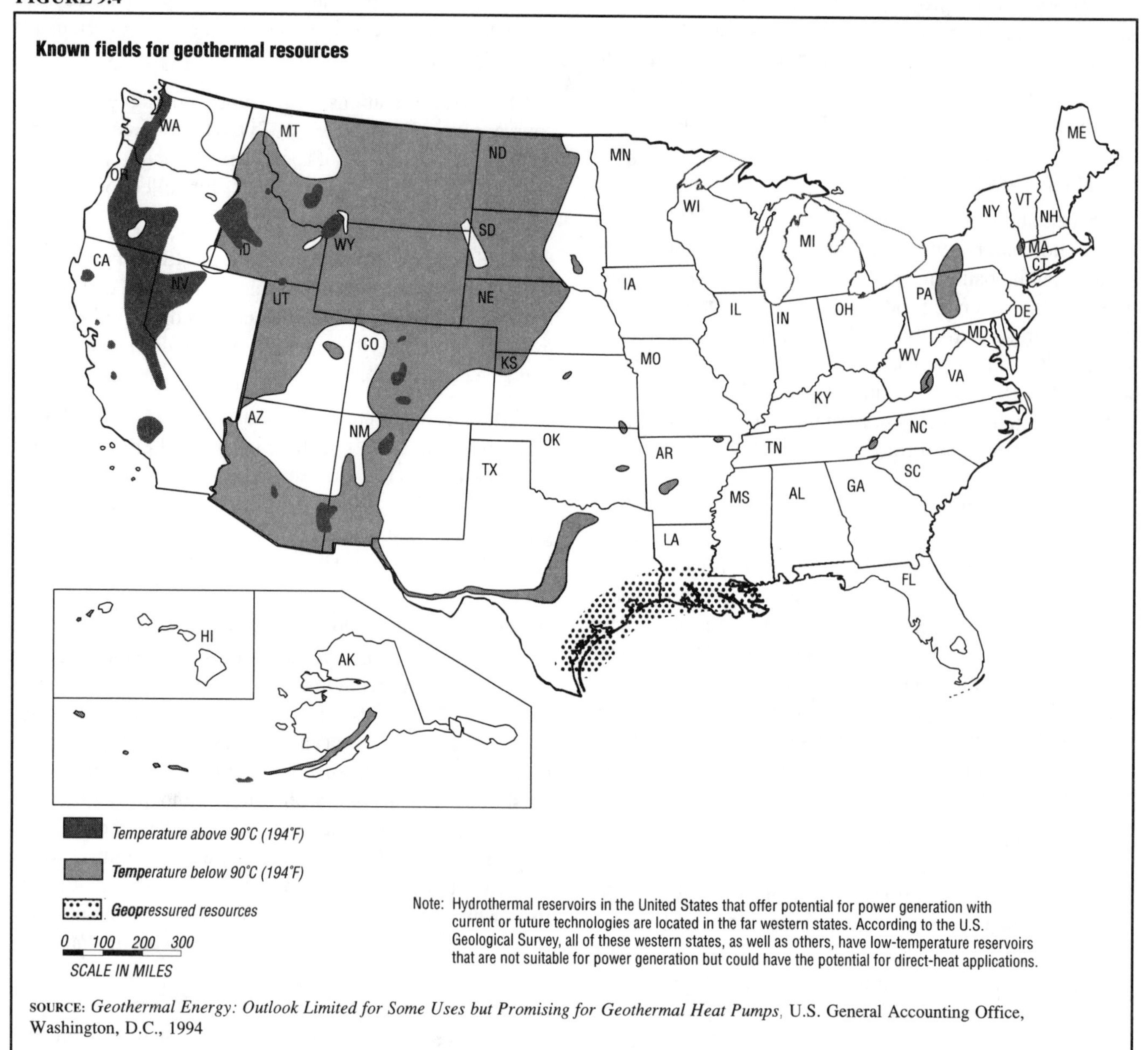

Note: Hydrothermal reservoirs in the United States that offer potential for power generation with current or future technologies are located in the far western states. According to the U.S. Geological Survey, all of these western states, as well as others, have low-temperature reservoirs that are not suitable for power generation but could have the potential for direct-heat applications.

SOURCE: *Geothermal Energy: Outlook Limited for Some Uses but Promising for Geothermal Heat Pumps*, U.S. General Accounting Office, Washington, D.C., 1994

International Production of Geothermal Energy

Since 1979 worldwide geothermal electrical generating capacity has nearly tripled. Nonetheless, it is still little more than the energy output of 10 average-size, coal-fired power plants. World geothermal reserves are immense, but unevenly distributed. These reserves fall mostly in seismically active areas at the margins or borders of Earth's nine tectonic plates. Currently, exploited reserves represent only a small fraction of the overall potential—many countries are believed to have in excess of 100,000 megawatts of geothermal energy available.

According to the Energy Information Administration, geothermal energy makes up less than 1 percent of world electrical production. The United States is the largest geothermal power producer, followed by the Philippines. The Philippine government has committed itself to the development of geothermal power by providing tax incentives and cooperation to the private sector; geothermal energy currently provides about one-fourth of the nation's electricity. New Zealand and Iceland both use their rich steam reserves to provide significant amounts of power—Iceland now heats more than 80 percent of its buildings with geothermal energy. Italy, Japan, and Mexico are the other major geothermal powers.

A few nations in the developing world, including El Salvador, Kenya, Bolivia, Costa Rica, Ethiopia, India, and Thailand, have considerable steam reserves available for power generation. Debt-ridden developing nations are especially eager to use available geothermal reserves instead of relying on fossil fuel imports for their energy needs.

WIND ENERGY

Winds are created by the uneven heating of the atmosphere by the sun, the irregularities of the earth's surface, and the rotation of the earth. Winds are strongly influenced by local terrain, water bodies, weather patterns, vegetation, and other factors. Wind flow, when "harvested" by wind turbines, can be used to generate electricity.

Wind machines have changed dramatically from those that were common in the 1800s. Early windmills produced mechanical energy to pump water and run sawmills. In the late 1890s, Americans began experimenting with wind power to generate electricity. Their early efforts produced enough electricity to light one or two modern light bulbs.

Compared to the pinwheel-shaped farm windmills that can still be seen dotting the American rural landscape, today's state-of-the-art wind turbines look more like airplane propellers. Their sleek, high-tech fiberglass design and aerodynamics allow them to generate an abundance of electricity, while they also produce mechanical energy and heat.

Over the past decade industrial and developing countries alike have started using wind power as a source of electricity to complement their existing power sources and to bring electricity to remote regions. Wind turbines cost less to install per unit of kilowatt capacity than coal or nuclear facilities. After installing a windmill, there are few additional costs, as the fuel (wind) is free.

Wind speeds are generally highest and most consistent in mountain passes and along coastlines. Europe has the greatest coastal wind resources, and clusters of wind turbines, or wind farms, are being developed there and in Asia. Denmark, the Netherlands, China, and India are especially interested in fostering the development of domestic wind industries. In the United States it is estimated that sufficient wind energy is available to provide more than one trillion kilowatt hours of electricity annually, or about one third of the total used in 1999. Currently, electricity-producing wind turbines operate in 95 countries.

Energy Production by Wind Turbines

Wind is the world's fastest-growing renewable energy source. Although wind power has not been adopted widely in the United States, U.S. companies export turbines to Spain, the Netherlands, Great Britain, India, and China. Following a slow period in the late 1980s when the U.S. government discontinued tax credits for wind installations, the market for wind turbines has grown.

The wind industry in California began in 1981 with the erection of 144 relatively small turbines capable of generating a combined total of seven megawatts of electricity. Within a year the number of turbines had increased 10 times, and by 1986 they had multiplied 100-fold. The 1980s saw an explosion of wind technology in California, where about 95 percent of the installed wind capacity in the United States is located. In 1995 California produced enough wind power to supply all of San Francisco's residents.

Experts point out that California's dominance has less to do with wind availability than with tax incentives that were offered by the state. Developers are in the process of building wind energy farms in the Midwest. Studies show that several states, especially the plains states, have wind speeds sufficient to supply electricity to those states. Twelve states—North Dakota, South Dakota, Texas, Kansas, Montana, Nebraska, Wyoming, Oklahoma, Minnesota, Iowa, Colorado, and New Mexico—contain 90 percent of the U.S. wind energy potential. Refinements in wind-turbine technology may enable a substantial portion of the nation's electricity to be produced by wind energy. Recent wind installations have been constructed in Texas and Minnesota. Other wind farm projects exist in Hawaii, Montana, New York, Oregon, and Wyoming. In 1998 and 1999 U.S. wind electrical generating capacity increased by 860 megawatts, as companies scrambled to take advantage of federal tax incentives that ended in June of 1999.

Growth in the U.S. wind industry has slowed in the last decade. Crude oil prices fell for most of the 1990s, making oil and gas the lowest-cost fuel sources. Federal budget cuts resulted in a change in federal policy toward renewable energies. Construction costs of wind farms has increased. Furthermore, the uncertainty involving electric utility deregulation has decreased investment in new energy sources.

International Development of Wind Energy

During the decade following the 1973 oil embargo, more than 10,000 wind machines were installed worldwide, ranging in size from portable units to multi-megawatt turbines. In Third World villages, small wind turbines recharge batteries and provide essential services. In China small wind turbines allow people to watch their favorite television shows, which has increased demand. In fact, five of the world's ten largest manufacturers of small wind turbines are Chinese.

Global wind power generating capacity exceeded 7,200 megawatts in 1997 from more than 25,000 operating wind turbines, more than double the 3,000 megawatts produced in 1993. Most of the growth was in northern Europe. Although wind power supplies less than 0.1 percent of the world's electricity, it is growing the fastest of renewable sources. The largest producers of wind energy are, in order, Germany, the U.S., Denmark, and India.

During the 1990s, Europe and India have displaced California as the main user of wind power. European countries had 63 percent of the world's wind energy capacity in 1997, with major increases in Germany, Denmark, and Spain, compared to 22 percent for the United States. India, the largest wind user of developing countries with 12 percent of the world's total, had an ambitious wind energy program planned. The Indian Ministry of Energy promoted the installation of enough windmills to produce 5,000 megawatts by 2000, which would have been enough electrical power to serve five million customers, but a slowed economy hampered those efforts. India is expected to be the most rapidly growing market for wind turbines, and if the planned wind energy program is successful, wind may supply more energy for India than the country's nuclear program. Other countries planning to install wind turbines include Australia, Belgium, Israel, Italy, and the United Kingdom.

Interest in wind energy has been driven in part by the declining cost of capturing wind energy. From more than 25 cents per kilowatt-hour in 1980, wind energy prices have declined to 5 cents per kilowatt-hour for new turbines today at sites with strong winds. Decreasing costs could make wind power competitive with gas and coal power plants, even before considering wind's environmental advantages.

Advantages and Disadvantages of Wind Energy

The main problem with wind energy is that the wind does not always blow. Some people object to the whirring noise of wind turbines or do not like to see wind turbines clustered in mountain passes and along shorelines because they interfere with scenic views. Environmentalists have charged that wind turbines are responsible for the loss of thousands of endangered birds that fly into the blades, as birds frequently use windy passages in their travel patterns.

On the other hand, generating electricity with wind offers many environmental advantages. Windfarms do not emit climate-altering carbon dioxide, acid-rain-forming pollutants, respiratory irritants, or nuclear waste. Because windfarms do not require water to operate, they are especially well-suited to semi-arid and arid regions. Windfarming also offers the added benefit of reducing soil loss on land prone to wind erosion because turbines capture the wind and decrease its potential for downwind destruction. Ironically, the winds that once created the Dust Bowl and contributed to the Great Depression may someday be harnessed to provide electricity.

SOLAR ENERGY

Ancient Greek and Chinese civilizations used glass and mirrors to direct the sun's rays to start fires. Solar energy (from the sun) is a renewable, widely available energy source that does not generate greenhouse gases or radioactive waste. Solar-powered cars have already competed in long-distance races, and solar energy has been used routinely for many years to power spacecraft. Although many people consider solar energy a product of the space age, the Massachusetts Institute of Technology built the first solar house in 1939.

Solar radiation is nearly constant outside Earth's atmosphere, but the amount of solar energy, or insolation, reaching any point on Earth varies with changing atmospheric conditions, such as clouds and dust, and the changing position of the earth relative to the sun. In the United States, insolation is greatest in the West and Southwest regions. Nevertheless, almost all U.S. regions have solar resources that can be used.

Passive and Active Solar Energy Collection Systems

Passive solar systems, such as greenhouses or windows with a southern exposure, use heat flow, evaporation, or other natural processes to collect and transfer heat. They are considered the least costly and least difficult systems to implement. (See Figure 9.5.)

Active solar systems use mechanical methods to control the energy process. They require collectors and storage devices as well as motors, pumps, and valves to operate the systems that transfer heat. Collectors consist of an absorbing plate that transfers the sun's heat to a working fluid (liquid or gas), a translucent cover plate that prevents the heat from radiating back into the atmosphere, and insulation on the back of the collector panel to further reduce heat loss. Excess solar energy is transferred to a storage facility so it may be used to provide power on cloudy days. In both active and passive systems, the conversion of solar energy into a form of power is made at the site where it is used. The most common and least expensive active solar systems are used for heating water.

Solar Thermal Energy Systems

A solar thermal energy system uses intensified sunlight to heat water or other fluids, reaching temperatures of over 750 degrees Fahrenheit. Mirrors or lenses constantly track the sun's position and focus its rays onto solar receivers that contain fluid. Solar heat is transferred to the water that in turn powers an electric generator. In a distributed solar thermal system, the collected energy powers irrigation pumps, providing electricity for small communities or capturing normally wasted heat from the sun in industrial areas. In a central solar thermal system, the energy is collected at a central location and used by utility networks for a large number of customers.

Other solar thermal energy systems include solar ponds and trough systems. Solar ponds are lined ponds filled with water and salt. Because salt water is denser than fresh water, the salt water on the bottom absorbs the heat, and the fresh water on top keeps the salt water contained and traps the heat. Trough systems use U-shaped mirrors to concentrate the sunshine on water or oil-filled tubes.

Photovoltaic Conversion Systems

The photovoltaic (PV) cell solar energy system converts sunlight directly into electricity without the use of

FIGURE 9.5

Solar house types

roof overhang

light

heat

storage

S N

Passive solar house

solar collector

light

heat

S N

Actives solar house

storage

SOURCE: U.S. Department of Energy

FIGURE 9.6

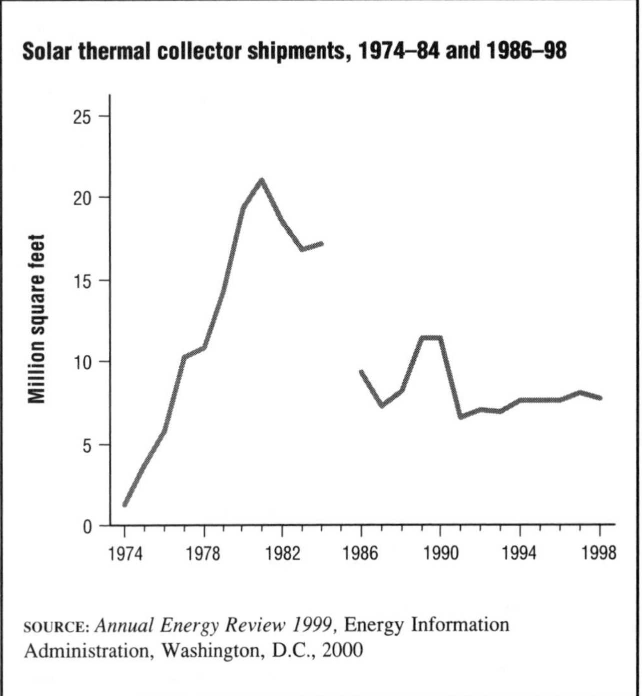

Solar thermal collector shipments, 1974–84 and 1986–98

SOURCE: *Annual Energy Review 1999,* Energy Information Administration, Washington, D.C., 2000

mechanical generators. PV cells have no moving parts, are easy to install, require little maintenance, do not pollute the air, and can last up to 20 years. PV cells are commonly used to power small devices, such as watches or calculators. They are also being used on a larger scale to provide electricity for rural households, recreational vehicles, and businesses. Solar panels using photovoltaic cells have generated electricity for space stations and satellites for many years.

Since PV systems produce electricity only when the sun is shining, a backup energy supply is required. PV cells produce the most power around noon, when sunlight is the most intense. A photovoltaic system typically includes storage batteries that provide electricity during cloudy days and at night.

The use of photovoltaic technology is expanding both in the United States and abroad. PV systems have advantages such as low operating costs because there are no turbines or other moving parts and maintenance is minimal. PV cell systems are nonpolluting and silent and can be operated by computer. Above all, the fuel source (sunshine) is free and plentiful. The main disadvantage of a photovoltaic cell energy systems is the initial cost. Although the price has fallen considerably, PV cells are still too expensive for widespread use. PV systems also use some toxic materials, which may cause environmental problems.

Solar Energy Usage

Because it is difficult to measure solar energy directly, shipments of solar equipment can be used as an indicator of activity. From a high of 84 low-temperature solar collector manufacturers in 1979, the number dropped to 12 manufacturers in 1998. Total shipments of solar thermal collectors peaked in 1981 at over 21 million square feet and declined to approximately 7.8 million square feet by 1998. (See Figure 9.6.)

Most of the solar thermal collector market is for residential purposes (mostly in the Sunbelt states), with only a small proportion for commercial purposes. Some state and municipal power companies have added solar systems as adjuncts to their regular power sources during peak hours. (See Figure 9.7.) In 1998 most of the solar-thermal collectors shipped were used for heating swimming pools, and a smaller percentage for hot water. (See Figure 9.8.) The market for solar energy space heating has virtually disappeared.

Advantages and Disadvantages of Solar Energy

The primary advantage of solar energy is its inexhaustible supply, while its primary disadvantage is its reliance on a consistently sunny climate to provide continuous electrical power, which is only possible in limited areas of the country. In addition, a large amount of land area is necessary for the most efficient collection of solar energy by electricity plants. Experts estimate that a new thermal energy plant would have a 60 percent higher cost of production than a conventional coal-fired plant.

Future Development Trends

Interest in photovoltaic solar energy systems is particularly high in rural and remote areas where it is impractical to extend traditional electrical power lines. In some

FIGURE 9.7

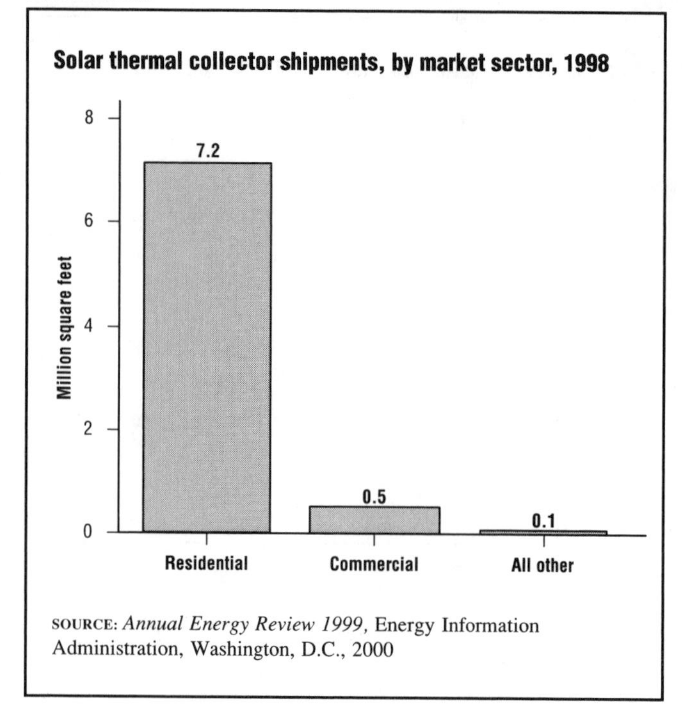

SOURCE: *Annual Energy Review 1999,* Energy Information Administration, Washington, D.C., 2000

FIGURE 9.8

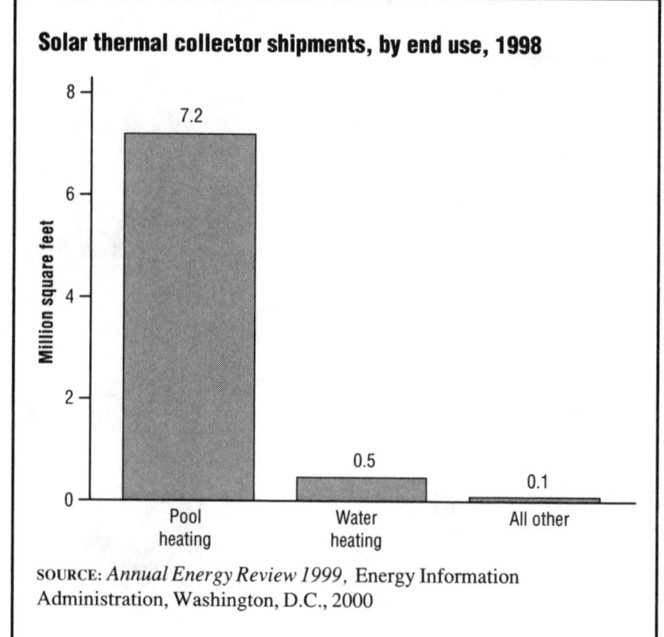

SOURCE: *Annual Energy Review 1999,* Energy Information Administration, Washington, D.C., 2000

remote areas, PV cells are used as independent power sources for communications or for the operation of water pumps or refrigerators.

Although solar power still costs more than three times as much as fossil fuel energy, utilities could turn to solar energy to provide "peaking power" on extremely hot or cold days. In the long run, some people believe that building solar energy systems to provide peak power capacity would be cheaper than building new and expensive diesel fuel generators which are now used.

POWER FROM THE OCEAN

The potential power of the world's oceans is unknown. Because the ocean is not as easily controlled as a river or as water that is directed through canals into turbines, unlocking that potential power is far more challenging. Three ideas being considered are tidal plants, wave power, and ocean thermal energy conversion (OTEC).

Tidal Power

The tidal plant uses the power generated by the tidal flow of water as it ebbs, or flows back out to sea. A minimum tidal range of three to five yards is generally considered necessary for an economically feasible plant. Canada has built a small 40-megawatt unit at the Bay of Fundy, with its 15-yard tidal range, the largest range in the world, and is considering building a larger unit there. The largest existing tidal facility is the 240-megawatt plant at the La Rance estuary in northern France, built in 1965. Russia has plans to construct a 400-megawatt plant.

Wave Energy

Norway has two operating wave power stations at Tostestallen on its Atlantic coast. The arrival of a wave forces water up a hollow 65-foot tower, displacing the air already in the tower. This air rushes out of the top through a turbine. The rotors of that turbine then spin, generating electricity. When the wave falls back, and the water level falls, air is sucked back in through the turbine, again generating electricity.

The second type of ocean power plant uses the overflow of high waves. As the wave splashes against the top of a dam, some of the water goes over and is trapped in a reservoir on the other side. The water is then directed through a turbine as it flows back to the sea. These two kinds of plants are experimental. Several projects are underway in Japan and the Pacific region to determine a way to use the potential of the huge waves of the Pacific.

Ocean Thermal Energy Conversion (OTEC)

OTEC uses the temperature difference between the ocean's warm surface water and the cooler water in its depths to produce heat energy that can power a heat engine to produce electricity. OTEC systems can be installed on ships, barges, or offshore platforms with underwater cables that transmit electricity to shore.

HYDROGEN: A FUEL OF THE FUTURE?

Hydrogen, the lightest and most abundant chemical element, is the ideal fuel from the environmental point of view. Its combustion produces only water vapor, and it is entirely carbon-free. Three-quarters of the mass of the universe consists of hydrogen. However, the combustible form of hydrogen is a gas, and is not found in nature. The many compounds containing hydrogen, for example water, cannot be converted into pure hydrogen without the expenditure of energy. The amount of energy that would be required to make gas is about the same as the amount of energy that would be obtained by the combustion of the hydrogen. Therefore, with today's technology little or nothing could be gained from an energy point of view.

Scientists are researching ways to economically produce hydrogen. Whether that will come from fusion, solar energy, or elsewhere, is not now possible to predict. Scientists have considered the possibility of a transition to hydrogen for more than a century, and today many see hydrogen as the logical "third-wave" fuel, with hydrogen gas following oil, just as oil replaced coal decades earlier. For now, however, wide-spread use of hydrogen as fuel is purely theoretical.

INTERNATIONAL RENEWABLE ENERGY USAGE

Wind Energy

In wind power Europe has already overtaken the United States, the nation that led the drive to wind energy in the 1980s. Europe's 1997 capacity of 4,500 megawatts was over two and one-half times that of the United States (1,600 megawatts), with Germany alone (1,900 megawatts) surpassing the output of American wind farms. European wind production has been encouraged by improved technology, falling costs, and government incentives such as tax credits and guaranteed prices. With windmills springing up from the coasts of Sweden to the tip of Spain, Europe's wind industry already employs 20,000 people. Close to 100,000 Danes own shares in hundreds of small cooperatives that operate 4,700 windmills. Furthermore, they are making a profit. Wind power generates 6 percent of Denmark's electricity, the highest per capita output of wind energy in the world.

The European Union wants to diversify its energy sources while clamping down on pollution. Almost no country supports the expansion of nuclear power, and in many areas wind power is becoming economically viable. New wind turbines already generate electricity less expensively than solar panels, biomass, or other nontraditional sources. The International Energy Agency, a research organization based in Paris, reports that wind energy is now competitive with electricity from Europe's oil- and coal-fired power plants.

Wind energy is produced differently in the United States than Europe. American entrepreneurs, seeing wind energy as a potentially profitable business, built large wind farms with huge numbers of turbines. When oil prices fell and tax credits were cut, growth stalled. In Denmark, Germany, Sweden, and the Netherlands, wind energy began as a grassroots movement with small groups of politically motivated investors installing one or a few machines at a time. In Spain, Britain, and Greece the clusters are larger because money has been provided by governments and utilities. The Netherlands, which had approximately 11,000 windmills less than a century ago, has 1,120 modern turbines today, many of them standing beside old-style windmills. The latest trend in Europe is to build wind farms offshore, where there is more wind and fewer complaints that they clutter the landscape.

Solar Power

Rural areas are more expensive to serve than cities, and electrification has been slow to reach many people in rural areas of developing countries. In the United States, it was only after the Rural Electrification Administration in 1935 provided low-cost financing to rural electric cooperatives that most farmers received power. The World Bank estimates that in places such as western China, the Himalayan foothills, or the Amazon basin, the cost of connecting new rural customers is seven times that in cities. Furthermore, state-owned power systems have been badly managed in many countries. This has left many national power systems all but bankrupt and blackouts have become common.

In India blackouts are so common that many factories and other businesses have at great expense set up their own private systems, using natural gas, propane, or fuel oil. Although rural families do not have access to those systems, they do have sunlight. In most tropical countries, considerable sunlight falls on rooftops. Electricity produced by solar photovoltaic cells was initially too expensive, as much as a thousand times more than that from conventional plants, but has continually fallen in price.

During the 1990s a different approach to solar electrification developed, driven less by government planners and more by the desire of individual families to meet their own needs for electricity. In more than a dozen countries, solar power is reaching thousands of families one by one, avoiding the delays of government planners. In Kenya, where the state power company has been on the verge of bankruptcy, eight domestic companies merged to market, install, and maintain solar home systems. With little state or international assistance, these companies managed to electrify 20,000 rural households between 1987 and 1992, 3,000 more than the state power system.

Financial Obstacles to International Renewable Energy Use

Despite advances, only one of every thousand potential electricity customers has yet been served. Part of the problem is lack of credit. Consumer credit is one of the most momentous financial advances of the twentieth century, leading to wide ownership of homes, automobiles, and appliances that the average person cannot afford outright. Millions of families in the developing world might afford solar energy if credit were available.

FIGURE 9.9

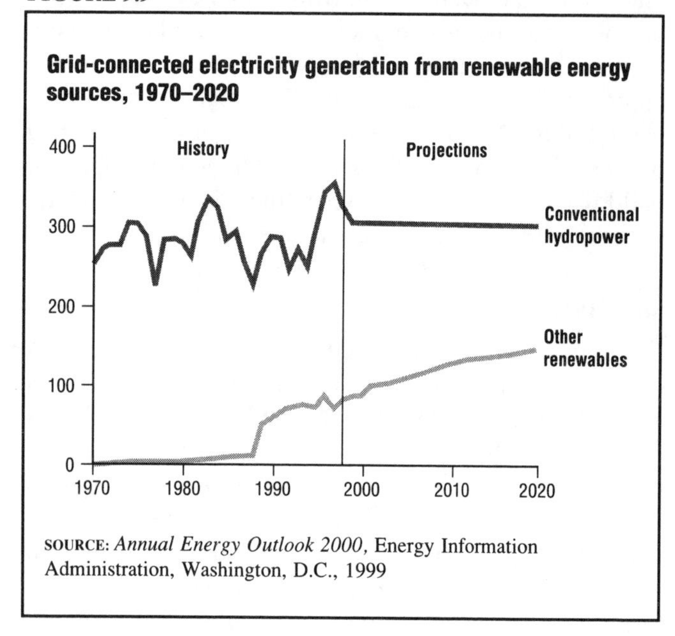

SOURCE: *Annual Energy Outlook 2000,* Energy Information Administration, Washington, D.C., 1999

FIGURE 9.10

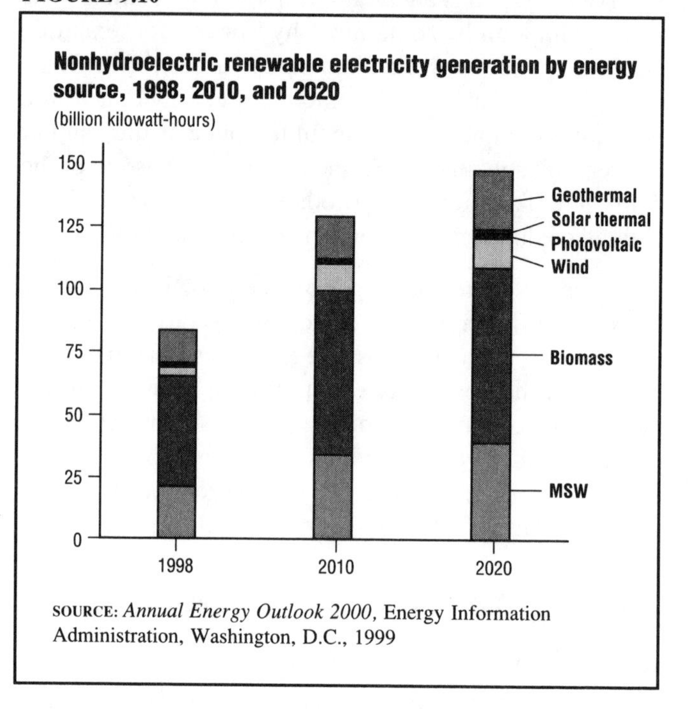

SOURCE: *Annual Energy Outlook 2000,* Energy Information Administration, Washington, D.C., 1999

To fill this gap, international agencies are working to set up revolving credit funds to finance solar, wind, and other renewable energy systems in many countries, including Vietnam, India, Indonesia, Uganda, Swaziland, and the Dominican Republic. A number of nonprofit agencies, private foundations, investment firms, and the World Bank have gotten involved. Many observers hope that newly developing nations can avoid many of the damaging environmental practices of industrialization, including a heavy dependence on fossil fuels.

FUTURE TRENDS IN U.S. RENEWABLE ENERGY USE

In its *Annual Energy Outlook 2000* (1999), the Energy Information Administration of the U.S. Department of Energy forecasts that for renewable energy sources, including ethanol for transportation, consumption will increase by 0.8 percent per year from 1998 to 2020. Renewable fuels used for electricity generation are projected to grow by 0.5 percent per year. In 1998 renewable energy provided 408 billion kilowatt-hours of electricity, which is projected to increase to 452 billion kilowatt-hours in 2020. Renewable energy is expected to decrease its share of total electricity generation, from 11.3 percent of electricity in 1998 to 9.5 percent in 2020, as low projected fossil fuel prices limit investment and expansion of renewable sources.

Hydropower production is projected to decline slightly by 2020, as 620 megawatts of new capacity will not offset declines in production from existing facilities due to changing operating conditions. Other renewables should increase. (See Figure 9.9.) Biomass and municipal solid waste energy production increase the most in the projections, from 65 billion kilowatt-hours to 109 billion in 2020. Biomass alone may increase by 26 billion kilowatt-hours between 1998 and 2020. (See Figure 9.10.)

Wind power is expected to increase in capacity by 2,800 megawatts by 2020, but high construction costs, lowered incentives, and unpredictability of wind supplies will limit growth. Geothermal energy is projected to increase by 860 megawatts, which will provide an extra 10 billion kilowatt-hours of electricity in 2020. Solar energy is not expected to contribute much to centrally generated electricity, but off-grid applications with PV cell technology will continue to grow.

Government-sponsored programs for renewable energy could increase its use significantly by 2020. The Clinton Administration's Comprehensive Electricity Competition Act (CECA), if approved by Congress, could mandate a Renewable Portfolio Standard (RPS), which requires a set percentage of electricity production from renewable sources. (See Chapter 8 for further discussion of CECA.) Individual states could also mandate more renewable energy. In a best-case scenario by the EIA, non-hydroelectric sources could contribute 72 billion kilowatt-hours more than projected amounts by 2020. If tax incentives that were cancelled in 1999 were extended to 2020, wind energy could increase by 46 percent and biomass by 10 percent more than projected.

CHAPTER 10

ENERGY CONSERVATION

ENERGY CONSERVATION AND EFFICIENCY

Energy conservation is the more efficient use of energy, without necessarily curtailing the services that energy provides. Conservation occurs when societies develop more efficient technologies that reduce energy needs. Environmental concerns, such as acid rain and the potential for global warming, have increased public awareness about the importance of energy conservation.

Energy efficiency can be measured by two indicators. The first is energy consumption per person (per capita) per year. Annual per person energy consumption in the U.S. topped 361 million Btu in 1979, dropped to 314 million Btu in 1983, and rose to 354 million Btu per capita in 1999. (See Figure 10.1.)

The second indicator of efficiency is energy consumption per dollar of gross domestic product (GDP). GDP is the total value of goods and services produced by a nation. When a country becomes more efficient, it uses less energy to produce the same amount of goods and services. In 1970, 19,000 Btu of energy were consumed to produce each dollar of GDP, dropping to 11,000 Btu per dollar in 1999. (See Figure 10.2.)

ENERGY CONSERVATION, PUBLIC HEALTH, AND THE ENVIRONMENT

In the coming decades, global environmental issues could significantly affect patterns of energy use around the world.—International Energy Outlook 2000, Energy Information Administration.

The connection between energy policy and the health of both people and the environment has become clearer in recent years. People living in cities with high levels of pollution have higher risks of mortality and certain diseases than those living in less polluted cities. Energy-related

FIGURE 10.1

Energy consumption per person, 1949–99

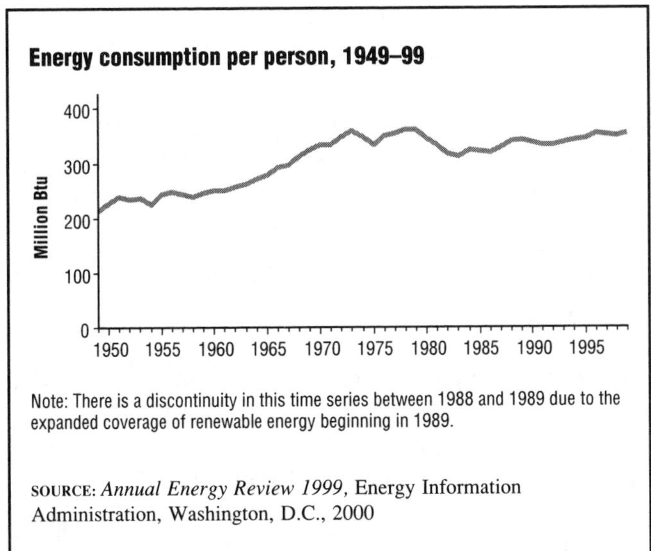

Note: There is a discontinuity in this time series between 1988 and 1989 due to the expanded coverage of renewable energy beginning in 1989.

SOURCE: *Annual Energy Review 1999,* Energy Information Administration, Washington, D.C., 2000

FIGURE 10.2

Energy consumption per dollar of gross domestic product, 1949–99

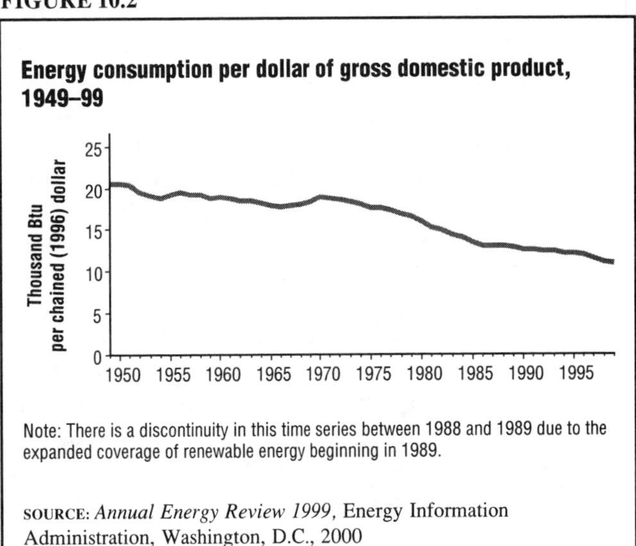

Note: There is a discontinuity in this time series between 1988 and 1989 due to the expanded coverage of renewable energy beginning in 1989.

SOURCE: *Annual Energy Review 1999,* Energy Information Administration, Washington, D.C., 2000

TABLE 10.1

Air pollutants, health risks, and contributing sources

Pollutants	Health risks	Contributing sources
Ozone[1] (O_3)	Asthma, reduced respiratory function, eye irritation	Cars, refineries, dry cleaners
Particulate matter (PM-10)	Bronchitis, cancer, lung damage	Dust, pesticides
Carbon monoxide (CO)	Blood oxygen carrying capacity reduction, cardiovascular and nervous system impairments	Cars, power plants, wood stoves
Sulphur dioxide (SO_2)	Respiratory tract impairment, destruction of lung tissue	Power plants, paper mills
Lead (Pb)	Retardation and brain damage, esp. children	Cars, nonferrous smelters, battery plants
Nitrogen dioxide (NO_2)	Lung damage and respiratory illness	Power plants, cars, trucks

[1] Ozone refers to tropospheric ozone which is hazardous to human health.

SOURCE: *Healthy People 2000—Statistical Notes,* Centers for Disease Control and Prevention, Atlanta, GA, 1995

FIGURE 10.3

Temperature increase for IPCC business-as-usual emissions and 222GtC carbon budget

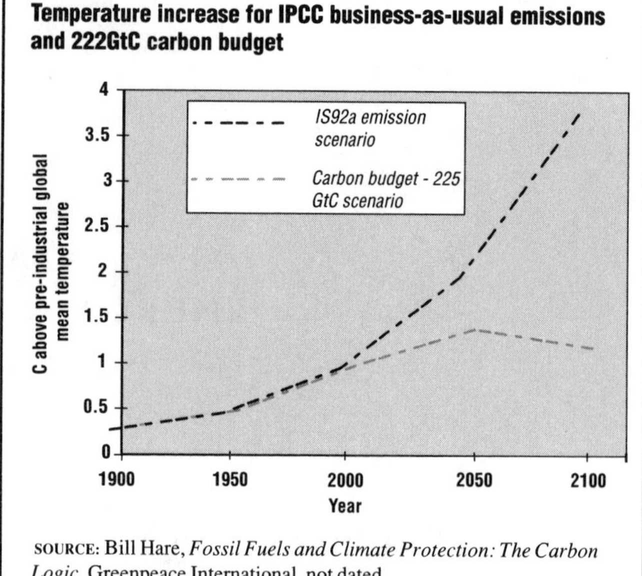

SOURCE: Bill Hare, *Fossil Fuels and Climate Protection: The Carbon Logic,* Greenpeace International, not dated

emissions generate a vast majority of these polluting chemicals. (See Table 10.1.) In the past few years a number of major studies have documented the growing body of evidence establishing the link between air pollution and public health. Air pollution has been related to diseases including asthma, bronchitis, emphysema, and lung cancer. The American Lung Association estimates the annual health costs of exposure to the most serious air pollutants at $40 to $50 billion. Clean and efficient energy technologies represent a cost-effective investment in public health.

The evidence of global warming and its consequences has also accumulated rapidly in recent years. The United Nations' Intergovernmental Panel on Climate Change (IPCC), a group of two thousand of the world's leading scientists, concluded in 1995 that global warming is real, serious, and accelerating. They determined that the most likely cause is primarily from humans burning coal, oil, and gasoline, which has increased the amount of carbon dioxide (CO_2) and other "greenhouse" gases in the earth's atmosphere, which trap heat. Deforestation is also contributing by reducing the earth's ability to absorb the carbon dioxide from burning fossil fuels.

The issues surrounding global warming are heavily debated. Industry officials claim that there is insufficient evidence of warming trends and that cleanup efforts will hamper economic growth. Environmentalists argue that climate change will be disastrous for people and for the planet, and that immediate actions must be taken to prevent it. Climate scientists are busy modeling predictions of global warming trends. Greenpeace International, an environmental group, claims that the earth's atmosphere has a limited capacity, or budget, for the amount of fossil fuels that can be burned before serious damage occurs to the environment. The carbon budget to keep the earth's temperature from increasing no more than one degree Centigrade per century, which they claim would still cause significant environmental and economic damage, is estimated at 225 gigatons of carbon (GtC, or billion tons). This is the amount of carbon that would be released by burning about one-fourth of the world's current known reserves of oil, coal, and natural gas. (See Figure 10.3.) From perspectives such as this, energy conservation becomes a very important issue.

The Kyoto Global Warming Treaty

In December 1997 the United Nations convened a conference of 160 nations on global warming in Kyoto, Japan, in hopes of producing a new treaty on climate change that would place legally binding limits on industrial emissions. However, wide gulfs existed between rich and poor countries. There was disagreement even among rich countries, with the European community contending that the United States, by far the world's largest polluter, had not done enough to reach its goals.

The treaty, called the Kyoto Protocol to the United Nations Framework Convention on Climate Change (UNFCCC), requires industrialized nations to reduce their emissions of six greenhouse gases—carbon dioxide, methane, nitrous oxide, hydrofluorocarbons, sulfur dioxides, and perfluorocarbons—below 1990 levels by 2012, with each country having a different target. The United States must cut emissions by 7 percent, most European nations by 8 percent, and Japan by 6 percent. Reductions must begin by 2008 and be achieved by 2012. Developing nations are not required to make such pledges.

The United States had proposed a program of voluntary pledges by developing nations, but that section was deleted, as was a tough system of enforcement. Instead, each country decides for itself how to achieve its goal. The treaty provides market-driven tools such as buying and selling credits for reducing emissions. It also sets up a Clean Development Fund to help poorer nations with technology to reduce their emissions. Countries would decide on their own whether to sign and ratify the treaty.

Although it is the first time nations have made such sweeping pledges, many sources expect difficulty in getting ratification. In the United States, President Clinton signed the Protocol but the Senate did not ratify it. Business leaders believe the treaty goes too far while environmentalists believe standards do not go far enough. Some experts doubt that any action emerging from Kyoto will be sufficient to prevent doubling of greenhouse gases. Representatives of the oil industry and business community contend the treaty will spell economic problems for the United States.

As of July 2000, 84 nations had signed the Kyoto treaty and 22 had ratified it, although none of these were major industrialized nations. The treaty becomes legally binding only when at least 55 countries have ratified. The UNFCCC met again in November 2000 to finalize issues relating to the Protocol and to work out some problems so that more nations will ratify it. In the meantime, the Clinton Administration developed the Climate Change Action Plan (CCAP), which consists of 44 actions the U.S. can voluntarily take to stabilize its carbon emissions to 1990 levels by 2000. The CCAP was created in response to an earlier United Nations climate convention in Rio de Janeiro in 1992. Of the CCAP's 44 actions, 31 are related to energy combustion or carbon dioxide emissions.

TABLE 10.2

Transportation conservation options

Improve the technical efficiency of vehicles
1. Higher fuel economy requirements—CAFE standards (R)
2. Reducing congestion: smart highways (E,I), flextime (E,R), better signaling (I), improved maintenance of roadways (I), time of day charges (E), improved air traffic controls (I,R), plus options that reduce vehicular traffic
3. Higher fuel taxes (E)
4. Gas guzzler taxes, or feebate schemes (E)
5. Support for increased R&D (EJ)
6. Inspection and maintenance programs (R)

Increase load factor
1. HOV lanes (I)
2. Forgiven tolls (E), free parking for carpools (E)
3. Higher fuel taxes (E)
4. Higher charges on other vmt trip-dependent factors (E): parking (taxes, restrictions, end of tax treatment as business cost), tolls, etc.

Change to more efficient modes
1. Improvements in transit service
   a. New technologies—maglev, high speed trains (EJ)
   b. Rehabilitation of older systems (I)
   c. Expansion of service—more routes, higher frequency (I)
   d. Other service improvements (I)—dedicated busways, better security, more bus stop shelters, more comfortable vehicles
2. Higher fuel taxes (E)
3. Reduced transit fares through higher US. transit subsidies (E)[a]
4. Higher charges on other vmt/trip-dependent factors for less efficient modes (E)—tolls, parking
5. Shifting urban form to higher density, more mixed use, greater concentration through zoning changes (R), encouragement of "infill" development (E,R,I), public investment in infrastructure (I), etc.

Reduce number or length of trips
1. Shifting urban form to higher density, more mixed use, greater concentration (E,R,I)
2. Promoting working at home or at decentralized facilities (EJ)
3. Higher fuel taxes (E)
4. Higher charges on other vmt/trip-dependent factors (E)

Shift to alternative fuels
1. Fleet requirements for alternative fuel-capable vehicles and actual use of alternative fuels (R)
2. Low-emission/zero emission vehicle (LEV/ZEV) requirements (R)
3. Various promotions (E): CAFE credits, emission credits, tax credits, etc.
4. Higher fuel taxes that do not apply to alternative fuels (E), or subsidies for the alternatives (E)
5. Support for increased R&D (EJ)
6. Public investment—government fleet investments (I)

Freight options
1. RD&D of technology improvements (E,I)

[a]U.S. transit subsidies, already among the highest in the developed world, may merely promote inefficiencies.

KEY: CAFE = corporate average fuel economy; E = economic incentive; HOV = high-occupancy vehicle; I = public investment; maglev = trains supported by magnetic levitation; R = regulatory action; RD&D = research, development, and demonstration; vmt = vehicle-miles traveled.

SOURCE: *Saving Energy in U.S. Transportation,* Office of Technology Assessment, Washington, D.C., 1994

EFFICIENCY IN THE TRANSPORTATION SECTOR

The U.S. transportation system plays a central role in the economy. Highway transportation is dependent on internal combustion engine vehicles fueled almost exclusively by petroleum. According to the Energy Information Administration and the Bureau of Transportation Statistics, the transportation sector accounted for 27 percent of all energy consumed in the United States in 1999. That year Americans used 25.9 quadrillion Btu for transportation, for which petroleum supplied 97 percent of the energy. Despite improvements in U.S. transportation efficiency in recent decades, the United States still consumes more than one-third of the world's transport energy. In 1996 the U.S. imported more oil than it produced for the first time in history, and has continued to do so in the years since then. American dependence on oil not only makes the economy vulnerable to the supply and price volatility of the world oil market, but also leads to air pollution problems.

Automotive Efficiency

Policy makers interested in transportation energy conservation have an array of conservation options. (See Table 10.2.) However, not all options are mutually supportive. For example, efforts to promote freer flow of automobile traffic, such as high-occupancy vehicle (HOV) lanes or free parking for carpools, may sabotage efforts to shift travelers to mass transit or to reduce trip lengths and frequency. Policy makers must consider how

FIGURE 10.4

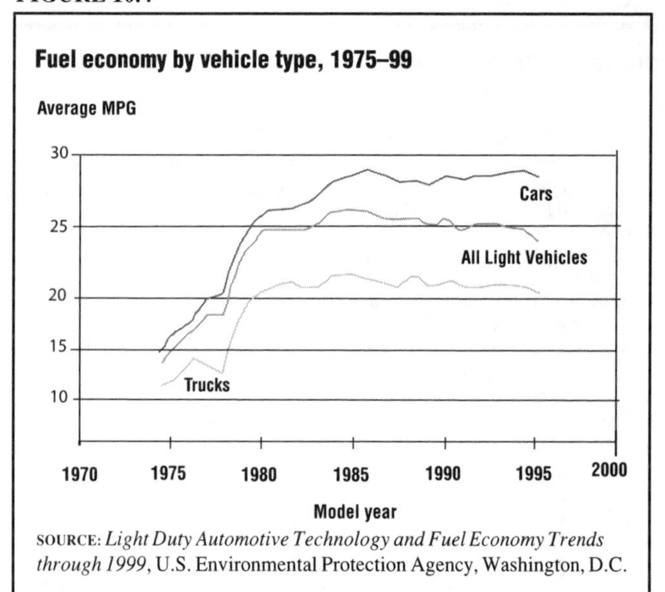

SOURCE: *Light Duty Automotive Technology and Fuel Economy Trends through 1999*, U.S. Environmental Protection Agency, Washington, D.C.

the implementation of one strategy will fit into an overall transportation plan. In the U.S. the automobile dominates the transportation sector, as light highway vehicles used 63 percent of all transportation energy in 1997. (See Figure 1.13 in Chapter 1.)

Motor gasoline, which is divided among passenger cars, light and heavy-duty trucks, aircraft, and miscellaneous other, consumes approximately 43 percent of the oil used in the United States. The major growth in fuel use over the past 10 years has been that consumed by trucks, due mainly to increased sales of sport utility vehicles (SUVs). In 1999 truck fuel use was nearly equal to automobile fuel use. Meanwhile, automobile fuel use has remained nearly constant due to fuel efficiency increases that have offset the growth in car miles traveled. Boosting truck efficiency will become increasingly important in holding down oil demand.

THE CORPORATE AVERAGE FUEL ECONOMY (CAFE) STANDARDS. The 1973 OPEC oil embargo painfully reminded America how dependent it had become on foreign sources of fuel. Although the United States makes up only 6 percent of the world's population, it consumes 26 percent of the world's supply of oil, much of which is imported from the Middle East. The oil embargo prompted Congress to pass the 1975 Automobile Fuel Efficiency Act (PL 96-426), which set the initial Corporate Average Fuel Efficiency (CAFE) standards.

The CAFE standards required domestic automakers to increase the average mileage of new cars sold to 27.5 miles per gallon (mpg) by 1985. Under CAFE rules, car manufacturers could still sell large, less efficient cars, but to meet the average fuel efficiency rates, they also had to sell smaller, more efficient cars. Automakers that failed to meet each year's CAFE standards were fined. Those that managed to surpass the rates earned credits that they could use in years when they fell below the CAFE requirements. Faced with the CAFE standards, the car companies became more inventive and managed to keep their cars relatively large and roomy, while increasing mileage with innovations such as electronic fuel injection and four valves per cylinder.

The effect of the CAFE regulations have been significant for fuel efficiency. In the decades since the first oil shock in 1973, the fuel economy of passenger cars (all cars currently on the road) increased from 13.4 miles per gallon (mpg) to 21.4 mpg in 1998. (See Table 10.3.) Greater gains have been made in the economy of new cars. In 1974, just after the oil embargo, vehicles averaged 14.2 mpg; in 1999, the average new car fuel economy was 28 mpg, although it has remained virtually unchanged for several years. (See Figure 10.4.) The total automobile fleet fuel economy is expected to increase as more fuel-efficient cars enter the market and older, less fuel-efficient autos drop out of the nation's fleet. However, since 1988 nearly all gains in automobile efficiency have been offset by increased weight and power in new vehicles.

The Persian Gulf War in 1991 was another strong reminder to the United States of its continuing heavy dependence on foreign oil, causing some members of Congress to want to raise the CAFE standards to 45 mpg for cars and 35 mpg for light trucks. Those in favor of raising CAFE standards claimed that this would save about 2.8 million barrels of oil per day. They also noted that if cars become even more fuel-efficient in the future, emissions of carbon dioxide would be significantly reduced. The domestic auto industry opposed the bill to raise CAFE standards, and actions to increase automobile efficiency failed in Congress.

SUVs and pickups are the fastest growing segment of the auto industry, accounting for 46 percent of the U.S market in 1999 and producing most of the profits for the major auto companies. (See Figure 10.5.) These types of vehicles fall under less stringent emissions standards than automobiles because they are classified as trucks, which have more lenient EPA regulations. In 1999 the EPA proposed new regulations tightening emissions standards on cars, mini-vans, SUVs under 8,500 pounds, and small pickup trucks. The regulations would not affect the rules for big SUVs and pickups, allowing these bigger vehicles to emit up to as much as five times more nitrogen oxides than cars. Automakers and buyers of trucks and SUVs have opposed tightening emissions for these vehicles, although critics contend that new SUVs are more like cars than trucks in design and should fall under the same rules. As of July 2000 no new restrictions were enacted, and these vehicles could still by law emit three times more nitrogen oxides, the main cause of smog.

The potential for savings from increased fuel economy in large trucks is huge, since their current fuel econo-

TABLE 10.3

Motor vehicle mileage, fuel consumption, and fuel rates, 1950–1998

	Passenger cars			Vans, pickup trucks, and sport utility vehicles [1]			Trucks [2]			All motor vehicles [3]		
Year	Mileage (miles per vehicle)	Fuel consumption (gallons per vehicle)	Fuel rate (miles per gallon)	Mileage (miles per vehicle)	Fuel consumption (gallons per vehicle)	Fuel rate (miles per gallon)	Mileage (miles per vehicle)	Fuel consumption (gallons per vehicle)	Fuel rate (miles per gallon)	Mileage (miles per vehicle)	Fuel consumption (gallons per vehicle)	Fuel rate (miles per gallon)
1950	[4]9,060	[4]603	[4]15.0	([5])	([5])	([5])	[6]10,316	[6]1,229	[6]8.4	9,321	725	12.8
1960	[4]9,518	[4]668	[4]14.3	([5])	([5])	([5])	[6]10,693	[6]1,333	[6]8.0	9,732	784	12.4
1970	[4]9,989	[4]737	[4]13.5	8,676	866	10.0	13,565	2,467	5.5	9,976	830	12.0
1975	[4]9,309	[4]665	[4]14.0	9,829	934	10.5	15,167	2,722	5.6	9,627	790	12.2
1980	[4]8,813	[4]551	[4]16.0	10,437	854	12.2	18,736	3,447	5.4	9,458	712	13.3
1981	[4]8,873	[4]538	[4]16.5	10,244	819	12.5	19,016	3,565	5.3	9,477	697	13.6
1982	[4]9,050	[4]535	[4]16.9	10,276	762	13.5	19,931	3,647	5.5	9,644	686	14.1
1983	[4]9,118	[4]534	[4]17.1	10,497	767	13.7	21,083	3,769	5.6	9,760	686	14.2
1984	[4]9,248	[4]530	[4]17.4	11,151	797	14.0	22,550	3,967	5.7	10,017	691	14.5
1985	[4]9,419	[4]538	[4]17.5	10,506	735	14.3	20,597	3,570	5.8	10,020	685	14.6
1986	[4]9,464	[4]543	[4]17.4	10,764	738	14.6	22,143	3,821	5.8	10,143	692	14.7
1987	[4]9,720	[4]539	[4]18.0	11,114	744	14.9	23,349	3,937	5.9	10,453	694	15.1
1988	[4]9,972	[4]531	[4]18.8	11,465	745	15.4	22,485	3,736	6.0	10,721	688	15.6
1989	[4]10,157	[4]533	[4]19.0	11,676	724	16.1	22,926	3,776	6.1	10,932	688	15.9
1990	[R]10,504	[R]520	[R]20.2	11,902	738	16.1	23,603	3,953	6.0	11,107	677	16.4
1991	[R]10,571	[R]501	[R]21.1	12,245	721	17.0	24,229	4,047	6.0	11,294	669	16.9
1992	[R]10,857	[R]517	21.0	12,381	717	17.3	25,373	4,210	6.0	11,558	683	16.9
1993	[R]10,804	[R]527	[R]20.5	12,430	714	17.4	26,262	4,309	6.1	11,595	693	16.7
1994	[R]10,992	[R]531	[R]20.7	12,156	701	17.3	25,838	4,202	6.1	11,683	698	16.7
1995	11,203	530	21.1	12,018	694	17.3	26,514	4,315	6.1	11,793	700	16.8
1996	[R]11,330	534	21.2	11,811	685	17.2	26,092	4,221	6.2	11,813	700	16.9
1997	[R]11,581	[R]539	21.5	12,115	703	17.2	27,032	4,218	6.4	[R]12,107	711	17.0
1998[P]	11,725	548	21.4	12,061	704	17.1	27,064	4,257	6.4	12,183	719	17.0

[1] Includes a small number of trucks with 2 axles and 4 tires, such as step vans.
[2] Single-unit trucks with 2 axles and 6 or more tires, and combination trucks.
[3] Includes buses and motorcycles, which are not shown separately.
[4] Includes motorcycles.
[5] Included in "Trucks."
[6] Includes vans, pickup trucks, and sport utility vehicles.
R=Revised. P=Preliminary.

SOURCE: *Annual Energy Review 1999*, Energy Information Administration, Washington, D.C., 2000

FIGURE 10.5

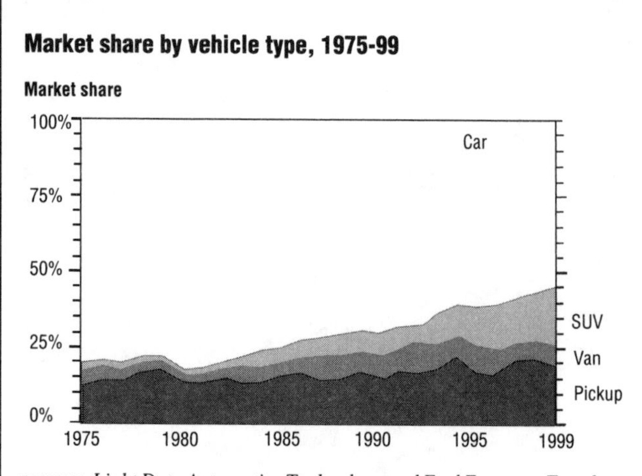

SOURCE: *Light Duty Automotive Technology and Fuel Economy Trends through 1999*, U.S. Environmental Protection Agency, Washington, D.C.

TABLE 10.4

Alternative fuel vehicle acquisition mandates for centrally fueled fleets of federal agencies, state governments, and alternative fuel providers

	Percentage of all acquisitions for groups mandated to acquire vehicles		
Year	Federal agencies	State governments	Alternative fuel providers
1996	25	N/A	N/A
1997	33	10	30
1998	50	15	50
1999	75	25	70
2000	75	50	90
2001 and beyond	75	75	90

Note: The act mandated that the federal government had to acquire 5,000 alternative fuel vehicles in 1993, 7,500 vehicles in 1994, and 10,000 vehicles in 1995. It did not require state governments and alternative fuel providers to acquire alternative fuel vehicles during these years. In addition, the states' and fuel providers' acquisition mandates for 1996 were postponed for 1 year.

SOURCE: *Energy Policy Act of 1992: Limited Progress in Acquiring Alternative Fuel Vehicles and Reaching Fuel Goals*, U.S. General Accounting Office, Washington, D.C., February 2000

my is so much lower than that of automobiles. The EIA projects a small increase in fuel efficiency for the heavy truck fleet and a larger increase for the small truck fleet. If the heavy truck fleet were to reach a fuel efficiency of 10 mpg through technological improvements, projected oil demand would drop by 300,000 barrels per day. Over the past few years, aerodynamically designed trucks have become more common on American roads.

Cheap gasoline prices throughout the 1990s virtually eliminated any sense of urgency with the issue of fuel efficiency, which was demonstrated by the high growth of large vehicle sales. In addition, when the federal 55-mph speed limit law was repealed, states allowed increased speed limits, which also lowered fuel efficiency. Carmakers have resisted building highly efficient cars, claiming that government mandates would saddle American motorists with car features they would not buy. In contrast, the European Commission has proposed an ambitious target of 47 mpg for gasoline-driven cars (compared to the current average of 29 mpg) and 52 mpg for diesel-powered cars by 2005. It must be noted that while European countries do not generally legislate fuel efficiency, the cost of gasoline is more than twice that in the United States. That serves as a powerful incentive to European drivers to buy fuel-efficient vehicles.

Mandating Alternative Fuel Vehicles

Over the past decade several laws have been passed to encourage or mandate the use of vehicles powered by fuels other than gasoline. The Clean Air Act Amendments of 1990 (PL101-549) required certain businesses and local governments with fleets of 10 or more vehicles in 21 metropolitan areas nationwide to phase in alternative fuel vehicles (AFVs) over time. Of those fleets, 20 percent were required to be AFVs by 1998. While great strides have been made in increasing the use of AFVs, there is no way to determine current compliance to the mandates because reporting and enforcement methods are inadequate.

The Energy Policy Act of 1992 (PL 102-486) was passed in the wake of the 1991 Persian Gulf War to conserve energy and increase the proportion of energy supplied domestically. It required the federal government to purchase 22,500 AFVs by 1995 and increase the percentage of such vehicles from 25 percent in 1996 to 75 percent in 1999 and thereafter. (See Table 10.4.) Agency budget cuts and inadequate enforcement have slowed compliance with these regulations. Many municipal governments and the U.S. Postal Service have put into operation fleets of natural gas vehicles, such as garbage trucks, transit buses, and postal vans.

Progress with Alternative Transportation Vehicles

In 1992, 251,352 AFVs were on U.S roads. By 1999, 418,128 AFVs were in use. (See Table 10.5.) These totals include vehicles originally manufactured to run on alternative fuels as well as converted gasoline or diesel vehicles. The manufacture of new AFVs has been steadily increasing.

The fuels used in AFVs are liquefied petroleum gas (LPG), compressed natural gas (CNG), liquefied natural gas (LNG), methanol and ethanol blends, electricity, and biodiesel. LPG is a mixture of propane and butane. CNG is natural gas that is stored in pressurized tanks. CNG releases one-tenth the carbon monoxide, hydrocarbon, and nitrogen as gasoline. Methanol is a liquid fuel that can be produced from natural gas, coal, or biomass (plant material, vegetation, or agricultural waste). Ethanol is ethyl alcohol, a grain alcohol mixed with gasoline and sold as gasohol. Three-fourths of the AFVs in use operate on LPG; two-thirds of the remainder use CNG.

TABLE 10.5

Estimated number of alternative-fueled vehicles in use in the United States, by fuel, 1992–99

Fuel	1992	1993	1994	1995	1996	1997	1998	1999	Avg. annual growth rate (percent)
Liquefied petroleum gases (LPG)[a]	221,000	269,000	264,000	259,000	263,000	263,000	269,000	*274,000*	3.1
Compressed natural gas (CNG)	23,191	32,714	41,227	50,218	60,144	70,852	85,730	*96,017*	22.5
Liquefied natural gas (LNG)	90	299	484	603	663	813	1,358	*1,517*	49.7
Methanol, 85 percent (M85)[b]	4,850	10,263	15,484	18,319	20,265	21,040	21,578	*21,829*	24.0
Methanol, neat (M100)	404	414	415	386	172	172	378	*378*	-0.9
Ethanol, 85 percent (E85)[b,c]	172	441	605	1,527	4,536	9,130	11,743	*17,892*	94.2
Ethanol, 95 percent (E95)[b]	38	27	33	136	361	347	14	*14*	-13.3
Electricity	1,607	1,690	2,224	2,860	3,280	4,453	5,824	*6,481*	22.0
Non-LPG subtotal	30,352	45,848	60,472	74,049	89,421	106,807	126,625	*144,128*	24.9
Total	**251,352**	**314,848**	**324,472**	**333,049**	**352,421**	**369,807**	**395,625**	***418,128***	**7.5**

[a]Values are rounded to thousands. Accordingly, these estimates are not equal to the sum of federal fleet data (for which exact counts are available) and non-federal fleet estimates (rounded to thousands).
[b]The remaining portion of 85-percent methanol and both ethanol fuels is gasoline.
[c]Does not include recently announced plans of some major automakers to make available large numbers of vehicles capable of operating on E85 fuel in the near future.
Note: Estimates for 1997 are revised. Estimates for 1998 are preliminary and estimates for 1999, in italics, are based on plans or projections. Estimates for historical years may be revised in future reports if new information becomes available.

SOURCE: Available at: http://www.eia.doe.gov/cneaf/solar.renewables/alt_trans_fuel97/table1.html

The largest numbers of AFVs are located in the South, followed by the West, the Midwest, and the Northeast. California, Texas, Illinois, Ohio, and Michigan have the most AFVs—42 percent of the total. (See Table 10.6.) Transit buses are one type of heavy-duty vehicle that has seen much AFV activity. In 1999 approximately one in four of new transit buses on order had alternative fuel capabilities.

Alternative Fuel and the Marketplace

AFVs cannot become a viable transportation option unless a fuel supply is readily available. (See Table 10.7.) Ideally, the infrastructure for supplying alternative fuels would be developed simultaneously with the vehicles. There were 5,180 refueling sites in the United States as of December 2000. In 1998, 76 percent of alternative fuel was used by privately owned vehicles, 22 percent for state and local vehicles, and the remainder for federal vehicles. (See Table 10.8.)

Many state policies and programs encourage the use of alternative fuels. California, for example, requires the sale of electric vehicles (EVs) by 2003. This has caused vehicle manufacturers to expedite vehicle research and development. In fact, EVs are already selling there and some rental car agencies now offer EVs to customers at prices only slightly higher than gasoline-powered cars.

Chrysler Corporation stopped making natural-gas-powered vehicles after the 1997 model year because it lost money on the vehicles, selling only 4,000 natural-gas-powered autos since production began in 1992. General Motors, which had suspended sales of natural-gas vehicles in 1994, resumed sales in 1997. Ford began selling some natural-gas versions of its cars and trucks in 1995. Commercial fleets, not retail customers, are the main buyers of natural-gas vehicles.

Market success of alternative fuels and AFVs depends upon public acceptance. People are accustomed to using gasoline as their main transportation fuel and it is readily available. As federal and state requirements for alternative fuels increase, so will the fuels' visibility and acceptance by the general public.

Electric Cars: Promise and Reality

The electric car is not a new invention. Popular during the 1890s, the quiet, clean, and simple vehicle was expected to dominate the automotive market of the twentieth century. Instead, it quietly disappeared as automobile companies chose to invest billions of dollars in the internal combustion engine. It has taken a century, but the electric car has returned.

TABLE 10.6

Estimated number of alternative-fueled vehicles in use, by state and fuel type, 1997

State	Liquefied petroleum gases	Natural gas	Methanol	Ethanol	Electricity	Total
Alabama	2,550	476	0	2	29	3,057
Alaska	87	20	2	0	4	113
Arizona	2,495	3,072	20	0	217	5,804
Arkansas	965	521	0	2	5	1,493
California	32,793	12,419	15,442	406	1,915	62,975
Colorado	4,175	2,955	273	366	247	8,016
Connecticut	1,439	735	14	5	26	2,219
Delaware	273	235	4	0	1	513
District of Columbia	36	400	329	104	91	960
Florida	8,254	2,247	1	57	66	10,625
Georgia	8,041	2,660	168	43	166	11,078
Hawaii	296	0	6	6	81	390
Idaho	1,486	319	0	77	104	1,986
Illinois	15,399	1,518	254	1,398	22	18,591
Indiana	7,455	1,721	1	587	34	9,798
Iowa	4,744	200	42	1,032	1	6,019
Kansas	1,449	46	22	201	3	1,721
Kentucky	3,088	518	0	411	3	4,020
Louisiana	3,173	447	124	3	0	3,747
Maine	494	3	0	3	8	508
Maryland	3,266	1,091	566	57	11	4,991
Massachusetts	3,022	850	252	19	140	4,284
Michigan	13,929	1,232	361	503	255	16,280
Minnesota	1,798	392	0	458	9	2,657
Mississippi	4,347	111	0	3	3	4,464
Missouri	2,954	801	389	1,020	18	5,182
Montana	1,164	386	0	22	1	1,573
Nebraska	2,266	293	0	746	3	3,308
Nevada	1,358	1,886	2	3	21	3,270
New Hampshire	399	10	0	7	19	435
New Jersey	3,915	1,512	51	16	99	5,593
New Mexico	3,041	860	1	3	14	3,919
New York	8,023	5,101	431	49	101	13,705
North Carolina	8,810	132	1	15	37	8,995
North Dakota	559	392	0	25	9	985
Ohio	14,613	2,809	190	276	32	17,920
Oklahoma	10,982	3,636	1	147	82	14,847
Oregon	6,976	279	310	16	32	7,613
Pennsylvania	9,889	2,473	460	39	74	12,935
Rhode Island	491	204	0	3	5	703
South Carolina	3,663	141	0	6	21	3,831
South Dakota	884	102	1	165	0	1,152
Tennessee	8,258	427	6	23	35	8,749
Texas	30,459	8,118	615	21	83	39,296
Utah	1,770	3,049	7	66	35	4,927
Vermont	265	4	0	3	38	310
Virginia	3,776	1,322	50	96	64	5,309
Washington	4,993	1,449	666	9	158	7,275
West Virginia	672	1,099	10	10	2	1,793
Wisconsin	6,627	918	140	938	29	8,652
Wyoming	1,139	74	0	5	0	1,218
U.S. total	**263,000**	**71,665**	**21,212**	**9,477**	**4,453**	**369,807**

Note: Natural gas includes compressed natural gas (CNG) and liquefied natural gas (LNG). Methanol includes M85 and M100. Ethanol includes E85 and E95.

SOURCE: Available at: http://www.eia.doe.gov/cneaf/solar.renewables/alt_trans_fuel97/table4.html

Energy standards have created a market for alternative energy cars. The combination of government mandates to reduce emissions and to encourage market opportunities has altered the automobile industry. By 2000 most car manufacturers either had an electric car on the roads or in testing.

The primary difficulty with EVs lies in inadequate battery power. The cars must be recharged often. These vehicles use lead-acid or nickel-cadmium batteries and have a range of 70 to 100 miles on a single charge. The range is reduced by factors such as cold temperatures, the use of air conditioning, vehicle load, and steep terrain.

In addition, EVs are expensive, although prices are coming down. General Motors' EV1 leases for between $480 and $640 per month. This is less than the cost of luxury cars but more than the average mid-size American car. Toyota produces the RAV-4-EV, Nissan makes the electric Altra, Ford the Ranger EV, and DaimlerChrysler the EPIC. In 1997 car industry officials announced plans to produce hybrid vehicles, which operate on gasoline in conjunction with fuel cell batteries and electric engines, by mid-2000. Some manufacturers, including Honda, met that goal. Such vehicles may overcome the problem of fuel availability by use of conventional fuel and abundant refueling stops and yet burn fuel twice as efficiently as with current technology.

Despite their high price, electric cars have many advantages. They are relatively quiet and simple in design and operation. They cost less to refuel and service and have fewer parts to break down. Their owners are likely to spend less time on maintenance and, if they recharge at home, will rarely have to go to the service station. These time savings have real value in the busy world of the twenty-first century. Over time the cost gap between cars that pollute and EVs that do not will narrow. With advances in battery development the gap could close entirely. Automobile industry experts believe electric cars will assume a "second car" role for commuters, much like the microwave oven has become an addition to, not a replacement of, conventional ovens for cooking.

California has led in the development of EVs. In 1990 the California Air Resources Board (CARB), facing severe air pollution in Los Angeles and other cities, passed the toughest auto emissions standards in the world. Most notable was the requirement that 2 percent of cars sold in the state by the seven major carmakers in 1998 must be zero-emission, which would rise to 10 percent by 2003. Auto-industry lobbyists protested and the 1998 mandate was lifted, but the big automakers were still required to achieve the 10 percent target by 2003. If the automakers actually achieve the challenging goal, approximately 800,000 zero-emission cars will be on California roads by 2010, up from 2,000 in use in the entire country in 1997. In 1998 New York joined California in EV requirements, when a federal judge there ruled that the state could mandate EV sales.

Air Travel Efficiency

The average American flew 1,739 miles in 2000. Europeans, though they flew fewer miles, had the world's most crowded skies, while the most rapid growth in flying was in Asia. Most air travel is done by a small portion of the world's people.

Flying carries an environmental price: it is a very energy-intensive form of transport. In much of the industrialized world air travel is replacing more energy-efficient rail or bus travel. Despite a rise in fuel efficiency of jet engines, jet fuel consumption has risen 65 percent since 1970.

Another problem with air travel is its impact on global warming. Airplanes spew nearly four million tons of nitrogen oxide into the air, much of it while cruising in the tropospheric zone five to seven miles above the earth where ozone is formed. The EPA estimates that air traffic accounts for about 3 percent of all global greenhouse warming. The Intergovernmental Panel on Climate Change (IPCC) notes that emissions deposited directly into the atmosphere do greater harm than those released at the earth's surface. (See Figure 10.6.)

In 1998 Pratt and Whitney, a designer and manufacturer of high-performance engines, announced plans to introduce a radical new engine design that would be cleaner, more efficient, and more reliable than conven-

TABLE 10.7

U.S. refueling site counts by state and fuel type as of 12/08/2000

STATE	M85	CNG	E85	LPG	ELEC	LNG	ALL
Alabama		15		76	35	2	128
Alaska	0	0	0	9	0	0	9
Arizona	0	28	1	103	46	3	181
Arkansas	0	7	0	68	0	0	75
California	32	207	0	346	337	9	931
Colorado	0	44	1	70	0	1	116
Connecticut	0	25	0	34	1	0	60
Delaware	0	6	0	4	0	0	10
Dist. of Columbia	0	3	0	0	1	0	4
Florida	1	38	0	149	3	1	192
Georgia	0	67	0	55	27	2	151
Hawaii	0	0	0	7	3	0	10
Idaho	0	8	1	34	1	0	44
Illinois	0	21	8	56	2	0	87
Indiana	0	34	1	46	1	3	85
Iowa	0	5	5	41	0	0	51
Kansas	0	5	1	68	0	1	75
Kentucky	0	6	2	25	0	0	33
Louisiana	0	14	0	33	0	0	47
Maine	0	0	0	20	0	0	20
Maryland	0	25	0	29	1	2	57
Massachusetts	0	15	0	37	3	0	55
Michigan	0	31	4	132	6	1	174
Minnesota	0	12	49	61	0	1	123
Mississippi	0	3	0	32	0	0	35
Missouri	0	7	4	130	0	0	141
Montana	0	10	1	42	0	1	54
Nebraska	0	5	7	28	0	0	40
Nevada	0	18	0	32	0	0	50
New Hampshire	0	1	0	29	1	0	31
New Jersey	0	22	0	28	0	0	50
New Mexico	0	14	1	90	0	1	106
New York	4	59	0	98	6	0	167
North Carolina	0	9	0	77	7	0	93
North Dakota	0	4	2	14	0	0	20
Ohio	0	48	0	75	1	1	125
Oklahoma	0	53	0	39	0	0	92
Oregon	0	15	0	51	0	1	67
Pennsylvania	0	53	0	106	1	1	161
Rhode Island	0	6	0	7	0	0	13
South Carolina	0	4	0	60	1	0	65
South Dakota	0	4	7	26	0	0	37
Tennessee	0	4	0	59	0	0	63
Texas	0	69	0	443	2	7	521
Utah	0	62	0	18	0	1	81
Vermont	0	0	0	17	7	0	24
Virginia	0	27	1	60	8	3	99
Washington	0	26	0	89	6	1	122
West Virginia	0	39	0	10	0	0	49
Wisconsin	0	22	1	82	0	0	105
Wyoming	0	18	0	32	0	1	51
Totals:	37	1218	97	3277	507	44	5180

SOURCE: Alternative Fuels Data Center. Available at http://www.afdc.nrel.gov/refuel/state_tot.shtml

TABLE 10.8

Estimated consumption of alternative transportation fuels in the United States, by vehicle ownership, 1995, 1997, and 1999
(thousand gasoline-equivalent gallons)

	1995				1997				1999			
Fuel	Federal	State and local	Private	Total	Federal	State and local	Private	Total	Federal	State and local	Private	Total
Liquefied petroleum gases (LPG)	42	25,092	207,567	232,701	55	26,814	211,487	238,356	*202*	*29,119*	*221,001*	*250,322*
Compressed natural gas (CNG)	4,250	12,340	18,572	35,162	4,394	29,770	30,131	64,295	*4,541*	*40,148*	*42,700*	*87,389*
Liquefied natural gas (LNG)	17	2,658	84	2,759	94	3,074	546	3,714	*228*	*5,692*	*968*	*6,888*
Methanol, 85 percent (M85)[a]	829	310	884	2,023	207	351	996	1,554	*37*	*288*	*976*	*1,301*
Methanol, neat (M100)	0	2,150	0	2,150	0	347	0	347	*0*	*1,923*	*0*	*1,923*
Ethanol, 85 percent (E85)[a]	49	128	13	190	286	510	484	1,280	*1,000*	*617*	*626*	*2,243*
Ethanol, 95 percent (E95)[a]	0	975	20	995	0	1,136	0	1,136	*0*	*59*	*0*	*59*
Electricity	25	281	357	663	48	332	630	1,010	*100*	*521*	*793*	*1,414*
Total	**5,212**	**43,934**	**227,497**	**276,643**	**5,084**	**62,334**	**244,274**	**311,692**	***6,108***	***78,367***	***267,064***	***351,539***

[a]The remaining portion of 85-percent methanol and both ethanol fuels is gasoline. Consumption data include the gasoline portion of the fuel.

Notes: Fuel quantities are expressed in a common base unit of gasoline-equivalent gallons to allow comparisons of different fuel types. Gasoline-equivalent gallons do not represent gasoline displacement. Gasoline equivalent is computed by dividing the lower heating value of the alternative fuel by the lower heating value of gasoline and multiplying this result by the alternative fuel consumption value. Lower heating value refers to the Btu content per unit of fuel excluding the heat produced by condensation of water vapor in the fuel.

Totals may not equal sum of components due to independent rounding.

Estimates for 1995 and 1997 are revised. Estimates for 1999, in italics, are based on plans or projections. Estimates for historical years may be revised in future reports if new information becomes available.

SOURCE: Available at: http://www.eia.doe.gov/cneaf/solar.renewables/alt_trans_fuel97/table13.html

tional designs. The new engine, expected to enter service in 2002, would reduce emissions by 40 percent and exceed noise restrictions set to take effect in 2000. The engine is designed for use on single-aisle planes carrying 120 to 180 passengers, such as the Boeing 737 or the Airbus A320.

Although each generation of airplane engine gets cleaner and more fuel-efficient, there seems to be little that can be done about the increased amount of flying. There is, however, a movement toward doing something about other engines in the airline industry—those in trucks, cars, and carts that service airplane fleets. Electric utility companies, including the Edison Electric Institute and the Electric Power Research Institute, launched a program in 1993 to electrify airports. By converting terminal transport buses, food trucks, and baggage-handling carts to electricity, airports could reduce air pollution considerably.

CONSERVATION IN THE RESIDENTIAL AND COMMERCIAL SECTORS

Total building energy use in the United States has increased: there are more people, more households, and more offices. Energy use per unit area (commercial) or per person (residential) has roughly stabilized over the past 10 to 12 years due to a variety of efficiency improvements. The sources of energy in buildings have changed dramatically. Use of fuel oil has dropped, and natural gas has largely made up the difference. At the same time, other energy demands have risen. Electronic office equipment has sharply increased electricity loads in commercial buildings, caused by computers, fax machines, printers, and copiers. Energy use in buildings accounts for an increasing share of total U.S. energy consumption: 27 percent in 1950, 33 percent in 1970, and 35 percent in 1999. The residential and commercial sectors used roughly 65 percent of U.S. electricity in 1999.

Building Efficiency

There are several potential areas for research and development in energy conservation in buildings. (See Table 10.9.) Among the techniques useful in reducing energy loads are advanced window designs, daylighting (letting light in from the outside by using high windows,

FIGURE 10.6

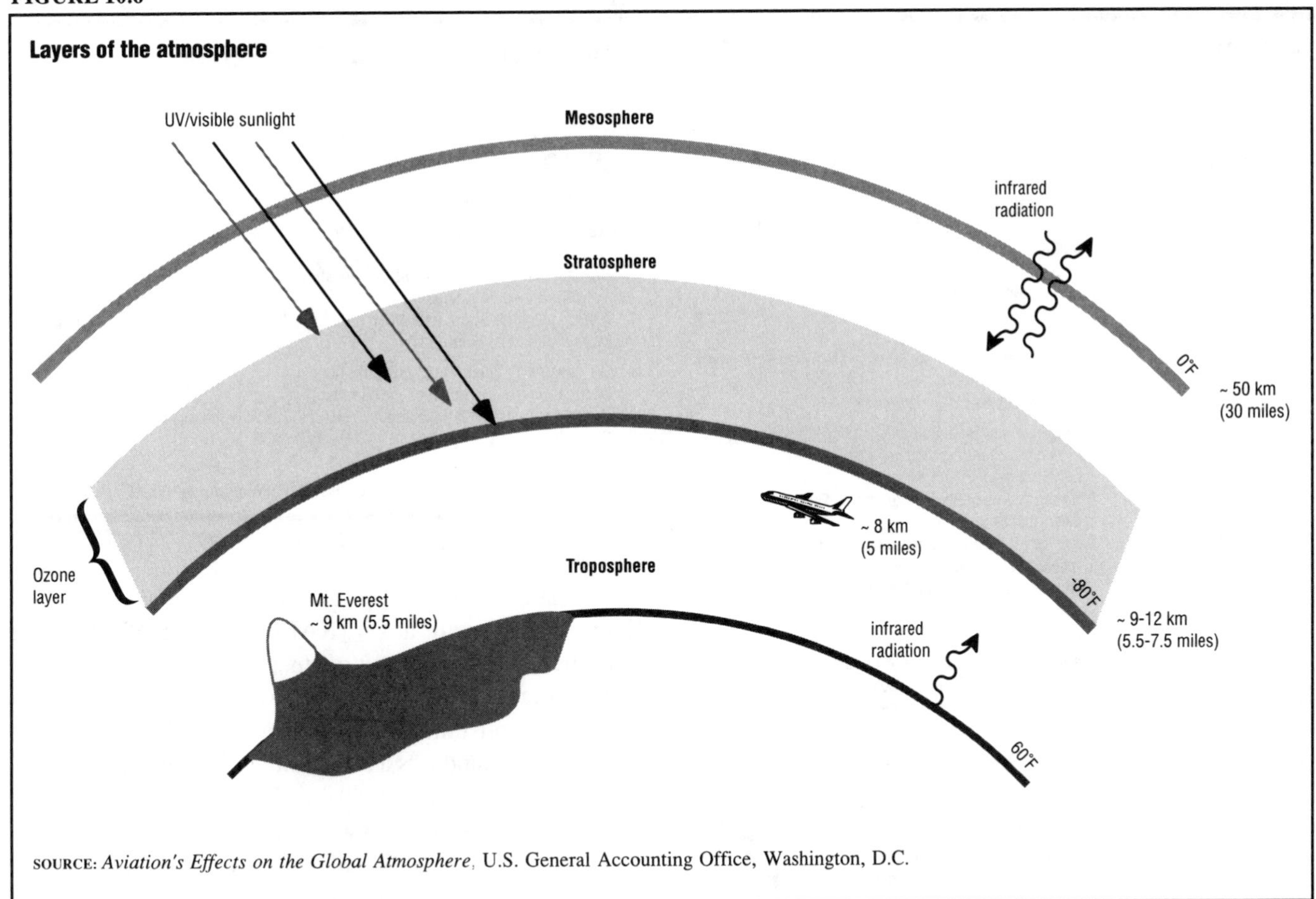

SOURCE: *Aviation's Effects on the Global Atmosphere*, U.S. General Accounting Office, Washington, D.C.

skylights, and atria in the center of large buildings), solar water heating, landscaping, and tree planting. Energy conservation efforts in the building sector have been substantial over the past 20 years.

Residential energy consumption can be reduced by introducing more efficient new housing and appliances, improving the existing stock of housing, and by building more multiple-family units. Residential energy consumption is reduced as well when people migrate to the South and West. (See Table 10.10.)

In the residential sector, the amount of energy used in newer homes is dramatically less than that used in older homes. The houses built since 1980 use about half of the energy consumed by older houses. The largest share of energy savings is due to better construction, higher quality insulation, and more energy-efficient windows and doors. The Office of Technology Assessment reported that roughly one-fourth the energy used to heat and cool buildings is lost through poor insulation and poorly insulated windows. Before the 1973 energy crisis, 70 percent of new windows sold were single-glazed. By 1990, due to changes in building codes and public interest, 80 percent of windows sold were double-glazed with double insulating ability, cutting energy loss in half.

FIGURE 10.7

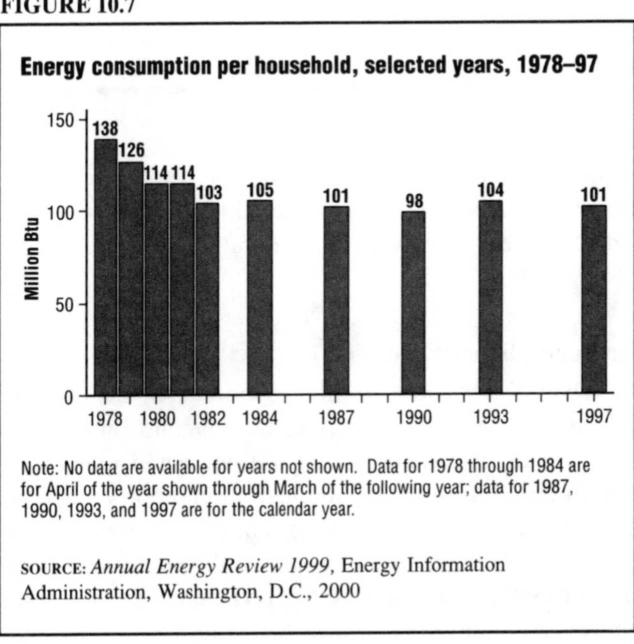

Note: No data are available for years not shown. Data for 1978 through 1984 are for April of the year shown through March of the following year; data for 1987, 1990, 1993, and 1997 are for the calendar year.

SOURCE: *Annual Energy Review 1999*, Energy Information Administration, Washington, D.C., 2000

Energy consumption per household has remained steady in the past decade, as technology gains have been offset by the increase in size of new homes and demand for energy services. (See Figure 10.7.) As in the residential sector, improved technology has helped

TABLE 10.9

Research development needs

Materials	Insulants, particularly transparent insulants such as aerogels Electronically adjustable spectrally selective windows Improved lighting controls for integrating daylighting and artificial lights Improved and longer life gaskets and sealants Phase-change materials Desiccants for cooling systems Selective surfaces Improved catalysts for small-scale biomass combustion emissions control Air-to-air heat exchanger materials
Building physics	Passive cooling techniques, including radiant cooling Perimeter daylighting systems, allowing deeper penetration of perimeter spaces Atria design for better daylighting and thermal performance Basic heat transfer and natural convection air-flow research to improve performance and comfort Moisture absorption and desorption in building materials Duct design
Whole buildings	Testing advanced concepts in buildings Performance monitoring of solar buildings Model land-use controls to encourage proper subdivision/site design
Human comfort research	Determining what makes people comfortable or uncomfortable with respect to temperature, humidity, lighting, and other factors within a building
Design tools	Improved residential and commercial building design tools that perform integrated analysis, including daylighting and window design, space heating, space cooling, and utility demand-side management Development of simplified design tools for the design and construction community Validation of design tools

SOURCE: *Renewing Our Energy Future,* Office of Technology Assessment, Washington, D.C., 1995

to slow the growth in commercial building energy use. Commercial buildings constructed after 1980 use considerably less energy than those built in the early part of the 1900s.

Home Appliance Efficiency

Overall, the number of households in the U.S. is increasing, which is increasing the demand for energy-intensive services such as air conditioning and appliances. The Energy Information Administration reports that residential energy use accounts for about 21 percent of the total national energy use. For household energy consumption in 1997, space heating used 51 percent of the total energy, appliances 26 percent, water heating 19 percent, and air conditioners 4 percent.

The number of appliances in households has been increasing steadily over the past several decades. (See Figure 10.8.) By 1997, 99 percent of American homes had color televisions, 47 percent had central air conditioning, and 35 percent had personal computers.

TABLE 10.10

Major factors influencing residential energy use

Factors causing an increase in consumption:
- Larger population–more households
- Fewer people per household–more households
- Increased demand for energy–intensive services

Factors causing a decrease in consumption:
- New housing more efficient than existing stock
- New appliances more efficient than existing stock
- Retrofits to existing housing
- Migration to the South and West
- More multifamily units

Factors causing fluctuations in consumption:
- Occupant behavior, changes in thermostat settings
- Fuel shifts–more electricity and less oil, changes in wood use
- Price changes

SOURCE: Office of Technology Assessment, Washington, D.C., 1992

In 1987 Congress passed the National Appliance Energy Conservation Act (NAECA; PL 95-629), which gave the U.S. Department of Energy (DOE) the authority to formulate minimum efficiency requirements for 13 classes of consumer products. It could also revise and update those standards as technologies and economic conditions changed. Table 10.11 shows the products and years in which standards were established or revised. In 1997 the DOE established an advisory committee to review and revise the standards.

Energy efficiency has increased for all major household appliances but most dramatically for refrigerators and freezers. Since 1972 new refrigerators and freezers have almost doubled in energy efficiency due to better insulation, motors, compressors, and accessories such as automatic defrost. These improvements have been accomplished at relatively low cost to manufacturers. In addition, efficiency labels for consumers are now required, which makes purchasing efficient models easier.

Air conditioners and heat pumps, another major group of appliances, has shown a 35 percent improvement over the past two decades. Although this improvement in energy efficiency is less than the improvements in refrigerators and freezers, it is significant because these appliances are large energy users.

Water heaters and furnaces have improved efficiency between 5 percent and 20 percent in two decades. The technological improvements in water heaters and furnaces are relatively costly compared to the overall price of the product. This means that the more energy-conserving models have a higher retail price, which discourages many consumers from purchasing efficient models, although the more efficient models will save money in the long run. In addition, many of the purchases of water heaters and furnaces are made by builders, who have little incentive to pay more for the most efficient models, or by homeowners

FIGURE 10.8

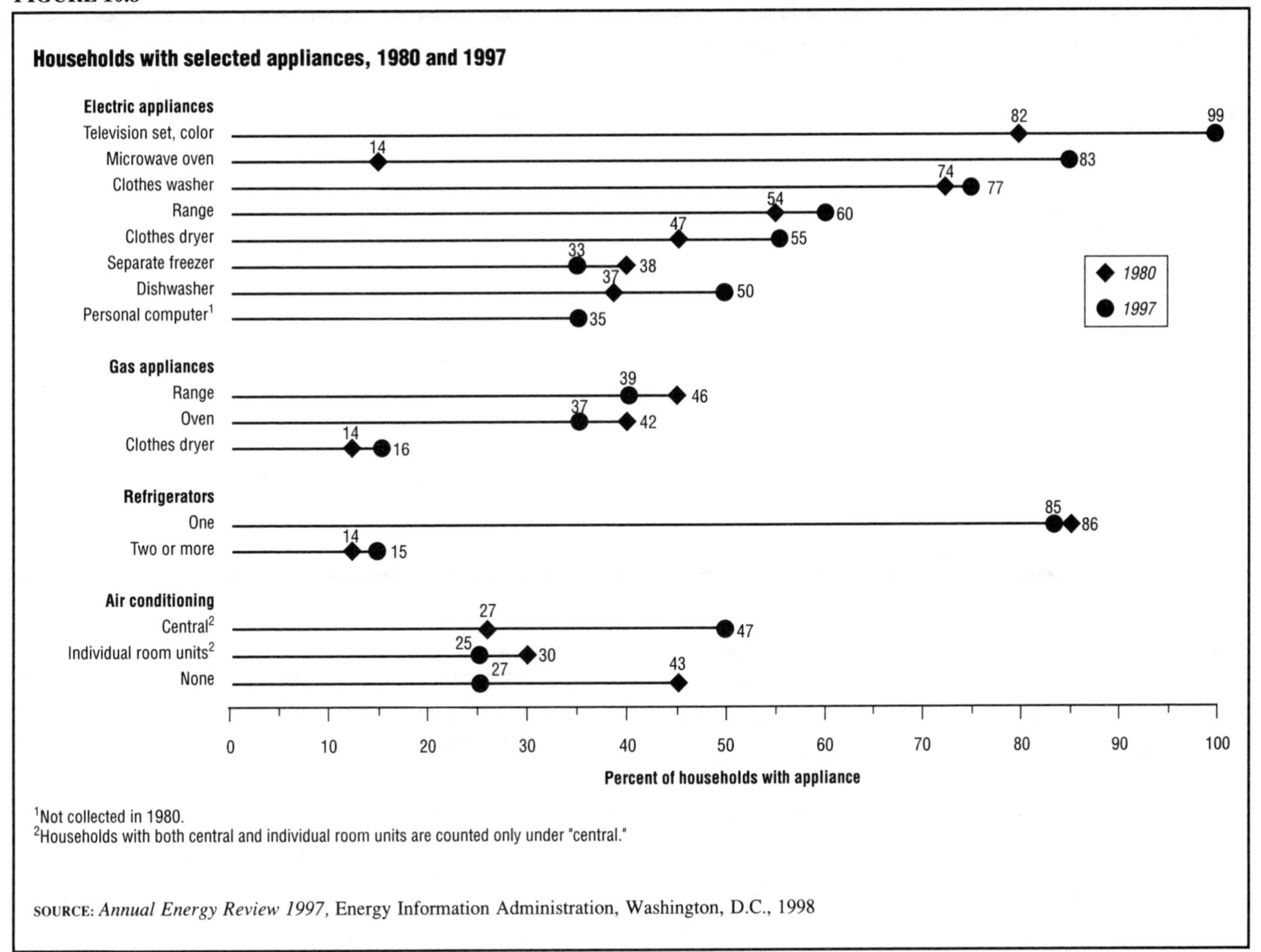

[1]Not collected in 1980.
[2]Households with both central and individual room units are counted only under "central."

SOURCE: *Annual Energy Review 1997,* Energy Information Administration, Washington, D.C., 1998

in emergencies, when fast availability and installation seem much more important than energy efficiency.

Nonetheless, consumers are generally willing to purchase more expensive, energy-efficient models of air conditioners, refrigerators, and lights if the devices can save them enough money in the long run on their electricity bills to offset the higher purchase costs. According to a U.S. General Accounting Office study (*Energy Conservation: Efforts Promoting More Efficient Use,*Washington, D.C., 1992), consumers will purchase such devices if the "payback period" is two years or less.

In addition to concerns about efficiency, appliance makers, especially of refrigerators and air conditioning systems, are striving to develop alternative cooling techniques as substitutes for chlorofluorocarbons (CFCs), which are ozone-damaging chemicals that can no longer be sold legally in the United States. Current technology is temporarily substituting CFCs with somewhat less dangerous HCFCs (hydro-chlorofluorocarbons). In Europe, refrigeration units using other substances, such as propane "greenfreeze" technology, are rapidly replacing HCFCs.

Lawn and Garden Equipment

In 1994 the EPA reported that as much as 10 percent of the nation's air pollution was generated by gasoline-powered lawn and garden equipment, including lawn mowers, chain saws, and golf carts. Former EPA Administrator Carol Browner estimated that Americans use 89 million pieces of such equipment, with lawn mowers alone accounting for 5 percent of the nation's air pollution.

Under Browner the agency established engine label and warranty requirements, exhaust emissions standards, and test procedures, requiring that engine makers meet the new requirements by 1996. Effective that year, new products offered for sale were equipped with improved carburetion systems, and additional standards were scheduled for subsequent years. Agency officials predict the new regulations will reduce smog-forming hydrocarbon emissions by 32 percent and carbon monoxide by 14 percent by 2003.

INTERNATIONAL COMPARISONS OF CONSERVATION EFFORTS

Compared to other industrialized countries, the United States is lagging behind in energy efficiency and conserva-

TABLE 10.11

Effective dates of appliance efficiency standards, 1988-2001

Product	1988	1990	1992	1993	1994	1995	2000	2001
Clothes dryers	X				X			
Clothes washers	X				X			
Dishwashers	X				X			
Refrigerators and freezers		X		X				X
Kitchen ranges and ovens		X					X	
Room air conditioners		X						
Direct heating equipment		X						
Fluorescent lamp ballasts		X						
Water heaters		X						
Pool heaters		X						
Central air conditioners and heat pumps			X					
Furnaces								
Central (>45,000 Btu per hour)			X					
Small (<45,000 Btu per hour)			X					
Mobile home		X						
Boilers			X					
Fluorescent lamps, 8 foot					X			
Fluorescent lamps, 2 and 4 foot (U tube)						X		

SOURCE: *Annual Energy Outlook 2000*, Energy Information Administration, Washington, D.C., 1999

tion efforts. According to EIA figures for energy consumption per dollar of GDP in 1998, the U.S. consumed 13,400 Btu per dollar of GDP compared to 7,400 for France, 7,300 for Germany, and 6,600 for Japan. This implies that other countries are making products more efficiently. If fuel prices increase in the future, the U.S. may face economic challenges from more efficient countries.

The U.S. also falls behind other industrialized countries in controlling carbon emissions. The United States is the world's largest producer of greenhouse gases, and its per capita emissions are also significantly higher than in other industrialized countries. For instance, EIA figures for carbon dioxide emissions from the consumption of fossil fuels in 1998 showed that the U.S. emitted 5.56 metric tons of carbon equivalent per person, compared to 2.8 metric tons per person in Germany, 2.3 in Japan, and 1.8 in France. If international environmental treaties are ratified, cleanup efforts would require the U.S. to become much more efficient with carbon emissions.

FIGURE 10.9

Energy use per capita and per dollar of gross domestic product, 1970–2020
(index, 1970=1)

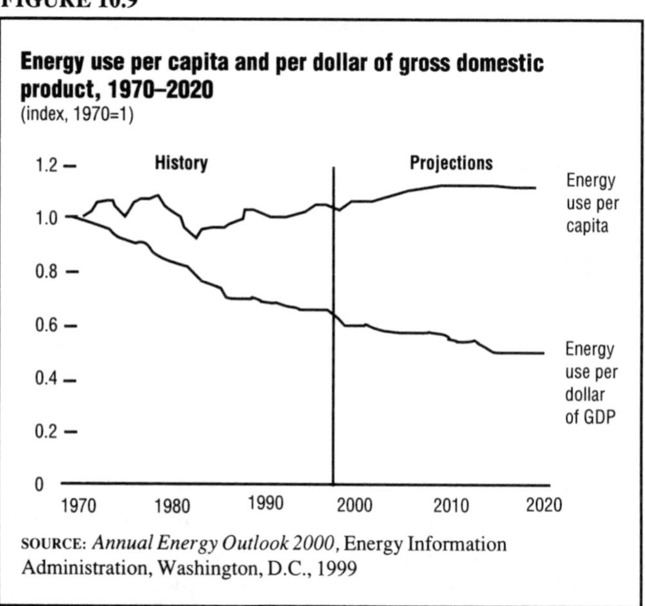

SOURCE: *Annual Energy Outlook 2000*, Energy Information Administration, Washington, D.C., 1999

FUTURE TRENDS IN CONSERVATION

The Energy Information Administration of the U.S. Department of Energy, in its *Annual Energy Outlook 2000* (1999), expects the United States to become more energy efficient over the next 20 years. Future gains in energy conservation, however, may be hampered or discouraged if the nation experiences continued low energy prices, which are also predicted. Low prices for fuel reduce the incentive for developing and implementing energy-saving technology. Nonetheless, the energy conservation trends of the past two decades are likely to continue as industries, homeowners, builders, businesses, and consumers replace older equipment and appliances with newer, more efficient models.

Per capita energy use is expected to increase slightly and then stabilize toward 2020, as increased demand for services is offset by technology gains that improve efficiency. Energy use per dollar of GDP is also projected to decrease as efficiency gains are made. (See Figure 10.9.)

The EIA projects that transportation fuel efficiency will grow more slowly from the present through 2020 than it did in the 1980s. (See Figure 10.10.) Light-duty vehicle efficiency is projected to remain essentially steady, because low fuel prices may lower the incentive for efficiency. For light-duty vehicles, including cars, the EIA projects that any gains in efficiency will by offset by consumer preferences for larger, more powerful vehicles.

FIGURE 10.10

Transportation stock fuel efficiency by mode, 1998–2020
(index, 1998=1)

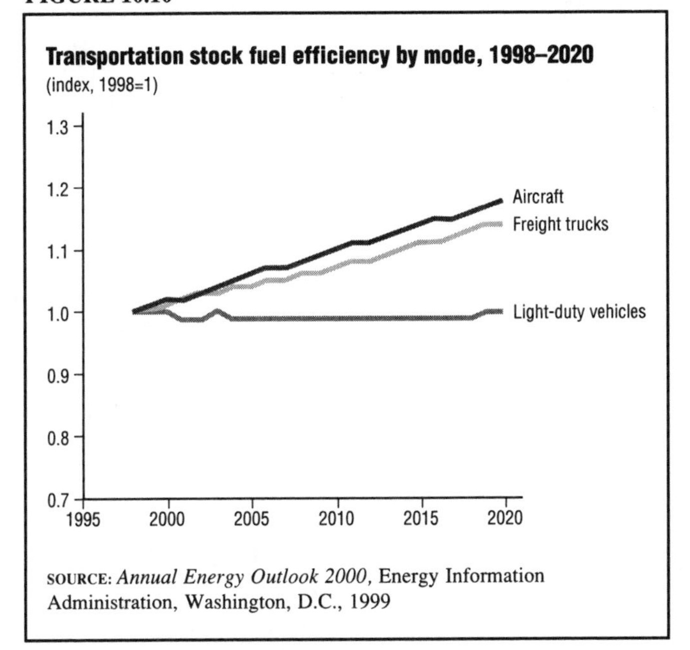

SOURCE: *Annual Energy Outlook 2000,* Energy Information Administration, Washington, D.C., 1999

FIGURE 10.11

Advanced technology light-duty vehicle sales by fuel type, 2010 and 2020
(thousand vehicles sold)

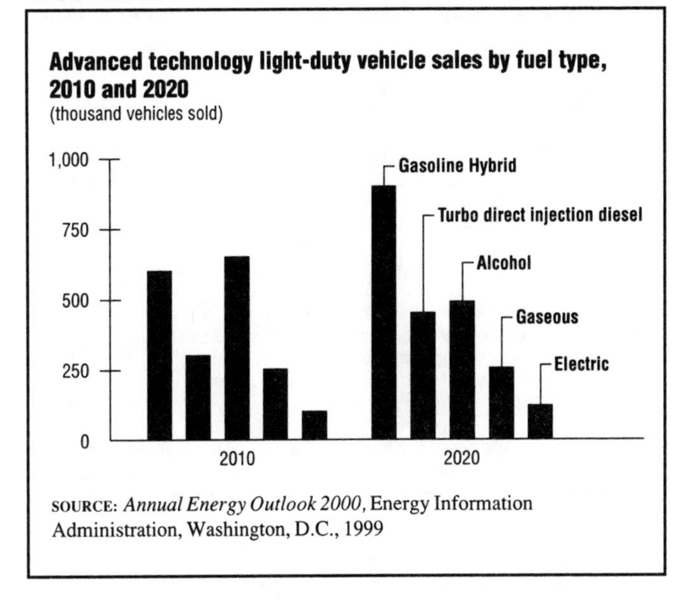

SOURCE: *Annual Energy Outlook 2000,* Energy Information Administration, Washington, D.C., 1999

FIGURE 10.12

Variation from reference case primary residential energy use in three alternative cases, 1999–2020

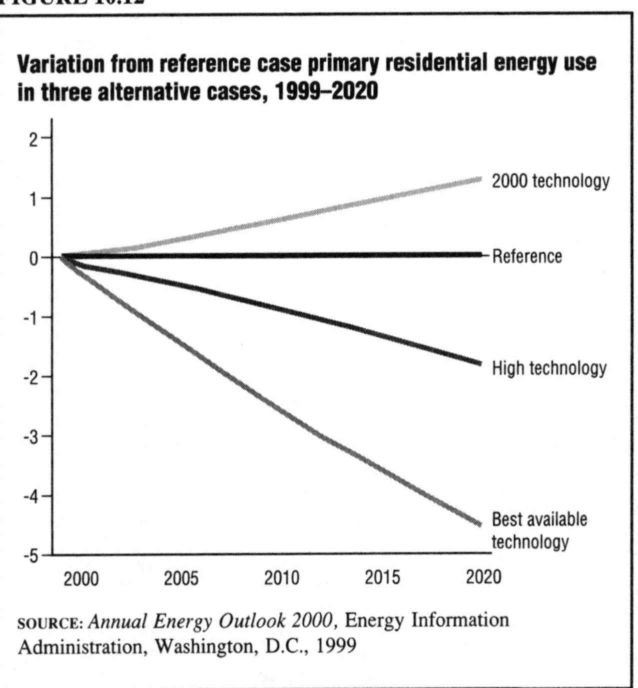

SOURCE: *Annual Energy Outlook 2000,* Energy Information Administration, Washington, D.C., 1999

For aircraft and large trucks, fuel efficiency is expected to increase by 15 to 20 percent by 2020.

The EIA predicts that the market for alternative fuel vehicles will grow as a result of federal and state mandates. By 2020, these vehicles are projected to account for 2.2 million vehicle sales, or 14.6 percent of total vehicle sales. (See Figure 10.11.) Alternative fuels could replace about 406,000 barrels of oil per day by 2020, or 4 percent of all light vehicle consumption. For residential energy consumption, if the most efficient technology is chosen in construction and operation, energy use could be 19.7 percent lower by 2020. (See Figure 10.2.)

The National Energy Policy Plan (NEPP) of the Department of Energy Organization Act of 1977 (PL 95-91) requires the president to submit to Congress every two years a national energy policy plan. This plan includes energy objectives and strategies as well as projections of energy supply, demand, and prices. The most recent plan, the Comprehensive National Energy Strategy (CNES), was produced by the Clinton Administration in April 1998. (See Table 10.12.) In 1999 the Clinton Administration also submitted to Congress the Comprehensive Electricity Competition Act (CECA), which aims to increase efficiency and reduce pollution by the nation's electricity system. (See Chapter 8 for further discussion of CECA.)

TABLE 10.12

The strategy at a glance

Goal I. Improve the efficiency of the energy system — making more productive use of energy resources to enhance overall economic performance while protecting the environment and advancing national security

Objective 1. Support competitive and efficient electric systems
Enact electric utility restructuring legislation, develop advanced coal/gas powerplants, improve existing nuclear powerplants

Objective 2. Significantly increase energy efficiency in the transportation, industrial, and buildings sectors by 2010
Develop more efficient transportation, industrial, and building technologies

Objective 3. Increase the efficiency of federal energy use
Adopt new/innovative energy-efficient and renewable technologies

Goal II. Ensure against energy disruptions — protecting our economy from external threat of interrupted supplies or infrastructure failure

Objective 1. Reduce the vulnerability of the U.S. economy to disruptions in oil supply
Stabilize domestic production, maintain readiness of Strategic Petroleum Reserve, diversify import sources, reduce consumption

Objective 2. Ensure energy system reliability, flexibility, and emergency response capability
Ensure reliable electricity/gas supply, refining and emergency response

Goal III. Promote energy production and use in ways that respect health and environmental values — improving our health and local, regional, and global environmental quality

Objective 1. Increase domestic energy production in an environmentally responsible manner; *increase domestic gas production, recover oil with less environmental impact, develop renewable technologies, maintain viable nuclear option*

Objective 2. Accelerate the development and market adoption of environmentally friendly technologies
Increase near-term deployment, expand voluntary efforts, design domestic greenhouse gas trading program, work with developing countries, design international trading/credit system

Goal IV. Expand future energy choices — pursuing continued progress in science and technology to provide future generations with a robust portfolio of clean and reasonably priced energy sources

Objective 1. Maintain a strong national knowledge base as the foundation for informed energy decisions, new energy systems, and enabling technologies of the future
Pursue basic research, including research on carbon/climate; support energy science infrastructure

Objective 2. Develop technologies that expand long-term energy options
Develop long-term options, such as fusion, hydrogen-based systems, and methane hydrates, that can have major impacts

Goal V. Cooperate internationally on global issues — developing the means to address global economic, security, and environmental concerns

Objective 1. Promote development of open, competitive international energy markets, and facilitate the adoption of clean, safe, and efficient energy systems
Encourage adoption of favorable legal/policy framework in other countries, promote clean/efficient energy systems and science/technology collaboration

Objective 2. Promote foreign regional stability by reducing energy-related environmental risks in areas of U.S. security interest
Prioritize concerns and develop cost-effective solutions

SOURCE: *Comprehensive National Energy Strategy,* Department of Energy, Washington, D.C., April 1998

IMPORTANT NAMES AND ADDRESSES

American Gas Institute
1515 Wilson Blvd.
Arlington, VA 22209
(703) 841-8647
FAX (703) 841-8697

American Petroleum Institute
1220 L St. NW
Washington, DC 20005
(202) 682-8000
FAX (202) 962-4730
URL: http://www.api.org

American Wind Energy Association
122 C St. NW, Fourth Floor
Washington, DC 20001
(202) 383-2500
FAX (202) 383-2505
URL: http://www.awea.org

Bureau of Land Management
Renewable Resources and Planning
1849 C St. NW, #5650
Washington, DC 20240
(202) 208-4896
FAX (202) 208-5016
URL: http://www.blm.gov

Council on Environmental Quality
722 Jackson Pl. NW
Washington, DC 20503
(202) 456-6224
FAX (202) 456-2710
URL: http://www.whitehouse.gov/ceq

Edison Electric Institute
701 Pennsylvania Ave. NW
Washington, DC 20004
(202) 508-5000
FAX (202) 508-5759
URL: http://www.eei.org

Electric Power Research Institute
2000 L St. NW, Suite 805
Washington, DC 20036
(202) 872-9222
FAX (202) 293-2697
URL: http://www.epri.com

Environmental Defense Fund
1875 Connecticut Ave., #1016
Washington, DC 20009
(202) 387-3500
FAX (202) 234-6049
URL: http://www.edf.org

Environmental Industry Associations
4301 Connecticut Ave. NW, #300
Washington, DC 20008
(202) 244-4700
FAX (202) 966-4818
URL: http://www.envasns.org

Environmental Protection Agency
401 M St. SW
Washington, DC 20460
(202) 260-4700
FAX (202) 260-0279
URL: http://www.epa.gov

Friends of the Earth
1025 Vermont Ave. NW, #300
Washington, DC 20005
(202) 783-7400
FAX (202) 783-0444
URL: http://www.foe.org

Greenpeace USA
702 H Street NW, Suite 300
Washington, DC 20001
(202) 462-1177
FAX (202) 462-4507
URL: http://www.greenpeaceusa.org

National Mining Association
1130 17th St. NW
Washington, DC 20036
(202) 463-2654
FAX (202) 833-9636

Natural Gas Supply Association
805 15th St. NW, #510
Washington, DC 20005
(202) 326-9300
FAX (202) 326-9330

Natural Resources Defense Council
1200 New York Ave. NW, #400
Washington, DC 20005
(202) 289-6868
FAX (202) 289-1060
URL: http://www.nrdc.org

Nuclear Energy Institute
1776 Eye St. NW, Suite 400
Washington, DC 20006
(202) 739-8000
FAX (202) 785-4019
URL: http://www.nei.org

Nuclear Regulatory Commission
11555 Rockville Pike
Rockville, MD 20852
(301) 415-2344
FAX (301) 415-2395
URL: http://www.nrc.gov

Public Citizen
1600 20th St. NW
Washington, DC 20009
(202) 588-1000
FAX (202) 588-7798
URL: http://www.citizen.org

Sierra Club
408 C St. NE
Washington, DC 20002
(202) 547-1141
FAX (202) 547-6009
URL: http://www.sierraclub.org

Solid Waste Association of North America
P.O. Box 7219
Silver Spring, MD 20907
(301) 585-2898

FAX (301) 589-7068
URL: http://www.swana.org

Union of Concerned Scientists
1616 P St. NW, #310
Washington, DC 20036
(202) 332-0900
FAX (202) 332-0905
URL: http://www.ucsusa.org

U.S. Department of Energy
1000 Independence Ave. SW
Washington, DC 20585
(202) 586-6151
FAX (202) 586-0956
URL: http://www.doe.gov

U.S. House of Representatives Subcommittee on Energy and Mineral Resources
1337 Longworth Bldg.
Washington, DC 20515
(202) 225-9297
URL: http://resourcescommittee.house.gov

U.S. Senate Committee on Energy and Natural Resources
364 Dirksen Bldg
Washington, DC 20510
(202) 224-4971
FAX (202) 224-6163
URL: http://www.senate.gov/~energy

Worldwatch Institute
1776 Massachusetts Ave. NW
Washington, DC 20036
(202) 452-1999
FAX (202) 296-7365
URL: http://www.worldwatch.org

RESOURCES

The Energy Information Administration (EIA) of the U.S. Department of Energy (DOE) is the major source of energy statistics in the United States and publishes weekly, monthly, and yearly statistical collections on most types of energy, available in libraries and online at http://www.eia. doe.gov. The *Annual Energy Review 1999* (2000) provided a complete statistical overview, while the *Annual Energy Outlook 2000* (1999) projected these findings into the future. The EIA's *International Energy Annual 1998* (2000) presented a statistical overview of the world energy situation, while the *Coal Industry Annual 1998* (2000) and *U.S. Crude Oil, Natural Gas, and Natural Gas Liquid Reserves* (1999) discussed reserves of coal, oil, and gas. *Conservation and Renewable Energy: Technologies for Transportation* (1990) discussed renewable energy use in U.S. transportation. The DOE's Office of Policy provided information on the National Energy Policy Plan in its *Comprehensive National Energy Strategy* (1998).

The DOE also provided *Spent Nuclear Fuel Discharges from U.S. Reactors* (2000), *Natural Gas Annual 1998* (2000), *Geothermal Energy in the Western United States and Hawaii* (1991), *Renewable Energy Annual 1999* (2000), *Petroleum: An Energy Profile* (1999), *Managing the Nation's Nuclear Waste* (1990), *Yucca Mountain Studies* (1990), *Alternatives to Traditional Transportation Fuels 1996* (1997), *Estimates of U.S. Biomass Energy Conservation 1992* (1994), and *Wind Energy Developments: Incentives in Selected Countries* (1998). The DOE also provided information on the Waste Isolation Pilot Project and on the development of alternative vehicles and fuels.

The U.S. General Accounting Office (GAO) published numerous helpful reports, including *Department of Energy: Problems and Progress in Managing Plutonium* (1998), *Nuclear Waste: Understanding of Waste Migration at Hanford Is Inadequate for Key Decisions* (1998), *Nuclear Waste: Uncertainties About Opening Waste Isolation Pilot Project* (1996), *Energy Security: Evaluating U.S. Vulnerability to Oil Supply Disruptions and Options for Mitigating Their Effects* (1996), and *Radioactive Waste: Status of Commercial Low-Level Waste Facilities* (1995). Also useful were *Air Pollution Allowance Trading Offers an Opportunity to Reduce Emissions at Less Cost* (1994), *Electric Vehicles: Likely Consequences of U.S. and Other Nations' Programs* (1994), *Geothermal Energy: Outlook Limited for Some Uses but Promising for Geothermal Heat Pumps* (1994), *Nuclear Safety: International Assistance Efforts to Make Soviet-Designed Reactors Safer* (1994), and *Nuclear Safety: Concerns with the Continuing Operation of Soviet-Designed Nuclear Power Reactors* (2000).

The U.S. Department of Transportation's Bureau of Transportation Statistics provided transportation information in its *Annual Report 1999* (1999). Greenpeace International provided information in its *Fossil Fuels and Climate Protection: The Carbon Logic* (1997).

The Department of the Interior's Mineral Management Service publishes information directly bearing on the nation's energy resources. *Federal Offshore Statistics: 1993* (1994) and *Managing Oil and Gas Operations on the Outer Continental Shelf* (1986) provided information on offshore oil.

The now-defunct Office of Technology Assessment (OTA) published a variety of materials on conservation. Of particular use in the preparation of this book were *Electric Power Wheeling and Dealing: Technological Considerations for Increasing Competition* (not dated), *Building Energy Efficiency* (1992), *Renewing Our Energy Future* (1995), *Saving Energy in U.S. Transportation* (1994), and *Improving Automobile Fuel Economy* (1991).

Information Plus thanks the National Conference of State Legislatures for use of information on nuclear energy from *Farewell to Arms* (Denver, Colorado, 1993). Its *Alternative Fuel Policies and Programs: A Legislator's Guide* (1997) discussed the use of alternative energy sources.

INDEX

Page references in italics refer to photographs. References with the letter t *following them indicate the presence of a table. The letter* f *indicates a figure. If more than one table or figure appears on a particular page, the exact item number for the table or figure being referenced is provided.*

D

E

F

S

T

U

V